A. Lee McAlester

Jim,

I want you
to have this
product of a
more peaceful era.

Lee

Physical Geology
Principles and Perspectives

A. Lee McAlester
Southern Methodist University
Dallas, Texas

Edward A. Hay
De Anza College
Cupertino, California

PRENTICE-HALL, INC.
Englewood Cliffs, New Jersey

Library of Congress Cataloging in Publication Data

McALESTER, ARCIE LEE, (date)
 Physical geology: principles and perspectives.

 Includes bibliographies and index.
 1. Physical geology. I. Hay, E. A., joint author.
II. Title.
QE28.2.M26   551   74-23840
ISBN 0-13-669523-X

10 9 8 7 6 5 4 3 2 1

Printed in the United States of America

PRENTICE-HALL INTERNATIONAL, INC., London
PRENTICE-HALL OF AUSTRALIA, PTY. LTD., Sydney
PRENTICE-HALL OF CANADA, LTD., Toronto
PRENTICE-HALL OF INDIA PRIVATE LIMITED, New Delhi
PRENTICE-HALL OF JAPAN, INC., Tokyo

For Colette, Martine, and Keven;
Barbara, Alison, Christopher, and Stuart

Contents

Preface

This book attempts to provide an introduction to physical geology that is brief and accessible, yet scientifically uncompromising. Every effort has been made to place the essential principles of the subject—minerals, rocks, geologic time, earth structure, plate tectonics, and land sculpture—in a rigorously up-to-date context while, at the same time, avoiding encyclopedic detail. In the Introduction we have also tried to set the stage by stressing the evolutionary nature of scientific conclusions and the philosophical foundations upon which rests our modern understanding of the earth sciences. Portions of the book have been adapted from appropriate chapters of *The Earth, An Introduction to the Geological and Geophysical Sciences* (Prentice-Hall, 1973). Much additional material has been added, however, to emphasize topics, such as rock origins and earth structure, that are particularly central to the study of physical geology.

We have profited greatly from discussions with colleagues during the early stages of writing this book, and wish to thank Sydney P. Clark, Jr., Brian J. Skinner, and Karl K. Turekian of Yale University, and Donald E. Buck, William R. Cotton, Duane E. Heinz, and David E. Williams of De Anza College for generously sharing both their time and knowledge. Reviews of early manuscript material provided important insights for improving our final product. Accordingly, we are indebted to Kathe Bertine, San Diego State University, Garniss H. Curtis, University of California at Berkeley, C. Earl Harris, Jr., Youngstown State University, Richard C. Robinson, Santa Monica College, Austin A. Sartin, Stephen F. Austin State University, and Mary Westerback, C. W. Post College. The efforts of Edythe Gramaglia to put handwritten manuscripts into readable form before they left De Anza College are much appreciated. Copy editing by William Green was done with considerable care, and we thank him for his efforts. Finally, we recognize the important roles played by William H. Grimshaw and Bruce Williams at Prentice-Hall in bringing this book to fruition; we are most anxious to extend our thanks to them for their skill, enthusiasm, and support.

A. L. McA.

E. A. H.

Physical Geology

introduction

Geology
in Perspective

Geology is one of the many fields of science, and science itself is but one of many broad areas of study. As an introduction to the main business of this book, geologic principles and perspectives, we shall first illustrate the manner in which sciences in general may be compared to other fields, and further, how geology in particular contains elements that are unique with respect to other sciences such as chemistry and physics.

I.1 Where Science Fits In

The difference between the sciences and other disciplines seems to reside not so much in their goals as it does in their *process of inquiry*. Some discussions of this topic even suggest that a rather special procedure exists called "the scientific method." Such a term is often misleading in its implication that a formula for the achievement of "truth" exists in the form of this "method." The observation Nobel Prize winner Percy W. Bridgeman made in 1945—that the method of science is "doing one's damnedest with one's mind, no holds barred"—is certainly a less pretentious statement of how scientists work. But the fact still remains that the scientist does proceed with his investigations often by using approaches and strategies for investigation that are different from those employed by other scholars. Why is this so? By virtue of the scientist's concern with questions about the workings of the material universe, he finds himself in the enviable position of being able to observe, measure, and experiment with objects of interest. He may engage in the exciting enterprise of prediction, and whether or not his predictions prove correct, the results of his experiments provide him with valuable input into the redesign of further experiments. Almost certainly, such chains of experimentation will lead to his more complete understanding.

Why is it, then, that scholars in other fields do not as successfully employ these same tools? We must remember that the tools useful to a given craftsman depend upon the task to be performed. Algebra and microscopes are as inappropriate to the historian in his search for understanding the causes of the Civil War as sledge hammers and jigsaws would be for the watch repairman. The scientist is thus seen to employ the tools best suited for deciphering how the material interactions of the universe work. He attempts to discover relationships between phenomena of the sort we think of as subject to so-called natural laws—gravity, relativity, and the like.

Because the interactions of matter and energy studied by the scientist are concerned with very *fundamental* properties, they lend themselves very nicely to controlled laboratory experiments and math-

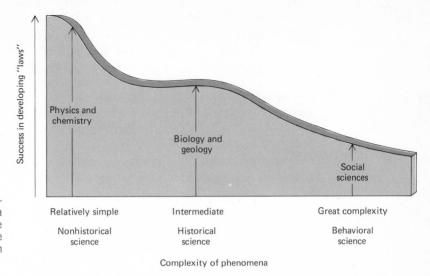

FIGURE I.1

Spectrum of complexity. As the phenomena being studied become more complex, there is greater difficulty in formulating "laws."

Figure labels:

Success in developing "laws" (vertical axis)

Physics and chemistry

Biology and geology

Social sciences

Relatively simple — Intermediate — Great complexity

Nonhistorical science — Historical science — Behavioral science

Complexity of phenomena

ematical descriptions. Consider, for example, the differences in complexity between considerations of how an object goes faster after falling longer as compared with trying to understand Hitler's rise to power in Germany after World War I. The physicist has only two variables to consider—time and velocity—whereas the historian must try to weigh and evaluate literally hundreds of events, social forces, and human interactions that took place over an extended period of time in many far-flung places. Clearly, the physicist will be able to experiment and numerically describe his investigations with dramatically greater success than the historian. It is thus little wonder that scientists have achieved more impressive results in their search for understanding: the problems they have chosen to attack are less complex, and therefore easier to solve. Although most people not familiar with mathematics may find it paradoxical, the simple truth is that the sciences are able successfully to employ sophisticated mathematical methods by virtue of the *relative simplicity* of what is being investigated! (Figure I.1.)

I.2   Geology's Scientific Uniqueness

Our example of how the historian and the physicist necessarily use different tools and strategies in their investigations describes what may be considered opposite ends of an academic spectrum. It now seems appropriate to ask where geology fits into these considerations. Is geology simply the discipline that applies chemistry and physics to problems concerning our planet? Or is there something about geology that makes it in some respects unique? Geologists have argued both points of view without reaching complete agreement about how much importance should be attached to the factors that make geology different, and we shall not enter into that debate here. We should, however, ask, "What is this difference and how has it come about?"

Geological sciences are concerned with the earth and how it has changed through time. Thus by their very nature they are historically oriented, and as such they are confronted with certain kinds of problems not encountered in other, nonhistorical sciences. We should not be surprised to find that geology occupies a sort of middle ground between chemistry or physics, on the one hand, and history or sociology on the other. When the changing conditions over long periods of time become an important factor, problems become significantly more difficult. The geologist can conduct experiments to learn about the *mechanisms* of erosion, for example, but he is faced with a completely different kind of problem in trying to understand the origin of Yosemite Valley or of the Grand Canyon, for these represent "one of a kind" sequences of erosional events occurring over millions of years, and these precise events are unrepeatable. In discussing this matter, George Gaylord Simpson (1963) has suggested that many of the concerns of the earth scientist (geologist) involve "postdiction" rather than prediction. In other words, because the earth has had such a long and complex history, understanding it requires methods of study beyond the approach employed so successfully in physics and chemistry, where controlled laboratory experiments are conducted to obtain *predictive* results.

Because of the complexity of the earth's history and the great difficulty in deciphering it, the history of geologic thought has seen a wide range of speculations important to the progress of geology. As Thomas S. Kuhn (1962) has observed, the scope of scientific speculations acceptable at any given time within a particular field of science (such as geology) will depend on the assumptions currently held by members of that scientific group. In reference to this, the term paradigm has recently begun to receive wide use. What are paradigms? Simply stated, they are models or generally agreed-upon "ground rules" that limit the ways a given investigation may proceed. Whether they realize it or not, people are constantly excluding many possible explanations for natural phenomena because they are not consistent with a currently accepted paradigm or assumption. During the time of Jesus, for example, who would have seriously proposed sailing westward from Rome in order to get to China? Such a suggestion would have been rejected during that era, even though today we know it to be reasonable and correct, because the earth was then widely believed to be flat. This speculation would have been "premature"—a concept set

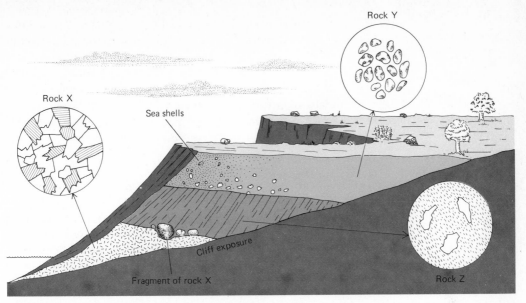

FIGURE 1.2
Different rocks exposed in a cliff. See text for discussion.

forth by Gunther S. Stent (1972)—because no appropriate paradigm existed then to legitimize it: going westward to end up in a land to the east could not have been related to the accepted knowledge of the day. The implications of such a proposal, had it been made then, would therefore be unexplored and unappreciated until people adopted—or facts forced them to adopt—a paradigm consistent with it. Perhaps even today we are failing to seize upon important ideas because they are not easily connected to currently held views.

Figure 1.2 and the following discussion will be used to depict two of many major paradigm changes (scientific revolutions) that have taken place in the early days of what we call the geological sciences. Three different explanations of the sketch will be presented, each derived from a once-acceptable and widely held view, although the specific quotations are hypothetical.

Statement 1 □ Greek poet, 600 B.C.

This remarkable cliff enables us to see the manner in which the earth was formed by the Gods out of chaos. It has remained essentially unchanged since its origin, with the exception that Poseidon [god of the sea] has flooded the surface with sea water so as to allow the sea shells to reside near the hill top.

This poet is clearly reflecting a popular attitude, or paradigm, of his day. The ancient Greeks explained all natural phenomena as functions of the whims and dictates of gods. Given such a philosophy of life, it would be totally unlikely—indeed unacceptable—to seek impersonal cause-and-effect relationships in natural processes as a means of discovering orderly interactions between matter and energy. The whole notion of scientific laws is dependent on very different attitudes from those expressed in Statement 1. If gods supply "purpose" to earth processes, then man cannot hope to discover reproducible relationships. Thus, ideas of invariable, impersonal cause-and-effect associations are essential to the emergence of modern science. It is apparent, therefore, that science could not have flourished under the paradigm of divine control.

Statement 2 □ A student of A. G. Werner (Werner was an influential professor at the Mining Academy of Freiberg), A.D. 1795

This exposure allows us to document the order of deposition of rock types that have settled-out from a world-encompassing ocean to form the crust of the earth. Rock X is the *oldest because it is granite*, and is therefore a primitive chemical precipitate from the primordial ocean, as is the basalt of Rock Z. The sandstone of Rock Y is a mechanical sediment deposited after an appreciable lowering of the ocean. It is obvious that the fossils [shells] were dissolved in the universal water and have been precipitated from it. This landscape has not changed significantly since the ocean receded to its present level.

Professor Werner was the major proponent of the widely held eighteenth-century view, called "Neptunism," that all rocks have precipitated from a world-encompassing ocean. The statement by Werner's student shows how once an investigator accepts a particular model, he automatically and subconsciously puts on "blinders" to a host of speculations and inferences that might otherwise have been considered. The view that order existed in the universe had been accepted by this time, but many investigators were too inclined to *pronounce* how things should be rather than carefully attempt to *discover* how they are. Werner was such a man. The model of Neptunism held sway for many years, but was eventually overthrown by a way of looking at the earth called uniformitarianism during the early decades of the nineteenth century.

Statement 3 □ A follower of Charles Lyell (Lyell is viewed by many as one of the founders of modern geology), A.D. 1835

By carefully observing the textures [grain relationships] of the three exposed rock types and the nature of the contacts between them, we can infer the genesis of each and the historical events involving their association with each other. The granite of Rock X is much older [by millions of years] than the others, and was formed by the slow crystallization of molten rock material at great depth beneath what was then the surface of the earth. In contrast to the large crystals formed by slow cooling in

Rock X, Rock Z has small crystals and, therefore, must have cooled quickly, perhaps as a lava flow on the earth's surface. Fragments of Rock X that have been incorporated within Rock Z provide clear *evidence* that Z is younger than X. A great period of erosion would necessarily have been required to cause Rock X, formed at great depth, to be exposed at the surface of the earth prior to the eruption of volcanic material of Rock Z. Rock Y, a sandstone, has resulted from the compacting and cementing of many abraded [worn] grains of sand, following transportation by streams into the sea. There they accumulated with shells of once living organisms, eventually to be formed into rock. This region has an involved and dynamic history. By using present geologic processes such as erosion, deposition, and volcanism as keys to our understanding of how the earth can change, given sufficient time, it is possible to reconstruct uplifting, prolonged erosion, subsidence, and many other subleties of earth history from a cliff exposure such as the one seen here. We need not invoke any supernatural or exceptional explanations to reconstruct the history of the earth.

This statement reflects the perspective of a modern geologist. It reflects a guiding paradigm that asserts validity for the principle of uniformitarianism. We shall return to a more complete discussion of this view shortly, but simply stated it asserts that the earth has an exceedingly long history during which complex geologic processes such as erosion, mountain building, and rock genesis have proceeded in accordance with fundamental laws (e.g., those of chemistry and physics).

The scientific revolution that gave rise to uniformitarianism was as intellectually labored as the American Revolution was hard-fought with the weapons of war. Let us briefly digress to review, at this point, how this exceedingly important scientific revolution came about.

I.3 The Beginning of Modern Geology

The well-known nineteenth-century American scholar Henry Adams wrote, "Anarchy is the law of nature, and order is the dream of man." Without entering into the philosophical debate invited by this statement as to whether man discovers or creates order, let us allow it to suggest a relationship that is fundamental to the rise of modern science: *if we are to understand nature, we must search for its orderliness.* All scientific inquiry begins with the assumption that there are discoverable orderly interactions going on within the universe. Simply stated, science rests fundamentally on a *faith* in the inductive and empirical processes and their ability to yield valid conclusions; that is to say, predictions about the future can be formulated reliably only from experiences in the past. Inasmuch as we cannot prove what will happen tomorrow, no scientist can ever be certain about his predic-

Statements explaining how phenomena are related may evolve through the following stages as experiments are conducted to test them:

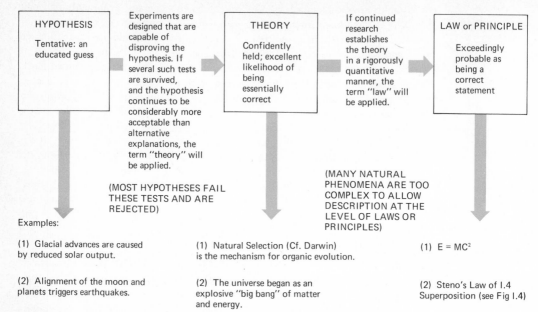

Examples:

(1) Glacial advances are caused by reduced solar output.

(2) Alignment of the moon and planets triggers earthquakes.

(1) Natural Selection (Cf. Darwin) is the mechanism for organic evolution.

(2) The universe began as an explosive "big bang" of matter and energy.

(1) $E = MC^2$

(2) Steno's Law of I.4 Superposition (see Fig I.4)

FIGURE I.3

Hierarchy of scientific confidence. The relationship between hypothesis, theory, and law.

tions, and hence no predictive statement (or law) can ever be wholly proved—even though such terminology is frequently applied to scientific assertions. The so-called law of gravity, for example, is a statement of a kind of relationship that exists between all forms of matter. We can have exceedingly great confidence in the accuracy of this statement, but in a philosophical sense it is clear that we *cannot be certain.* Consider the so-called "hierarchy of confidence" for scientific statements that is depicted in Figure I.3.

The extreme complexity of nature forces investigators to generate models to account for the complicated reality being studied. Scientific models are mental constructs of how nature *might* function. The particular models an investigator is likely to formulate are ultimately controlled by the paradigms of his science. His models may be mathematical, mechanical, or verbal, and typically they take the form of hypotheses, theories, and laws, serving us by providing suitably simple vehicles for our use in reasoning by analogy. We do not truly know what an atom is like, for example, but we have developed the so-called atomic theory which has provided a variety of models used to generate analogies helpful in gaining greater insights into many fields of science.

To return to the initial point, it is clear that modern science could

THE STRUCTURE OF SCIENCE

The following anecdote is told of the scholar William James. Whether it is true or not is less important than the comment it makes on various paths in the search for knowledge. Modern science could not begin its development until the **assumption** was made that the workings of nature are characterized by **discoverable** relationships.

After a lecture on cosmology and the structure of the solar system, James was accosted by a little old lady.

"Your theory that the sun is the center of the solar system, and that the earth is a ball that rotates around it, has a very convincing ring to it, Mr. James, but it's wrong. I've got a better theory," said the little old lady.

"And what is that, madam?" inquired James politely.

"That we live on a crust of earth which is on the back of a giant turtle."

Not wishing to demolish this absurd little theory by bringing to bear the masses of scientific evidence he had at his command, James decided to gently dissuade his opponent by making her see some of the inadequacies of her position.

"If your theory is correct, madam," he asked, "what does this turtle stand on?"

"You're a very clever man, Mr. James, and that's a very good question," replied the little old lady, "but I have an answer to it. And it's this: the first turtle stands on the back of a second, far larger, turtle, who stands directly under him."

"But what does this second turtle stand on?" persisted James patiently.

To this, the little old lady crowed triumphantly, "It's no use, Mr. James—it's turtles all the way down."

be nurtured only within an environment where the notion of order was accepted, thus allowing models to be devised for explaining nature. When did such mind-sets begin to appear with regard to matters of the earth? Although Aristotle and other Greek philosophers functioned under such a paradigm, an early example dealing more explicitly with geological considerations comes from the work of Nicolas Steno, a Danish scholar living in Florence in the seventeenth century. Steno observed exposures of layered rocks throughout the Florentine countryside. Reasoning that they must have formed in the past in much the same manner as the layered deposits he observed being formed from sediments along Mediterranean shores and Italian streams valleys, he formulated three important statements which he suggested

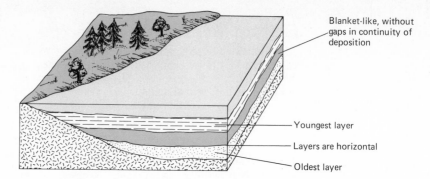

Blanket-like, without gaps in continuity of deposition

Youngest layer

Layers are horizontal

Oldest layer

LAW OF SUPERPOSITION: Lower layers of sediment must have achieved firmness before subsequent layers could settle upon them, hence for rocks forming by this sedimentary process the layers become progressively younger from bottom to top. (1 is older than 2, 2 is older than 3, etc.)

LAW OF ORIGINAL CONTINUITY: Sedimentary layers were deposited as blanket-like, continuous sheets of material that gradually thin to nothing at their edges, or have been terminated against a solid surface.

LAW OF ORIGINAL HORIZONTALITY: Although bottom layers may irregularly conform to the shape of the surface upon which they are deposited, most layers were deposited parallel to the horizon (horizontal) or very nearly so.

FIGURE I.4
Steno's laws.

could be applied to *all* layered rocks. Today we have such extreme confidence in their correctness that we refer to them as "laws" (Figure I.4).

The great historical and philosophical achievement of Steno's work is that (1) order in nature was assumed, (2) comparison of the present and the past (modern sedimentary environments and old layered rocks) was carefully undertaken, and (3) a general statement about *all* layered rocks was formulated. Furthermore, with the advent of such "laws" (then "theories," perhaps) it became possible, indeed necessary, to view the earth as undergoing dynamic change through time —change which could be deciphered by careful application of Steno's Laws (Figure I.5). Steno was probably the first person who seriously attempted to work out the geologic history of a portion of the earth. His work in the seventeenth century heralded the beginning of an important human undertaking.

With a new paradigm asserting orderliness in the universe, men eventually began to observe and carefully describe rocks exposed all around them at the earth's surface. They found that, indeed, order could be found there, although it was not always clear what the order was. The earth was found to be complex, and men, whose minds were unaccustomed to such ambiguity and uncertainty, frequently imposed their own order on nature. Such was the case with A. G. Werner and

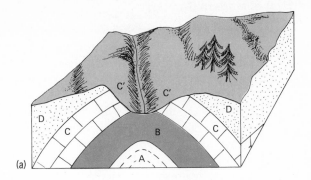

(a)

LAW OF SUPERPOSITION: A is the oldest and D is the youngest of these sedimentary layers.

LAW OF ORIGINAL CONTINUITY: Where edges of layers are exposed at the earth's surface, erosion must have occurred to destroy the original continuity (see regions C' on sketch)

LAW OF ORIGINAL HORIZONTALITY: All these layers have been folded since their time of origin, because they began as horizontal strata.

FIGURE 1.5
(a) Application of Steno's laws: Historical inferences such as sequence of deposition, folding, and erosion can result from Steno's laws. (b) Imagine the enthusiasm of Steno if he had been able to see the Grand Canyon! Superposition, original horizontality, and original continuity at their finest. (b: E. A. Hay.)

(b)

his "Neptunism." Well-meaning investigators often succumbed to temptation out of intellectual frustration and ascribed simple solutions (recall "turtles all the way down" on p. 11) to relationships of bewildering complexity. But always they were guided by a belief in discoverable orderliness. Albert Einstein's suggestion that while God is ingenious, he is not malicious, delightfully characterizes the spirit and essence of this most fundamental of all scientific assumptions.

I.4 The Arrival of Modern Geology

We have recognized that the assumption of orderliness in the universe provides the basis for scientific inquiry. However, the mechanism for achieving this order on earth has never been obvious. Following the time of Steno, an explanation for the evolution of the earth called "catastrophism" became dominant in western Europe. Subsequent challenges to this theory developed under the banner of "uniformitarianism," and in the late eighteenth and early nineteenth centuries a major scientific revolution took place within the field of geology.

Out of the conflict between catastrophism and uniformitarianism geology emerged as a modern science, with uniformitarianism as its paradigm. In a sense, there *was* no geology prior to the acquisition of this paradigm. Let us briefly consider the major points of philosophical disagreement between catastrophists and uniformitarians in this revolutionary battle.

Although no one man is ever responsible, all by himself, for a major school of thought, the catastrophist Georges Cuvier and the uniformitarian James Hutton were clearly the luminaries of their respective groups. Cuvier, a French scholar of great reputation, argued that the earth had repeatedly undergone incredibly violent upheavals of a sort that could not currently be observed during the normal day-by-day activities of our planet. Cuvier functioned within the intellectual framework, perhaps religiously inspired, that the earth is a relatively young (thousands of years old) entity. Consider his statement, excerpted from a treatise written in 1817:

> [Fossil seashells] have therefore once lived in the sea, and been deposited by it; the sea consequently must have rested in the places the deposition has taken place. Hence it is evident that the basin or reservoir containing the sea has undergone some change at least, either in extent, or in situation, or in both. Such is the result of the very first search, and of the most superficial examination. . . . This change was sudden, instantaneous not gradual, and that which is so clearly the case in this last catastrophe is not less true of those which preceded it. The dislocation and overturning of the older strata show without any doubt that the causes which brought them into the position which they now occupy, were sudden and violent.

Theories of sudden catastrophic change such as this were based on rock exposures similar to that depicted in Figure I.8. Cuvier, functioning within the framework of Steno's Laws, was particularly concerned with the significance of the surface separating older rocks from younger ones. Such a surface, called an "unconformity," was judged

FIGURE I.6
Georges Cuvier (1769–1832), leader of the
school of Catastrophism. (Brown Brothers.)

FIGURE I.7
James Hutton (1729–1797), the
"father" of modern geology.
(Dover Publications Inc., New
York.)

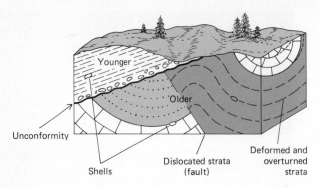

Younger

Older

Unconformity

Shells

Dislocated strata
(fault)

Deformed and
overturned
strata

FIGURE I.8
Block diagram showing an unconformity
and other rock relationships. See text for
discussion.

to have formed suddenly during the same brief interval when the older
rocks were undergoing "dislocation and overturning." Such an explanation is unavoidable for someone operating under the assumption
that the earth is quite young. It is not, however, the only interpretation.

Let us now consider, in contrast, how it was possible for James Hutton to view similar rock relationships and be led to significantly different explanations for their origin. Hutton, a Scot who was trained in
medicine and prospered as an agriculturist, was an avid student of
present-day earth processes. He became profoundly influenced by
strong similarities existing between modern and ancient features of the
earth. He observed that sedimentary features in beach deposits along
the coast of Scotland are essentially indistinguishable from those seen
in hard rocks located in the highlands many miles from the sea. Sediment-laden streams flowing to the sea, he proposed, clearly attest to
the slow destruction of land. Hutton reasoned that the processes he
was observing were capable of bringing about a sequence of changing
geographies whose record is frozen into rocks of the earth's crust—
capable, that is, if these present-day processes had been able to operate over inconceivably long periods. So impressed was Hutton by this
explanation that he quickly accepted the conclusion that the earth is
truly ancient. Indeed, in 1795 he wrote that, as far as he could determine, for the earth there is "no vestige of a beginning, no prospect of
an end."

Hutton's analysis of the unconformity separating the older and
younger rock strata in Figure I.8 would surely have been that it represented a very long period of time during which slow geologic process
such as folding and faulting (breaking) of layers caused the marine
sediments to rise above sea level and undergo prolonged erosion until
they became essentially featureless. Later subsidence of the region
then led to renewed deposition of sediment burying the eroded surface
(Figure I.9). The contrast between this interpretation and Cuvier's

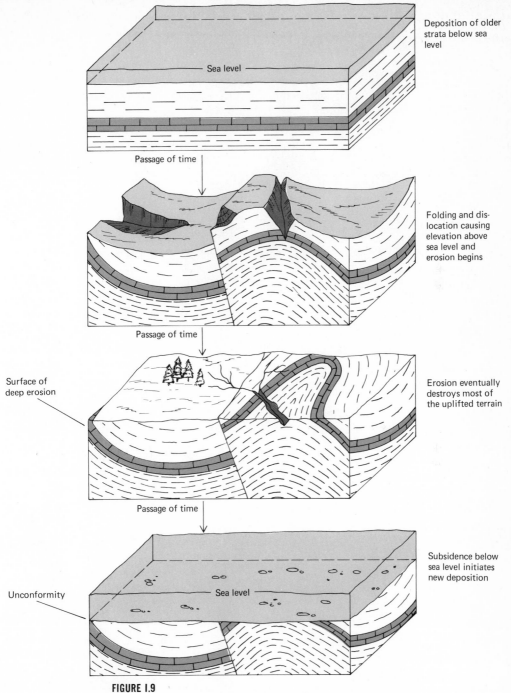

Deposition of older
strata below sea
level

Sea level

Passage of time

Folding and dis-
location causing
elevation above
sea level and
erosion begins

Passage of time

Surface of
deep erosion

Erosion eventually
destroys most of
the uplifted terrain

Passage of time

Subsidence below
sea level initiates
new deposition

Unconformity

Sea level

FIGURE I.9
Sequence of events leading to the relationships shown in Figure I.8.

(a)

FIGURE I.10
(a) Charles Lyell (1797–1875), author
of the first textbook of geology. (b)
Title page of "Principles of Geology"
(3 volumes), first published 1830–1833.
(a: Crown copyright, Geological Survey
photograph; b: G. Craven.)

(b)

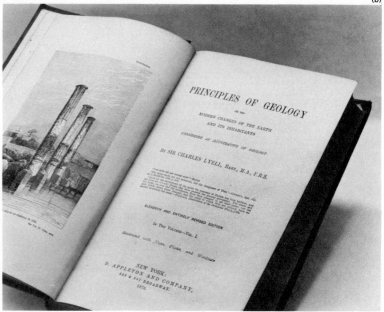

could hardly be more extreme. Hutton saw the unconformity as documenting a very old earth that had slowly changed through time by means of presently observable processes; in the words of Archibald Geikie (1905), he was guided by the view that "the present is the key to the past." That same unconformity and the strata surrounding it, when viewed from the perspective of a different paradigm, led Cuvier to conclude "without any doubt that the causes which brought them into the position which they now occupy, were sudden and violent." Both Hutton and Cuvier recognized orderly changes through time. Their great disagreement concerned the pace at which the changes occurred and the processes responsible for them.

The doctrine of uniformitarianism, with its insistence that the earth has great antiquity and its commitment to the assumption that matter and energy have always interacted in an unchanging and hence "lawful" manner throughout the earth's long history, was thus explicitly put forth by James Hutton during the late 1700s. The triumph of this view, however, was not immediate. Catastrophism did not easily suffer defeat, and intellectual battles raged on into the 1830s until, with the publication of Charles Lyell's three-volume *Principles of Geology* (1880–1833), this most important of revolutions in geology was, for all practical purposes, won. Lyell's *Principles* was, in effect, the first textbook in geology. The fact that such a book could be written was an affirmation that a new paradigm had emerged. It was destined to exert a profound influence on how men would view themselves and the universe. Charles Darwin, for example, carried the first volume of Lyell's *Principles* up the gangplank of the *H.M.S. Beagle* with him in 1831, at the beginning of the five-year voyage during which he developed the initial ideas of a mechanism to account for organic evolution. Darwin later commented: "I feel as if my books came half out of Sir Charles Lyell's brain."

Many new geologic insights have marked the remainder of the nineteenth century and the twentieth century till now. They have all flourished from research founded upon the paradigm of uniformitarianism, which itself has required significant clarification during this period. The term *uniformitarianism* did not, originally at least, have the same meaning for all geologists. As a consequence, there emerged a threefold ambiguity as to whether a person advocating uniformitarianism was (1) suggesting a uniformity of *rate* and *intensity* for geologic events and processes, (2) asserting that presently observable processes are *completely* sufficient to decipher earth history, or (3) simply assuming that matter and energy have always interacted in the same manner. Only the last of these interpretations has been retained as acceptable. The first two attitudes fail to account for the evolution of the earth's sources of energy and its physical and chemical systems. Inasmuch as the energy of the earth is conceived to be less now than it was in the past, we cannot reasonably expect all geologic processes to have maintained their intensity during several billions of years. We do, however, assume that these processes, whatever their intensities, have continued to act according to uniform laws throughout these long spans of time.

I.5   Geology in the 1970s

Until recently geologists continued their investigations of the earth in the spirit of Steno, Hutton, and Lyell, but were increasingly frustrated by their failure to discover principles capable of unifying many seemingly unrelated areas of study. Scientists intuitively seek order and simplicity, and for the geologists the earth had long been a very difficult object of study. But during the 1960s a new revolution in geology began to develop that we are still in the midst of today.

Later in this book (in Chapter 7) you will encounter discussions of "plate tectonics" as a model describing major rigid "plates" of crustal rock and the interactions that take place between them. Acceptance of this new paradigm represents the revolutionary overthrow of many long-entrenched geologic notions. Our ocean floors, for example, are now considered to be among the youngest major features of the earth, whereas just a few years ago, prior to the development of the plate tectonics model, they were universally acknowledged to represent rocks of fundamental crustal antiquity. Many other examples of dramatic changes could be cited, but at this point it will suffice to recognize that the unprecedented vitality and excitement currently characterizing the geological sciences are direct consequences of this most recent scientific revolution involving the paradigm of plate tectonics and the new perspectives it has provided. Many previously insoluble problems will become less difficult when now viewed within this framework.

ADDITIONAL READINGS

*Adams, F. D.: *The Birth and Development of the Geological Sciences*, The Williams & Wilkins Co., Baltimore, 1938 (Dover Publications paperback edition, 1954). A standard and detailed history of earth studies from classical times through the early nineteenth century.

*Albritton, C. C., Jr. (ed.): *The Fabric of Geology*, Freeman, Cooper & Co., Stanford, California, 1963. Advanced essays on the history and philosophy of the geological sciences, with extensive bibliographies.

*Geikie, A.: *The Founders of Geology*, The Macmillan Company, New York, 1905 (Dover Publications paperback edition, 1962). Classic history of Geology through the mid-nineteenth century.

*Fenton, C. L., and M. A. Fenton: *Giants of Geology*, Doubleday & Company, Inc., Garden City, New York, 1952. Well-written, popular biographies of influential nineteenth-century geologists.

*Kuhn, T. S.: *The Structure of Scientific Revolutions*, The University of Chicago Press, Chicago, 1962.

Stent, G. S.: "Prematurity and Uniqueness in Scientific Discovery," *Scientific American*, December 1972, pp. 84–93.

*Available in paperback.

chapter **1**

Matter
and Minerals

Physical geology is concerned with the origin of earth materials and how they have been changed through time by forces of mountain building and erosion. It is appropriate, therefore, to begin with an introduction to the materials that make up the solid earth.

Earth materials are traditionally and most conveniently considered on two different scales: (1) The units of smaller scale are **minerals**, natural substances of uniform chemical composition possessing a definite crystalline structure. Each mineral has a combination of composition and structure that characterizes it and no other. About 2,000 different minerals have been found, but only about 20 common ones make up the bulk of the earth. Some examples among these common minerals are mica, quartz, and gypsum. (2) The units of larger scale are **rocks**, usually aggregates of several different minerals that have been brought together by one of many rock-forming processes. Examples are granite, basalt, and sandstone. Unlike minerals, rocks are not rigidly defined by their composition and structure since they may be composed of endlessly varying mixtures of minerals. Nevertheless, only about 20 mineral combinations are really common; these 20 rocks make up most of the solid earth.

Rocks and minerals, like all other substances, or **matter**, in the universe, are composed of elemental particles assembled according to certain fundamental chemical principles. Before we look more closely at the common rocks and minerals, it will be useful to turn to this smaller scale and review some of these principles. In Chapters 2 and 3 we shall discuss the major categories of rocks found on earth.

MATTER

1.1 Elements and Compounds

The smallest units of matter that normally concern earth scientists are the **elements**, substances that cannot be broken down into other substances by chemical means. There are 88 such elements known to occur naturally; in addition, 15 others have been produced artificially by physicists using atomic reactors and particle accelerators. These 103 elements are the basic building blocks of *all* matter, living and nonliving, on earth and throughout the universe. The interrelationships of the elements are best shown by arranging them into a **periodic table**, which groups together elements with similar physical and chemical characteristics (Figure 1.1). Reviewing the complete periodic table would be a cumbersome introduction to the structure of earth materials were not most of the elements extremely rare in the earth. Indeed, the rocks and minerals of the solid earth are made up almost entirely of only 10 elements—those shown in color in Figure 1.1. Five additional elements, shown in gray in Figure 1.1, are of less importance

Group Period	I	II											III	IV	V	VI	VII	VIII
1	1 H Hydrogen																	2 He Helium
2	3 Li Lithium	4 Be Beryllium											5 B Boron	6 C Carbon	7 N Nitrogen	8 O Oxygen	9 F Fluorine	10 Ne Neon
3	11 Na Sodium	12 Mg Magnesium											13 Al Aluminum	14 Si Silicon	15 P Phosphorus	16 S Sulfur	17 Cl Chlorine	18 Ar Argon
4	19 K Potassium	20 Ca Calcium	21 Sc Scandium	22 Ti Titanium	23 V Vanadium	24 Cr Chromium	25 Mn Manganese	26 Fe Iron	27 Co Cobalt	28 Ni Nickel	29 Cu Copper	30 Zn Zinc	31 Ga Gallium	32 Ge Germanium	33 As Arsenic	34 Se Selenium	35 Br Bromine	36 Kr Krypton
5	37 Rb Rubidium	38 Sr Strontium	39 Y Yttrium	40 Zr Zirconium	41 Nb Niobium	42 Mo Molybdenum	43* Tc Technetium	44 Ru Ruthenium	45 Rh Rhodium	46 Pd Palladium	47 Ag Silver	48 Cd Cadmium	49 In Indium	50 Sn Tin	51 Sb Antimony	52 Te Tellurium	53 I Iodine	54 Xe Xenon
6	55 Cs Cesium	56 Ba Barium	57 La Lanthanum	72 Hf Hafnium	73 Ta Tantalum	74 W Tungsten	75 Re Rhenium	76 Os Osmium	77 Ir Iridium	78 Pt Platinum	79 Au Gold	80 Hg Mercury	81 Tl Thallium	82 Pb Lead	83 Bi Bismuth	84 Po Polonium	85* At Astatine	86 Rn Radon
7	87* Fr Francium	88 Ra Radium	89 Ac Actinium															

58 Ce Cerium	59 Pr Praseodymium	60 Nd Neodymium	61* Pm Promethium	62 Sm Samarium	63 Eu Europium	64 Gd Gadolinium	65 Tb Terbium	66 Dy Dysprosium	67 Ho Holmium	68 Er Erbium	69 Tm Thulium	70 Yb Ytterbium	71 Lu Lutetium
90 Th Thorium	91 Pa Protactinium	92 U Uranium	93* Np Neptunium	94* Pu Plutonium	95* Am Americium	96* Cm Curium	97* Bk Berkelium	98* Cf Californium	99* Es Einsteinium	100* Fm Fermium	101* Md Mendelevium	102* No Nobelium	103* Lw Lawrencium

FIGURE 1.1
Periodic table of the elements. The 10 abundant solid earth elements are shaded in color; the abundant fluid earth elements are shaded in gray. The steplike line separates metals (below) from nonmetals (above). Asterisk denotes a man-made element. Elements 104 and 105 have been tentatively created and identified in the Soviet Union.

in the solid earth but are abundant in the water and air of the fluid earth. Our primary concern throughout this book will be with these 15 elements that make up the bulk of the earth, although we shall have occasion to investigate several others of special interest.

The properties of the elements are expressed in the periodic table in the following way: the elements become progressively heavier from left to right and from top to bottom. Thus hydrogen (element 1, upper left) is the lightest element and lawrencium (element 103), the heaviest. Note that the most abundant elements are all relatively light; no element heavier than nickel (element 28) is among the 15. The periodic table also associates in vertical "groups" those elements having similar chemical properties. For example, the six elements in the extreme right-hand column (group VIII) are all very stable, nonreactive gases called the noble gases. For the abundant elements of the earth, this columnar grouping shows that sodium and potassium, although differing in weight, have similar chemical properties, as do the pairs magnesium-calcium, carbon-silicon, and oxygen-sulfur. Atlhough it is listed in group I, hydrogen has properties found in no other element.

Pure elements seldom occur in nature; instead, two or more elements are usually found chemically combined in substances called compounds, which are more stable than pure ("free") elements. Within the solid earth, only carbon, sulfur, copper, and iron are found in any abundance as free elements, and even these four are much more common in compounds such as *calcium carbonate*, a compound of calcium, carbon, and oxygen that makes up ordinary limestone, or *iron disulfide*, the familiar "fool's gold" that is a compound of iron and sulfur. With few exceptions, then, the bulk of the earth is composed of chemical compounds.

Why is it that elements typically have greater stability when in combination than in their pure form? To answer this question we must take a closer look at them and develop a *model* to explain how they function.

1.2 An Atomic Model

Our discussions of earth materials will deal primarily with elements and compounds, but in order to understand better the behavior of these materials, we shall now briefly consider the components of the elements themselves and the ways they combine to make compounds. The smallest particles of an element that still possess its characteristic properties are called atoms. Reactions between atoms form compounds. Just as elements are made up of individual atoms, so are compounds composed of combinations of elements bonded together in definite proportions. Common salt (sodium chloride), for example, is a combination of the elements sodium and chlorine. Strictly speaking, this combination involves particles of each called ions, rather than the more familiar "atoms." This distinction will be clarified shortly. That a difference truly exists between atoms and ions can be anticipated, though, by noting the contrast between table salt and its elemental components. If a sample of table salt were separated to yield uncombined sodium and chlorine, then the properties of the compound would be lost, for chlorine alone is a poisonous, greenish gas and sodium is a gray, explosively reactive metal.

Until the beginning of this century, scientists believed that the atoms of the chemical elements were indivisible units of matter. Then, in the early 1900s, a model for the atom emerged that was more helpful to scientists in their efforts to understand how elements interact with each other. Although this model is now known to be oversimplified, it will nevertheless be satisfactory for our purposes.* It proposes a central nucleus containing most of the mass of the atom, surrounded

*It is important to remember that *all* scientific models are mental constructs; they *are not* "truth." It is sensible to use whichever model is simplest, yet appropriate, for whatever task is at hand. Inasmuch as models are *tools* for thinking, we shall always attempt to use the most convenient one, so long as it adequately serves our purpose.

by one or more extremely tiny, rapidly moving, and electrically charged particles called electrons. Researchers also found that the manner in which the elements combine to make compounds is determined by the number of electrons surrounding the nucleus. Each element has a characteristic number of electrons, ranging from only 1 in hydrogen to 103 in lawrencium. The normal number of electrons corresponds to the atomic number of the element (atomic numbers are shown above the element symbols in Figure 1.1).

The positions of the electrons around the nucleus are not random; instead, the rapidly moving electrons occur in concentric, cloudlike clusters called shells (Figure 1.2). The innermost shell of a given atom always contains only two electrons (or one in hydrogen), and any additional electrons are arranged in complex cycles of eight to form additional shells. The electron arrangement is important because the manner in which elements combine to form compounds is largely determined by the number of electrons in the *outermost* shell. The number of these outer-shell electrons for the 15 most abundant elements are shown arranged around their element symbols in Figure 1.3.

The most stable elements are those with completely filled outer shells containing *eight electrons.* Examples are the previously mentioned noble gases (helium,* neon, argon, krypton, xenon, and radon) occupying group VIII in the periodic table. These gases almost never react with other elements to form compounds. In contrast, the most reactive elements are those that are only one or two electrons short of having a completely filled outer shell—that is, those having one, two, six, or seven electrons in that shell. These elements occur in groups I, II, VI, and VII of the periodic table. As was seen with the noble gases,

---

*Helium is an exception, having only two electrons—which nevertheless fill the first shell and thus yield a stable configuration.

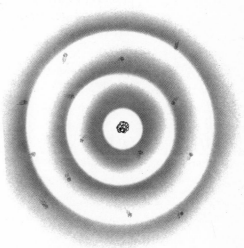

FIGURE 1.2
A schematic representation of the concentric electron shells of the silicon atom. Silicon, with atomic number 14, has 14 electrons distributed in three shells— two in the inner, eight in the middle, and four in the outer shell.

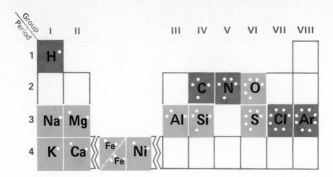

FIGURE 1.3
Electron-dot formulas of the 15 abundant elements. The dots show the number of electrons in the out- ermost electron shell.

the presence of eight electrons in the outer shell seems to be a highly stable configuration, so much so that elements not intrinsically posses- sing that structure react in a variety of ways to achieve it. These are the reactions that give rise to the formation of compounds. In general, elements with one or two electrons in the outer shell tend to react with those having six or seven electrons in the outer shell to form stable compounds having complete outer shells containing eight electrons. Thus two atoms of hydrogen, with one outer electron each, combine with one atom of oxygen, with six outer electrons, to form a single molecule of water (Figure 1.4).

1.3 Bonding to Form Compounds

There are two models frequently used to describe how the bonding between elements takes place in the formation of compounds.* One involves the *transferring* of electrons from one atom to another, thus forming electrically charged particles called ions, and the other envi- sions the *sharing* of one atom's electrons with another atom; com- pounds formed by these different mechanisms are said to possess ionic and covalent bonds, respectively (Figure 1.5). Why are there two dif- ferent types? The fundamental answer to this question is, "We do not know," but within the framework of our models there does seem to

*There are several other bonding models of lesser geologic significance not being considered here.

FIGURE 1.4
A schematic repre- sentation of the com- bination of two hydro- gen atoms with one oxygen atom to form a molecule of water sharing eight outer- shell electrons.

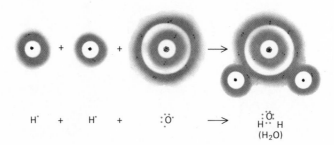

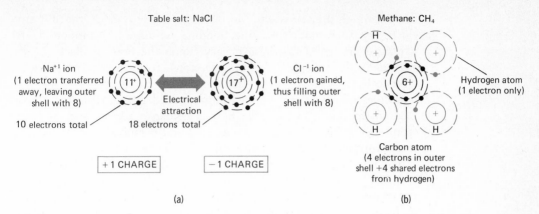

Table salt: NaCl

Na^{+1} ion
(1 electron transferred
away, leaving outer
shell with 8)

Electrical
attraction

10 electrons total

18 electrons total

Cl^{-1} ion
(1 electron gained,
thus filling outer
shell with 8)

+ 1 CHARGE   − 1 CHARGE

(a)

Methane: CH$_4$

Hydrogen atom
(1 electron only)

Carbon atom
(4 electrons in outer
shell +4 shared electrons
from hydrogen)

(b)

FIGURE 1.5
(a) Ionic bonding
(electron transfer).
(b) Covalent bonding
(electron sharing).

be an intuitive explanation. Elements that can achieve the stable eight-electron outer-shell configuration by giving away or receiving only one or two electrons tend to do just that, thus forming ions capable of bonding by electrical attraction with oppositely charged ions. On the other hand, elements that require the transfer of several electrons to become ions tend not to do so; instead, they tend to form covalent bonds. Carbon and silicon, for example, would each require the movement of four electrons to form ions. Apparently it is more efficient for them to achieve a stable configuration by sharing the electrons of other atoms than by moving so many of their own.

You may be thinking, "How do we know that such differences really exist?" Observing a grain of table salt and a grain of sugar, for example, what is the evidence supporting the assertion that salt is ionically bonded and that sugar is covalently bonded? Because we cannot directly see electrons, circumstantial evidence must be used in generating these two different models for chemical bonding. For example, a simple yet convincing test for salt and sugar involves dissolving both in separate beakers of water and observing the effects of attempting to pass an electrical current through each (Figure 1.6): the current

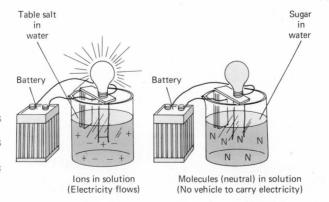

FIGURE 1.6
Ions versus molecules in solution. The electrically charged ions in solution facilitate the flow of electricity; neutral molecules cannot.

Table salt
in
water

Sugar
in
water

Battery

Battery

Ions in solution
(Electricity flows)

Molecules (neutral) in solution
(No vehicle to carry electricity)

flows through the salt water, but the sugar water fails to transmit any electricity. Our models for the bonding within each compound are wholly consistent with these observations: the flow of electricity is essentially the bumping along of electrons, requiring the presence of electrically charged particles (ions), whereas the presence of neutral molecules of covalently bonded sugar would not provide a suitable medium for electron movement. It is evidence of this sort, derived from observable properties of elements and compounds, that has led to the formulation of different models for chemical bonding.

The number of electrons gained, lost, or shared by an element when it forms an ion determines the *oxidation number* of that element. By convention, elements that tend to *receive* electrons during the process of forming compounds are given negative oxidation numbers, while those that tend to *donate* electrons are given positive numbers. All compounds contain elements having both positive and negative oxidation numbers.

The **empirical formula** for a compound gives the simplest proportion in which its elements combine. It can be deduced by remembering that the sum of all positive oxidation numbers must equal the sum of all negative oxidation numbers. Consider the following examples:

Example 1:   Silicon $+$ 2 Oxygen $=$ SiO_2
(mineral quartz) Si^{+4} $2 \times O^{-2}$

$$(+4) \underset{=}{=} (-4)$$

Example 2:   2 Aluminum $+$ 3 Oxygen $=$ Al_2O_3
(mineral corundum) $2 \times Al^{+3}$ $3 \times O^{-2}$

$$(+6) = \underset{=}{} (-6)$$

The normal oxidation numbers of the 15 most abundant elements are shown in Figure 1.7. Note that some elements may have more than one oxidation number, depending on how the electrons are gained, lost, or shared in forming a particular compound.

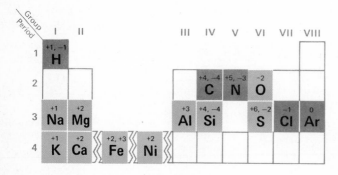

FIGURE 1.7
The normal oxidation numbers of the 15 abundant elements.

1.4   **The Atomic Nucleus**

Although its atoms' pattern of electrons determines most of an element's *chemical* properties, many *physical* properties, including the element's weight, are determined not by the electrons but by the massive nucleus around which the almost weightless electrons move. Atomic nuclei are themselves composed of two principal kinds of particles: protons and neutrons. Protons carry a positive electrical charge; an element always has the same number of protons as electrons, for the positively charged protons in the nucleus serve to balance the negative electrical charges of the surrounding electrons. Thus the atomic number of an element corresponds not only to the number of electrons in that element, but also the number of nuclear protons.

Neutrons, in contrast, have no electrical charge and may be thought of as being composed of both a proton and an electron that balance each other in charge, but are grossly unequal in weight, the much larger proton contributing almost all the mass of the neutron. The important point about neutrons is that in a random sample of atoms of a single element, the individual atoms may have *differing numbers of neutrons* in their nuclei, whereas each atom will have the same *fixed number of nuclear protons and surrounding electrons* (Figure 1.8. Because an element may have atoms with differing numbers of (heavy) neutrons, each element occurs in differing weight configurations called nuclides or isotopes. Nuclides of a single element differ in number of neutrons and in weight, but not in ordinary chemical properties.

In principle, any element might have any number of neutrons in its nucleus, thus leading to an infinite number of nuclides. In fact, however, most nuclides are *unstable*; that is, the nuclear forces of attraction are insufficient to hold the neutrons together. In these circumstances the neutrons "decay" by breaking into separate electrons, protons, or other particles; this decay process is called radioactivity. Many thousands of nuclides are known, but only about 270 are stable, an average of less than three stable nuclides per element. All other nuclides are radioactive and tend to decay at rates that are constant for a particular nuclide but vary greatly between nuclides. For most of these nuclides the decay rate is so rapid that they do not occur naturally on earth (any that were present early in earth history having long since decayed), but must be created artificially in nuclear reactors and particle accelerators.

Because almost all the weight of an atom comes from its neutrons and protons, nuclides are distinguished by the total number of *both* types of particles in the nucleus. This total is known as the mass number of the nuclide. All of the important naturally occurring nuclides and their mass numbers are shown in Figure 1.9. Most elements occur

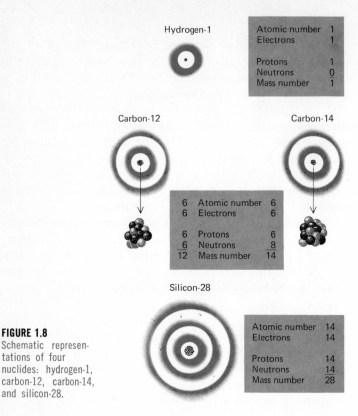

FIGURE 1.8
Schematic represen-
tations of four
nuclides: hydrogen-1,
carbon-12, carbon-14,
and silicon-28.

in nature not as pure nuclides, but as relatively uniform mixtures containing all the natural nuclides of that element. The proportions of
each nuclide in these mixtures are also shown in Figure 1.9.

1.5   Gases, Liquids, and Solids

Let us now turn from the structure of atoms and ions to a more general consideration of the three *states* or *phases*—gaseous, liquid, and
solid—in which all matter can occur. Most elements and compounds
can exist in each of the three phases, depending on the physical conditions, particularly temperature and pressure, to which they are subjected. At room temperature and normal atmospheric pressure, 90 of
the 103 elements are solids, 11 are gases, and only 2 (bromine and
mercury) are liquids. With changing temperature and pressure, however, the elements, and compounds made up of them, may pass from
one state to another. The most familiar example is water: at normal
atmospheric pressure it is a solid ice below 0°C(32°F), a liquid between 0 and 100°C (32 and 212°F), and a gas steam above 100°C

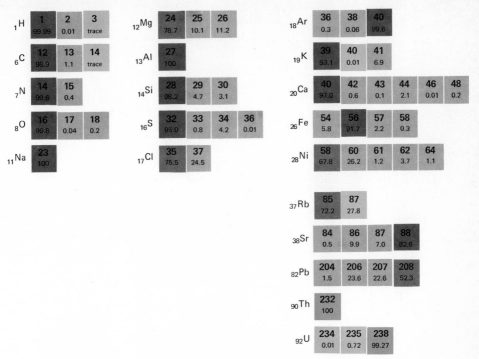

FIGURE 1.9
Natural nuclides of the 15 common elements and of 5 additional elements (lower right) important in earth chronology (see Chapter 4). Radioactive nuclides are shown in color. The lower numbers of each box show the percentage abundance of each nuclide in natural mixtures of the element; darker shading indicates dominant nuclides. (Data from Wedepohl, "Handbook of Geochemistry," vol. 1, 1969.)

(212°F). We are all aware of these temperature effects on water, but less familiar are the effects of changes in pressure. At high pressures water remains liquid at temperatures as great as 371°C (700°F). The combined effects of both temperature and pressure on water are shown in Figure 1.10, an example of a phase diagram.

At the atomic level, the principal differences among the three states of matter relate to the *degree of ordering* of the constituent atoms or ions. In typical solids, the atoms or ions have a very regular arrangement; most commonly they are joined in rigid geometrical frameworks (Figure 1.11). Such complete frameworks seldom exist in liquids, although the atoms are still held in close associations. In gases the atoms (or molecules, in the case of gaseous compounds) are completely separated from each other and move about freely. Many processes important to an understanding of the earth are concerned with transformations among these three states. Liquid-gas transitions are particularly crucial in the interaction of the ocean and atmosphere, whereas liquid-solid transitions, and the resulting solid state, are of primary importance in the minerals that make up the solid earth.

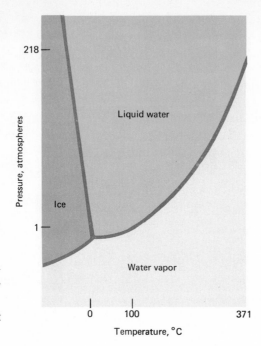

FIGURE 1.10
Phase diagram for water, showing the relation of the gaseous, liquid, and solid phases to pressure and temperature (not to scale).

PROPERTIES OF MINERALS

1.6 Mineral Structure

FIGURE 1.11
Solid, liquid, and gaseous states: a schematic representation of the degree of atomic ordering in the three states of matter.

Naturally occurring inorganic solids of fixed chemical composition are called minerals. There are, however, some materials that superficially seem to be minerals, such as glass and opal, that in a precise sense do not qualify because they lack the internal geometric regularity required of matter to be classified as solid. Glass, for example, has the slightly disordered structure of a liquid, and may best be thought of as a supercooled liquid rather than a true solid (Figure 1.11). Matter

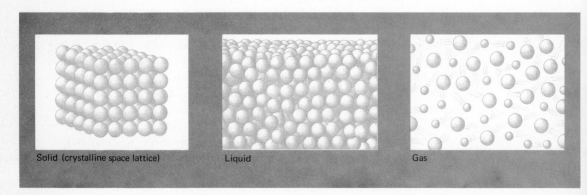

Solid (crystalline space lattice)   Liquid   Gas

having internal orderliness is said to be **crystalline,** and all minerals are by definition crystalline. An individual crystal is characterized by an extensive three-dimensional pattern of repeated atoms or ions called a **crystalline space lattice** (Figure 1.11). Most solid substances found in nature are aggregates of many such individual crystals. Inasmuch as we cannot directly look into an individual crystal and see its crystalline space lattice, you might legitimately ask, "How is it known that such regularity exists?" The evidence for internal orderliness is largely circumstantial but compelling. It derives from four major properties of minerals:

1. Crystal Form   Crystals are characteristically bounded by flat surfaces, called *crystal faces*, which are large-scale reflections of the internal arrangement of the atoms in the crystal. Long before the principles of atomic structure were understood, study of the arrangement of faces in natural crystals showed that there are six basic groups, each

Isometric   Three equal-length axes at right angles to each other

Garnet   Magnetite   Halite, Pyrite

Tetragonal   Two equal axes and a third either longer or shorter, all at right angles

Zircon

Orthorhombic   Three unequal axes, all at right angles

Olivine   Aragonite   Anhydrite   Goethite

Monoclinic   Three unequal axes, two at right angles and a third perpendicular to one but oblique to the other

Pyroxene, Amphibole   Mica, Clay minerals   Orthoclase   Gypsum

Triclinic   Three unequal axes meeting at oblique angles

Plagioclase   Kyanite

Hexagonal   Three equal axes in the same plane intersecting at 60° and a fourth perpendicular to the plane of the other three

Quartz   Calcite, Dolomite   Hematite

with a characteristic symmetry of the faces (Figure 1.12). These six groups of natural crystals, each of which can be divided into additional subgroups, provided the principal clues to the structure of solid matter. As early as 1669 Steno recognized that even though the arrangement of faces sometimes varies for a given mineral, *the angles between corresponding faces of a mineral always remain constant*. This statement has become known as the *law of constancy of interfacial angles* (Figure 1.13). Such external regularity could almost certainly not exist except as an effect of an internally regular crystalline space lattice.

2. Cleavage Many crystals fracture in a manner that yields smooth planar surfaces which develop only at fixed orientations with respect to their external form (Figure 1.14). It has long been concluded that there must be some internal control for this remarkably reproducible phenomenon. Our model suggests that there are internal planes within many crystals across which the bonding forces are weaker than across others. It is only natural, therefore, that these weaker planes yield to fracturing more readily than others, thus producing the phenomenon of cleavage.

3. Polarization and Double Refraction Ordinary light may be thought of as vibrating in many different planes, as diagrammatically indicated for the unpolarized light in Figure 1.15 before it enters the calcite crystal. The unpolarized light undergoes a very remarkable change as it passes through the crystal. Instead of one unpolarized light ray emerging from the other side, two separated rays are seen, each having been caused to vibrate in only one plane (hence plane-polarized). Furthermore, the two planes of polarization are exactly 90 degrees apart from one another. Many minerals affect light in this

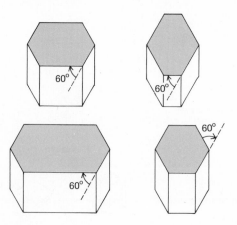

FIGURE 1.13
Law of constancy of interfacial angles, first noted by Steno in 1669.

FIGURE 1.14
Cleavage of mica.
(Walter R. Fleischer.)

manner, and with calcite the amount of this double refraction is so great that instruments are not necessary for its detection (Figure 1.16). This same experiment meets with failure when glass, or some other material lacking an orderly crystalline space lattice, is used.

4. X-ray Patterns The most recently discovered evidence of internal regularity in crystals has also proved to be the most useful in constructing precise models of their three-dimensional framework. In 1912 it was discovered that crystals alter X-ray beams passing through them in such a way that it becomes possible to determine precisely the internal arrangement of the atoms or ions (Figure 1.17). This X-ray technique quickly supplanted older methods of measuring crystal faces and became the principal tool for studying the structure of crystalline solids. X-ray methods now also serve as a quick and reliable means of

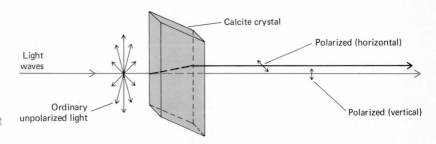

FIGURE 1.15
Polarization of light
by a mineral. See text
for discussion.

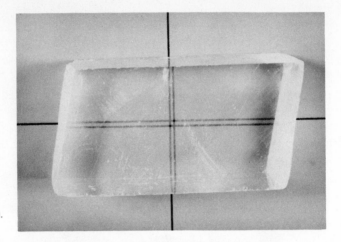

FIGURE 1.16
Calcite rhomb showing double refraction. (D. Ester.)

identifying even the smallest grains of a crystalline substance, for each element or compound has a distinctive pattern of X-ray transmission.

X-ray studies of atomic structure, combined with other lines of evidence, have shown that the arrangement of atoms and ions in crystals is largely determined by two properties: the *size* and *oxidation number* of the constituent atoms. In order to visualize how these properties affect crystal structure, it is helpful to think of individual atoms or ions as small spheres which become packed together to form crystals. These spheres tend to arrange themselves so that positive and negative oxidation numbers are balanced and so that there is a minimum distance between the spheres. When the positive and negative spheres differ greatly in size, a triangular packing arrangement is favored in which each element bonds with three adjacent atoms of opposite charge, giving rise to what is known as *threefold* coordination (Figure 1.18). As the difference in size between positive and negative elements decreases, *fourfold*, *sixfold*, and *eightfold* coordinations are favored. Finally, when the size of the positive and negative spheres are equal, a *twelvefold* coordination with a maximum density of closely packed spheres occurs. Figure 1.19 summarizes the normal oxidation numbers and comparative sizes of the ions that are common in rock-forming minerals. Sizes of *ions*, rather than *atoms*, have been shown in Figure 1.19 because the majority of common minerals may be most conven-

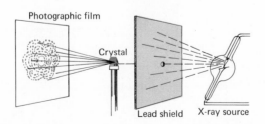

Photographic film

Crystal

Lead shield   X-ray source

FIGURE 1.17
X-ray determination of crystal structure. The X-ray pattern shown indicates a hexagonal arrangement of the internal ions.

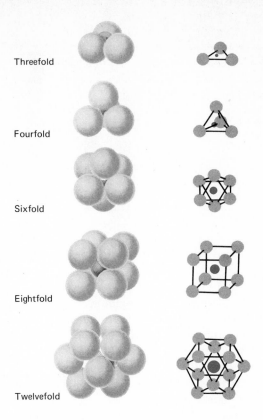

Threefold

Fourfold

Sixfold

Eightfold

Twelvefold

FIGURE 1.18
Five types of ionic coordination. As the positive and negative (color and gray) ions become more similar in size, additional negative (gray) ions can be packed around a single positive (color) ion.

iently treated "as if" they were the products of ionic bonding. This is not precisely true, but whatever the truth is, it is too complex for us easily to model; this simpler model nicely serves our present purposes.

1.7   Mineral Composition

As noted earlier, the relative abundances of the 103 chemical elements are such that only 10 elements make up most of the earth's minerals. This composition has been established by chemical analyses of the

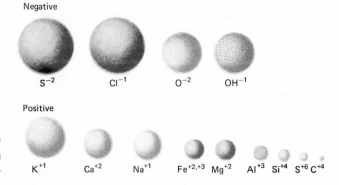

Negative

S^{-2} Cl^{-1} O^{-2} OH^{-1}

FIGURE 1.19
Oxidation numbers and relative sizes of ions found in common rock-forming minerals.

Positive

K^{+1} Ca^{+2} Na^{+1} $Fe^{+2,+3}$ Mg^{+2} Al^{+3} Si^{+4} S^{+6} C^{+4}

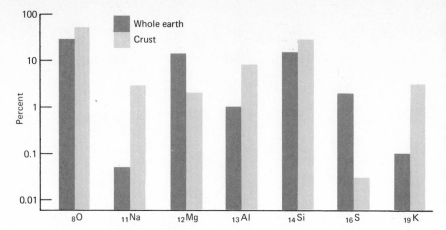

FIGURE 1.20
Abundances of elements (weight percent) in the whole earth and in the surficial crust. (Data from Mason, "Principles of Geochemistry," 1966.)

minerals exposed at the earth's surface in combination with less direct evidence about the materials of the earth's vast interior. The darker bars of Figure 1.20 present a modern estimate of the relative abundances of the elements most common in the entire solid earth. Note that only 8 elements constitute over 99 percent of the total earth; all the other 80 naturally occurring elements combine to make up less than 1 percent. Of these 8 elements, only 4—iron, oxygen, silicon, and magnesium—constitute over 90 percent of the solid earth.

Later we shall see that there is good evidence that the rocks exposed on the earth's surface are part of a thin "skin," called the **crust**, that formed at some earlier stage of the earth's development by a separation of lighter and more volatile elements from heavier materials of the interior. Because of this separation, the distribution of elements in the crust differs from that in the earth's deeper interior. (The crust makes up less than 1 percent of the earth's total mass, and thus the abundances of elements in the entire solid earth primarily reflect the composition of the interior, where heavy elements, particularly iron and nickel, dominate.) The most abundant elements in the minerals of the earth's crust are shown in the lighter bars of Figure 1.20. Once again, 8 elements constitute the bulk of the whole (in this case, 98 percent of the crust). Note that oxygen, sodium, aluminum, silicon, potassium, and calcium are enriched in the crust, whereas magnesium and sulfur, along with iron and nickel, are more abundant in the interior.

It is now possible to apply some of the chemical principles discussed earlier to predict the *combinations* of elements present in the minerals of the solid earth. Refer again to Figure 1.19, which shows the normal oxidation numbers and ionic sizes of the principal elements. The only really abundant element with a *negative* oxidation number is oxygen. This means that in most compounds making up the solid earth, oxygen supplies the negative charge and the other elements supply the bal-

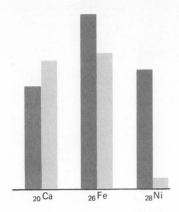

ancing positive charges. Note also the very large size of the oxygen ion relative to all the common positive ions (except potassium). This leads us to predict that minerals are, *by volume*, mostly composed of large spheres of ionic oxygen interspersed with smaller, positively charged ions of other elements. The character of this oxygen-positive ion interaction provides a useful means of classifying many of the 2,000 minerals that have so far been discovered in the rocks of the solid earth (Figure 1.21).

In the simplest case, oxygen combines directly with various positive ions to give **oxide** minerals such as *hematite* (iron oxide, Fe_2O_3). More commonly, the positive ions combine not with oxygen alone, but with combinations of oxygen and additional positively charged atoms that function together as a unit in carrying the negative charge of the mineral. These combinations, called **ion groups,** have an overbalance of negative charges and thus form compounds just as if they were single, negatively charged elements. Three such atom groups are of great importance in minerals: the **silicate group,** in which oxygen combines in various proportions with silicon; the **sulfate group,** in which four oxygen ions combine with one sulfur ion; and the **carbonate group,** in which three oxygen ions combine with one carbon.*

Sulfur is one of the few common elements that can have either a positive or negative oxidation number (see Figures 1.7 and 1.19). When it combines with oxygen in the sulfate atom group its oxidation number is positive. With a negative oxidation number it provides the sole negative charge in **sulfide** minerals.

The relatively rare element chlorine occurs concentrated as the sole

---

*Carbon is not one of the most abundant elements of the solid earth, although it makes up a small but significant fraction of the atmosphere. Even though it composes much less than 1 percent of the crust, carbon is locally concentrated into accumulations of carbonate minerals that are of great importance in earth chemistry.

MINERAL GROUP		MINERAL	GENERALIZED CHEMICAL COMPOSITION						Negative ions Group of ions
			Positive ions						
			Sodium	Magnesium	Aluminum	Potassium	Calcium	Iron	
Silicates	Isolated tetrahedra	Olivine		Mg				Fe	SiO_4^{-4}
		Aluminum silicates			Al				
		Garnets		Mg	Al		Ca	Fe	
	Chain	Pyroxenes	Na	Mg	Al		Ca	Fe	SiO_4^{-4}
		Amphiboles	Na	Mg	Al		Ca	Fe	$(OH)^{-1}$
	Sheet	Micas		Mg	Al	K		Fe	SiO_4^{-4}
		Clay minerals			Al	K			$(OH)^{-1}$
	Framework	Quartz	[Si only]						SiO_4^{-4}
		Orthoclase feldspar			Al	K			
		Plagioclase feldspar	Na		Al		Ca		
Carbonates		Calcite					Ca		
		Aragonite					Ca		$(CO_3)^{-2}$
		Dolomite		Mg			Ca		
Sulfates		Gypsum					Ca		
		Anhydrite					Ca		$(SO_4)^{-2}$
Oxides		Hematite						Fe	
		Goethite						Fe	
		Magnetite						Fe	O^{-2}, $(OH)^{-1}$
		Aluminum oxides			Al				
Halide		Halite	Na						Cl^{-1}
Sulfide		Pyrite						Fe	S^{-2}

FIGURE 1.21
Classification and generalized chemical composition of the principal rock-forming minerals.

negative ion in the mineral *halite* (sodium chloride, NaCl). Such minerals in which chlorine or other closely related elements provide the only negative charge are called halides.

These six principal kinds of negative chemical charges define six principal groups of rock-forming minerals: silicates, carbonates, sulfates, oxides, halides, and sulfides (Figure 1.21). A seventh mineral group can be added for those few elements—primarily carbon, iron, nickel, and sulfur—that may occur not only as compounds, but also as free, uncombined elements. Such relatively rare minerals are called native elements. These seven mineral groups contain most of the 2,000

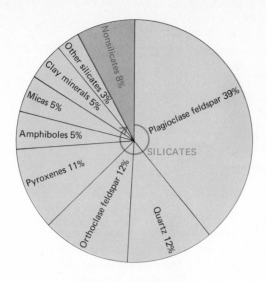

FIGURE 1.22
Percentage (by volume) of common minerals in the earth's crust. (Data from Ronov and Yoroshevsky, "American Geophysical Union Monograph 13," 1969.)

known minerals, of which only about 20 are really common in the solid earth (see Figure 1.22); the others are very rare and restricted in occurrence. These common rock-forming minerals are those that we shall consider here and be referring to in later chapters.

MINERAL DIVERSITY

1.8 Silicate Minerals

Of the principal mineral groups, the silicates are by far the most important because they make up about 92 percent of the earth's crust (Figure 1.22). In addition to being an extremely abundant element, silicon also has unique chemical properties that permit it to link with oxygen to form large structural networks of atoms. Silicon has an oxidation number of $+4$ and a very small ionic size (Figure 1.19); a silicon ion is always surrounded in crystal structures by four much larger oxygen ions in a fourfold coordination that resembles a geometric tetrahedron (Figure 1.23). This structural unit, the silicon-oxygen tetrahedron, is the *fundamental building block* of all silicate minerals. We shall continue our discussion of this important mineral group by treating these tetrahedra as large ionic units possessing a negative charge of -4 each. Inasmuch as minerals are without charge, we must consider the various ways these negatively charged tetrahedra can be bonded together to yield electrically neutral crystalline space lattices.

In the simplest case, these silicon-oxygen tetrahedra occur as isolated units, bound together only by positive metallic ions. More commonly, silicon ions share a linkage to one or more oxygen ions so that

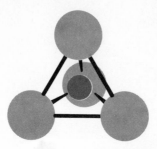

FIGURE 1.23
The silicon–oxygen tetrahedron.

more complex structures—chains, sheets, or completely interlocking frameworks—are formed; these are illustrated in Figure 1.24. These structural differences define four principal groups of silicate minerals: *silicates with isolated tetrahedra; chain silicates; sheet silicates;* and *framework silicates.** The minerals within each of these groups may be further classified by the positive ions that combine with the negative silicon-oxygen tetrahedral groups. These are principally ions of sodium, magnesium, aluminum, potassium, calcium, and iron, the dominant positive elements, other than silicon, found in the earth's crust.

*Several minor groups of silicates, whose minerals are relatively uncommon, have been omitted.

(a)

— 4 charge remains

— 2 charge remains

(b)

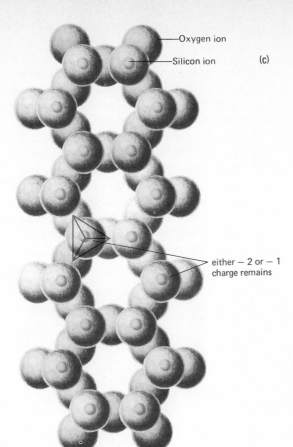

Oxygen ion

Silicon ion

(c)

either − 2 or − 1 charge remains

FIGURE 1.24
Linking patterns of silicon—oxygen tetrahedra in silicate minerals. (a) Isolated tetrahedra; no sharing of O. (b) Single chain: each tetrahedron shares two O's. (c) Double chain; adjacent tetrahedra share two or three O's (alternating). (d) Sheet; each tetrahedron shares three O's. (e) Framework; each tetrahedron shares all four O's. A single tetrahedral unit is outlined in dark color on each pattern.

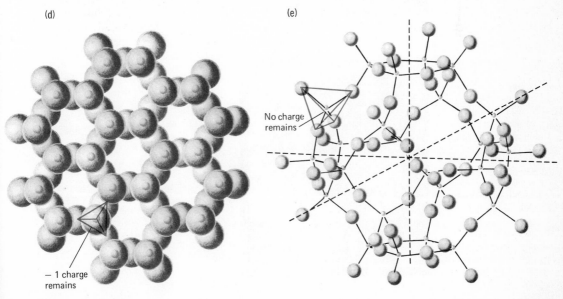

(d)

− 1 charge remains

(e)

No charge remains

Silicates with isolated tetrahedra   In this group the ratio of silicon to oxygen atoms is 1:4 (SiO_4) because none of the oxygen atoms is shared among adjacent silicon atoms (Figure 1.24). The basic oxidation number of the silicon-oxygen group is -4 since the oxidation number of the silicon is $+4$ and that of each of the four oxygen ions is -2. This large excess negative charge serves to bind a high proportion of other positive atoms into the structure so that the charges are balanced. Three important mineral subgroups have this structure: the olivines, in which magnesium and iron provide the principal additional positive ions; the aluminum silicates, in which aluminum is the principal positive ion; and the garnets, which have various complex combinations of positive ions, principally magnesium, aluminum, calcium, and iron (Figure 1.25).

(a)

(b)

(c)

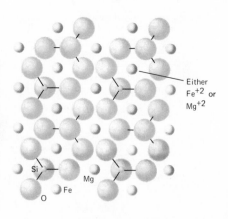

(d)

FIGURE 1.25
Isolated tetrahedra silicate minerals. (a) Olivine. (b) Aluminum silicate. (c) Garnet. (d) Diagrammatic structure of olivine. Isomorphism: Fe^{+2} and Mg^{+2} ions can substitute for each other within this space lattice in any proportion because they have the same charge and nearly the same size. (a-c: Yale Peabody Museum.)

Either Fe^{+2} or Mg^{+2}

Si

Mg

Fe

O

Chain silicates   In this structure oxygen ions are shared between silicon ions in such a way that long silicon-oxygen chains of two kinds are formed. In one, each silicon ion shares two of its four oxygen ions with adjacent silicon ions; this is the *single-chain structure*. The principal minerals with this structure are the pyroxenes (Figure 1.26), important rock-forming minerals that may contain all the common positive atoms except potassium, which is prevented from entering the chain structure by its large ionic size. The other group of chain silicates has a *double-chain* structure in which adjacent silicon ions alternately share two and then three of their oxygen ions. The principal double-chain silicates are the amphiboles (Figure 1.26). Like the closely related pyroxenes, these may contain various combinations of all the common positive atoms except potassium. Because of their struc-

(a)

(b)

(c)

FIGURE 1.26
Chain silicate minerals. (a) Pyroxene. (b) Amphibole. (c) Diagrammatic structure of a calcium- and magnesium-bearing pyroxene. (a, b: Yale Peabody Museum.)

ture, amphiboles and pyroxenes tend to crystallize as elongate minerals whose external shape reflects their internal atomic arrangement.

Sheet silicates   In this structure each silicon ion shares three of its oxygen ions to form a sheetlike, two-dimensional network. Because of this arrangement, minerals with this structure tend to split into thin parallel sheets. The micas, one of two important groups of sheet silicates, are familiar examples (Figure 1.27). Micas all contain aluminum ions, some of which replace some silicon ions in the structural tetrahedron. They are also characterized by potassium ions whose large size permits them to fit *between* the sheetlike layers of silicon-aluminum-oxygen tetrahedra (Figure 1.27c. In addition, some micas contain iron and magnesium. The other important group of sheet silicates, the clay minerals, are similar, but differ in structural detail. Clay minerals never occur as large crystals but are always found in earthy masses of extremely tiny crystalline particles—particles that are usually so

(a)

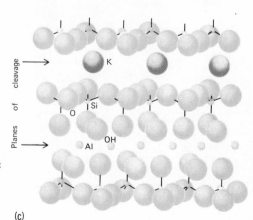

(b)

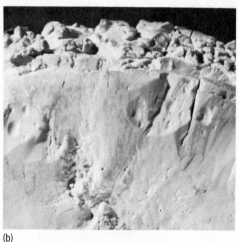

(c)

FIGURE 1.27

Sheet silicate minerals. (a) Mica. (b) Clay mineral [earthy mass about 2 cm (¾ in.) across; the extremely tiny individual crystals are not visible]. (c) Diagrammatic structure of a mica. The large potassium atoms fit between the structural sheets of silicon, oxygen, and aluminum. Note that cleavage occurs only between the sheets of silicon–oxygen tetrahedra.

small that they can be seen only with an electron microscope. Both the clay materials and micas have water-derived oxygen-hydrogen (OH) ionic groups in their structures in addition to the principal oxygen-silicon groups.

Framework silicates   In this last group every silicon ion shares each of its four oxygen ions with adjacent silicon ions to give a three-dimensional network with a ratio of one silicon for every two oxygens (SiO_2). They are thus the most silicon-rich of the silicate minerals. Note that the basic charge of the SiO_2 is neutral, since silicon's oxidation number of $+4$ is balanced by the two oxygen ions, each of which has an oxidation number of -2. Framework silicates have the most stable structural arrangement and are by far the most abundant of all the minerals in the earth's crust. The two principal types are quartz, which is composed *only* of silicon and oxygen, and the feldspars, a group of minerals in which sodium, aluminum, potassium, and calcium occur in the framework in varying combinations and proportions (Figure 1.28). The bulk of the earth's crust is composed of feldspar (51 percent) and quartz (12 percent) (Figure 1.22).

Because the basic electrical charge on tetrahedra in framework sili-

(a)

(b)

FIGURE 1.28
Framework silicate minerals. (a) Quartz. (b) Orthoclase feldspar. (c) Diagrammatic structure of orthoclase feldspar. The unbalanced charges caused by the substitution of aluminum for some silicon ions serve to hold potassium in the structure. (a, b: Yale Peabody Museum.)

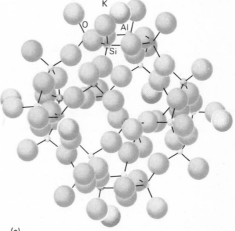

(c)

cates is neutral, we need to explain how additional positive ions—such as sodium, aluminum, potassium, and calcium—can occur in the structure of the feldspars. The reason for this is related to the presence of aluminum, which is unique among all the common elements in being able to substitute for silicon in some of the silicon-oxygen tetrahedra. This is possible because the small size of the aluminum ion is close to that of silicon (see Figure 1.19). Aluminum, however, has an oxidation number of $+3$, whereas silicon's is $+4$; thus when an aluminum ion substitutes for a silicon ion, it leaves an unbalanced negative charge which serves to bind other positive ions into the structure (Figure 1.28d). Two groups of feldspars are defined by these extra ions: in orthoclase feldspars potassium is the dominant additional ion; in plagioclase feldspars varying proportions of sodium and calcium dominate. The presence of these ions explains why the feldspars have cleavage while quartz, a pure SiO_2 framework silicate, does not: planes of weakness occur where the metallic ions are located.

A summary of the compositions of all the major silicate minerals is given in Figure 1.21. Note again the predominance of only a few elements in the ionic structure of these common minerals; these elements are the principal constituents of the solid earth.

Perhaps the most important aspect of our study of different silicate minerals is the recognition that physical properties of a mineral are manifestations of its crystalline space lattice structure and its chemical composition. Consider, for example, the properties of *color* and *density* for the olivine mineral group. As shown diagrammatically in Figure 1.25d, olivine has a structure of isolated silicon-oxygen tetrahedra ionically bonded together by $+2$ ions of either iron or magnesium. Because iron and magnesium ions both have a $+2$ change and approximately the same size (see Figure 1.19), they may substitute for each other in any proportion within the lattice structure of olivine. Olivine is said to be an isomorphic (*iso* = equal; *morphic* = form) mineral because of this. Olivine may be a pure iron silicate (Fe_2SiO_4), or a pure magnesium silicate (Mg_2SiO_4), or any mixture of iron and magnesium in between—but always with the same lattice structure and hence isomorphic. What differences would you predict between the iron-rich olivine and the magnesium-rich variety? Remembering that the *only* difference between them is compositional, we might expect iron ions with a mass number of 56 to impart a greater density than mass-number-24 magnesium ions (see Figure 1.9). Indeed such is the case: Fe_2SiO_4 weighs 4.4 grams per cubic centimeter, and Mg_2SiO_4 only 3.3 grams. What about color? Here again the presence of iron seems to be important. The iron-rich variety is black while the magnesium-rich olivine is a pale green. A similar contrast exists in the mica group of silicates: biotite (iron-bearing mica) is black, while muscovite, devoid of iron, is colorless.

Cleavage is another physical property that can be easily related to a mineral's space lattice pattern. The one perfect direction of cleavage characterizing the micas, for example, results from the weak bonding between tough sheets of silicon-oxygen tetrahedra (see Figures 1.14 and 1.27c). The ionic bonding between adjacent sheets is not particularly strong because the sheets themselves do not possess a great deal of negative charge to be satisfied—only −1 per tetrahedron.

Both types of chain silicates provide another instructive comparison. They display two directions of cleavage parallel to their chains of tetrahedra, because the weakest parts of their structure are where positive ions are bonding the chains together. But pyroxenes and amphiboles differ in the angles between their two cleavage directions. Figure 1.29 provides an explanation for this quantitative difference.

These examples should serve to clarify an important generalization: the physical properties displayed by a mineral are reflections of its chemical composition (atoms or ions present) and its pattern of packing (crystalline space lattice) at the atomic level.

1.9 Nonsilicate Minerals

Although only 8 percent of the earth's crust is composed of minerals that lack the characteristic silicon-oxygen tetrahedra of the silicates,

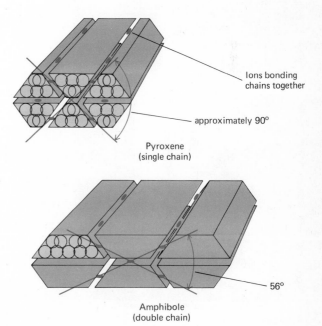

FIGURE 1.29
Cleavage differences for chain silicates. Weakest planes are between chains of tetrahedra, therefore wider chains (amphibole) have different cleavage angles from narrower chains (pyroxene).

several of these nonsilicate minerals are common enough in certain geologic settings to require consideration.

Carbonates and sulfates   In carbonate and sulfate minerals the negative charges are provided by ionic groups of oxygen combined with either carbon or sulfur, rather than with silicon as in the silicates. Carbon, like silicon, has a normal oxidation number of +4; unlike silicon, however, it combines with oxygen in only one arrangement in forming carbonate minerals. Because of the extremely small size of the carbon, it has only three attached oxygen ions arranged in a triangular, threefold coordination (see Figures 1.18 and 1.30).

There are three common carbonate minerals. Two of these, calcite and aragonite, have the same chemical composition, in which the only positive atom is calcium ($CaCO_3$). The two minerals differ only in

(a)

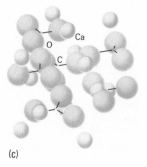

(c)

(b)

FIGURE 1.30

Carbonate minerals. (a) Calcite. (b) Dolomite. (c) Diagrammatic structure of aragonite. Each carbon atom is linked to three oxygen atoms. (a, b: Yale Peabody Museum.)

crystal structure (minerals which differ only in structure but not in chemical composition are termed *polymorphs* of each other). In calcite the ions are arranged with hexagonal symmetry; in aragonite the ions have orthorhombic symmetry (see Figures 1.12 and 1.30). The other important carbonate mineral, dolomite [CaMg (CO$_3$)$_2$], is similar to calcite except that its crystal structure contains varying amounts of magnesium in addition to calcium.

In sulfate minerals, four oxygen ions combine with one positive sulfur ion to form tetrahedral atomic groups with an overall oxidation number of −2 (Figure 1.31). There are only two common sulfate minerals, and both have calcium as the only additional positive ion: gypsum (CaSO$_4$·2H$_2$O) has water bound into the crystal structure, while anhydrite (CaSO$_4$) lacks water and has a different structural arrangement.

Oxides   A few minerals have oxygen *alone* as the negative ion, rather than oxygen combined into ion groups with silicon, carbon, or sulfur. The most important such minerals are the iron oxides. Iron is one of several elements that may exist with either of two different oxidation numbers, depending on how the electrons in the outer shell combine with other elements. For iron these oxidation numbers are +2 and +3; because of them, iron combines directly with oxygen in a variety of arrangements. In one the iron has an oxidation number of +3 and forms either the mineral hematite (Fe$_2$O$_3$) or the mineral goethite [FeO(OH)], which also contains an oxygen-hydrogen ionic group (from water) in its crystal structure (Figure 1.32). In addition, iron of both oxidation numbers mixes and combines with oxygen in the ratio of Fe$_3$O$_4$ to make the important iron oxide mineral magnetite.

FIGURE 1.31
Sulfate minerals. (a) Gypsum. (b) Diagrammatic structure of anhydrite. Each sulfur atom is linked to four oxygen atoms. (a: Yale Peabody Museum.)

(a)

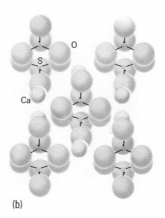

(b)

A second important group of oxide minerals contains aluminum as the only positive element. Like the silicon-bearing clay minerals, these aluminum oxides occur mostly as earthly masses in which the tiny individual crystals can be seen only under an electron microscope (Figure 1.32).

Halides and sulfides   Two other common minerals that need to be considered are the only ones made up of compounds which lack oxygen in their structure. In halite (NaCl), the negative charge is carried by the relatively rare element chlorine, with sodium as the positive ion (Figure 1.33). The other important oxygen-free mineral is pyrite

(a)

(b)

FIGURE 1.32
Oxide minerals. (a) Hematite. (b) Goethite. (c) Diagrammatic structure of magnetite. (a, b: Yale Peabody Museum.)

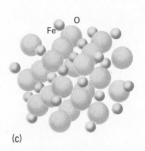

(c)

(FeS₂), in which sulfur supplies the negative charge and iron the positive (Figure 1.33). Although pyrite (iron disulfide) is the only *common* mineral in the group, many *rare* sulfide minerals are of great economic importance as ore minerals for lead, copper, and zinc.

Native elements   The final group of nonsilicate minerals comprises the elements that are *not* combined into compounds but occur instead as uncombined "native" elements (Figure 1.34). Such elements are rare in the earth's crust, although some—particularly gold, silver, copper, sulfur, and carbon (as either *graphite* or the high-pressure polymorph *diamond*)—are found in local concentrations that are of great eco-

(a)

(b)

FIGURE 1.33
Halide and sulfide minerals. (a) Halite. (b) Pyrite. (c) Structure model of halite. (a, b: Yale Peabody Museum.)

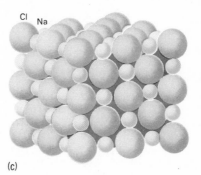

Cl   Na

(c)

(a)   (b)

FIGURE 1.34
Native element min-
erals. (a) Copper.
(b) Diamond (carbon).
(a: American Museum
of Natural History;
b: DeBeers Consoli-
dated Mines.)

nomic importance. Uncombined iron, although very rare in the crust, is believed to be present in large quantities deep within the interior of the earth, where it occurs mixed with nickel and other uncombined metallic elements.

SUMMARY OUTLINE

Matter

1.1 *Elements and compounds:* the bulk of the solid earth is composed of only 10 elements chemically combined in compounds.

1.2 *An atomic model:* an element is most stable when it has eight electrons in its outer shell (helium is an exception, having only two).

1.3 *Bonding to form compounds:* ionic bonding (electron transfer) and covalent bonding (electron sharing) are models widely used to explain the formation of most compounds.

1.4 *The atomic nucleus:* atomic nuclei are composed of protons and neutrons, the sum of which equals an element's atomic weight.

1.5 *Gases, liquids, and solids:* transformations among these three fundamental states of matter continuously occur among the elements and compounds making up the earth's rocks, water, and air.

Properties of Minerals

1.6 *Mineral structure:* minerals have internal orderliness, as indicated by their crystal form, cleavage, and their ability to polarize light and produce X-ray patterns.

1.7 *Mineral composition:* most of the earth's solid crust is composed of crystalline minerals dominated by oxygen-silicon groups; a small but significant proportion of crustal minerals lack silicon; only a very few common minerals lack oxygen.

Mineral Diversity

1.8 *Silicate minerals:* adjacent silicon-oxygen tetrahedra share oxygen ions in differing ratios to make four principal groups: framework, sheet, and chain silicates, and silicates with isolated tetrahedra; the physical properties of silicate minerals derive from these groups and their chemical composition.

1.9 *Nonsilicate minerals:* the most important mineral groups which lack silicon are carbonate, sulfate, and oxide minerals, all of which contain oxygen; and halide and sulfide minerals, which lack oxygen.

ADDITIONAL READINGS

*Ahrens, L. H.: *Distribution of the Elements in Our Planet*, McGraw-Hill Book Company, New York, 1965. A brief, intermediate survey of the distribution of the elements, stressing their occurrence in igneous rocks.

Desautels, P. E.: *The Mineral Kingdom*, Grosset & Dunlap, Inc., New York, 1968. A popular introduction to gems, minerals, and mineral collecting, with sumptuous illustrations.

*Ernst, W. G.: *Earth Materials*, Prentice-Hall, Inc., Englewood Cliffs, N.J., 1969. A comprehensive introduction to minerals and rocks.

Mason, B.: *Principles of Geochemistry*, John Wiley & Sons, Inc., New York, 1966. A clearly written and accessible introduction to all aspects of the chemistry of the earth.

Mason, B., and L. G. Berry: *Elements of Mineralogy*, W. H. Freeman and Co., San Francisco, 1968. An outstanding intermediate text on minerals.

*Turekian, K. K.: *Chemistry of the Earth*, Holt, Rinehart and Winston, Inc., New York, 1972. A brief but authoritative introduction.

Vanders, I., and P. F. Kerr: *Mineral Recognition*, John Wiley & Sons, Inc., New York, 1967. A guide to mineral identification, with many fine color photographs.

*Zim, H. S., and P. R. Shaffer: *Rocks and Minerals*, Golden Press, New York, 1957. An excellent elementary guide to mineral and rock identification.

*Available in paperback.

chapter **2**

Igneous and Metamorphic Processes

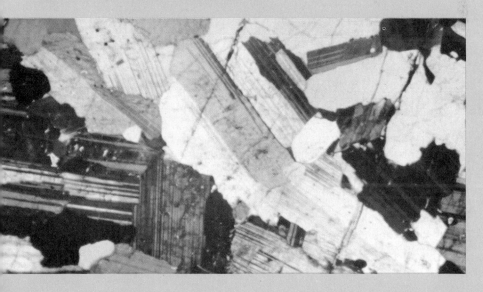

Just as the 10 most abundant elements of the earth's crust are combined to make only about 20 common minerals, so these minerals in turn are associated into a relatively few kinds of common rocks, the next larger structural units of the solid earth. In general, rocks are aggregates of several minerals; that is, rock-forming processes do not ordinarily completely segregate different minerals but, instead, lead to characteristic associations of minerals. Granite, for example, is a rock made up primarily of three minerals: mica, quartz, and feldspar. Under special circumstances *monomineralic rocks* (composed of only one mineral) may be formed. Ordinary limestone is such a rock; it is composed mostly of the mineral calcite.

There are three universally recognized classes of rocks, each of which has a different origin: igneous rocks, which form by crystallization from a hot molten mass called a magma; sedimentary rocks, which form on the earth's surface when fragmentary particles or dissolved substances from preexisting rocks are transported and deposited by water, ice, or wind; and metamorphic rocks, which are made up of either igneous or sedimentary rocks that have been profoundly modified, but not completely melted, by high temperatures and pressures.

We can directly observe only two of these three rock types in the making. Modern volcanic (igneous) and sedimentary processes can be seen at the earth's surface, but for those rocks requiring extreme conditions of temperature and pressure to form (metamorphic rocks), we can only reason by analogy from laboratory expeirments and field relationships. In this chapter we shall begin with observations of volcanic rocks and then consider how crustal materials of the earth respond when subjected to the elevated temperatures and pressures existing at depths of tens or even hundreds of kilometers beneath the surface. In Chapter 3 we shall return to a consideration of sedimentary rocks.

IGNEOUS ROCK DIVERSITY

2.1 Igneous Textures

When molten magmas erupt onto the earth's surface, two factors contribute to their rapid crystallization: (1) the temperature of the surface environment, which is dramatically lower (even on a hot day) than the temperature of the magma and the crustal interior from which it came; and (2) the reduced (atmospheric) pressure of the earth's surface, which allows any gases dissolved in the magma to effervesce, thus cooling it even more. Sometimes these cooling effects are so rapid that the silicon-oxygen tetrahedra and other ionic constituents of the magma are frozen into relatively rigid positions before they achieve orderly crystalline space lattice orientations. Volcanic glass such as obsidian (Figure 2.1) is formed when this occurs.

More frequently, however, a magma erupting onto the earth's surface forms crystalline assemblages primarily consisting of two or three silicate minerals. Most of the crystals are quite small and form a substance referred to as **groundmass**; but typically there are much larger crystals, called **phenocrysts** (*pheno* = apparent, *cryst* = crystals), scattered throughout the rock (Figure 2.2). We refer to this crystalline relationship as **porphyritic texture**. Like *all* rock **textures** (the relationships between the grains composing a rock), porphyritic textures provide valuable clues to the genesis (mode of origin) of the rock. As an example, consider the following interpretation for the porphyritic texture shown in Figure 2.3. Imagine a completely liquid magma, formed at a depth of several kilometers, beginning an assent toward the earth's surface (Figure 2.3, position 1). As this magma rises through somewhat cooler and lower-pressure environments, a few crystals, usually of only one or two mineral types, begin to grow. These are typically minerals of high melting points, so the bulk of the magma remains liquid throughout its movement toward eventual eruption (Figure 2.3, position 2). Immediately prior to the eruption of magma its condition is typically that depicted at position 2: a few large phenocrysts surrounded by liquid silicate material. Upon exposure to the earth's surface the liquid phase of this magma undergoes rapid crystallization. Because the silicate liquid is extremely viscous (sticky), the ions within it cannot move great distances, so they must quickly organize themselves into *many small crystals*, thus forming the fine-grained groundmass that typifies volcanic rocks (Figure 2.3, position 3).

FIGURE 2.1
Obsidian: volcanic glass resulting from eruption and quick cooling of magma. (Robert Navias.)

(a)

(b)

FIGURE 2.2
(a) Photo of volcanic eruption. (b) Porphyritic texture.
(a: R. S. Fiske, U. S. Geological Survey; b: D. Ester.)

This analysis argues that, since (1) the formation of porphyritic texture requires an *increased* rate of crystal growth at a time when most of the magma is still liquid; and since (2) modern lava flows are observed to have undergone such an increase upon eruption, and they typically possess porphyritic texture; therefore, (3) any rock having

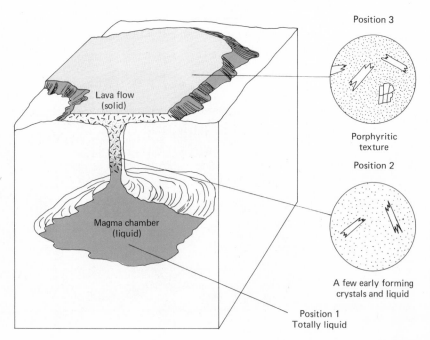

FIGURE 2.3
Formation of porphyritic texture. Changes in magma as it solidifies from a melt during the process of eruption.

porphyritic texture and a fine-grained groundmass probably formed from a magma on the earth's surface. Thus, porphyritic texture is genetically significant as an indicator of volcanic or extrusive igneous origin. Although some very old rocks may show no indication of volcanic forms—indeed, they may even be observed as stream-worn pebbles—if their texture is *fine-grained* and *porphyritic*, we will nevertheless conclude that their ultimate origin was an extrusive igneous process. Under some conditions the phenocrysts are too few or too small to be apparent, in which case the very small crystal size of the rock will still serve as evidence of rapid crystallization and hence of an extrusive origin.

Now comes the time for some geological detective work. Consider the rock material shown in Figure 2.4. This rock seems to have originated by crystallization of a silicate magma—but where? Surely not at the earth's surface! Such large crystals, all of approximately the same size, suggest two characteristics of crystallization that are distinctly different from what happens for volcanic rocks: (1) the large size indicates rather slow growth, and (2) the size equality suggests a more-or-less constant rate of growth. Because geologists have not discovered any place on earth where such coarse-grained silicate rocks having this granular texture (intergrowth of equal-sized crystals) are forming, we must conclude that their origin takes place in some region not accessible to direct observation. But where? Where might magma crystallize slowly and at a constant rate? What if magma that was

(a)

FIGURE 2.4
Plutonic rocks. (a) Batholithic terrane, Sierra Nevada. (b) Photomicrograph of granular texture. (a: E. A. Hay; b: D. Ester.)

(b)

formed at depth within the earth did not migrate to the surface, but simply moved into another deep, but somewhat cooler, environment? In such a setting it could slowly crystallize over a very long interval of time, eventually becoming solid rock. Surely its texture would be *coarse-grained* and *crystalline granular* (Figure 2.5a). Indeed, the logic of this analysis is so attractive that we have come to believe that this texture indicates intrusive igneous—or, as it is often called, "plutonic"—origin.

One important problem still surrounds our interpretation of plutonic rocks. If they originate deep within the earth's crust, why do we frequently observe them at the surface? If our inferences about their site of origin are correct, then their widespread exposure at the surface, particularly in mountainous regions, demands another inference of no less geologic significance: deep erosion must have stripped away huge volumes of rock from above them. *Whenever exposures of plutonic rocks are encountered, we infer conditions of crustal uplifting and subsequent deep erosion in the geologic past* (see Figure 2.5b and c).

2.2   Classification of Igneous Rocks

Igneous rocks fall naturally into two broad categories on the basis of their textures and the different modes of origin their textures suggest: glassy and fine-grained porphyritic textures indicate volcanic origin, and coarse-grained granular texture suggests plutonic origin. The description and classification of igneous rocks would thus be very simple, if their variability were limited to only these three textures. It turns out, however, that one additional variable of great importance must be taken into consideration: *mineral composition*. Some volcanic rocks for example, contain quartz, others do not; some are rich in dark silicate minerals, while others contain mostly light-colored feldspars. A host of other variations could be cited. How should we respond to this complexity that nature has forced upon us? Geologists have reacted in a manner consistent with the pattern of all scientific activity: they have developed systems of *classification* to subdivide the entire spectrum of igneous rocks into a relatively few categories. This is simply the process of imposing *order* on nature, and it is absolutely necessary before further understanding can proceed. Any system of classification must be concerned with *similarities* and *differences,* and all systems are characterized by some degree of compromise to determine which factors should have the greater emphasis. Consider, for example, attempts to classify the materials shown in Figure 2.6. We would accept a three-year-old's classification as highly satisfactory if he differentiated B, D, F, G, and H as rocks and A and I as flowers. For him to recognize the rocks as different from flowers, flies, and shoes and similar to each other seems far more important than noting the variations that exist between the different rocks. A professional geologist, on the other hand, would hardly settle for such an unsophisticated

FIGURE 2.5
Events to be inferred from the exposure of plutonic rocks. (a) Crystallization of plutonic rocks at depth. (b) Uplifting that causes erosion. (c) Erosion eventually exposes plutonic rock.

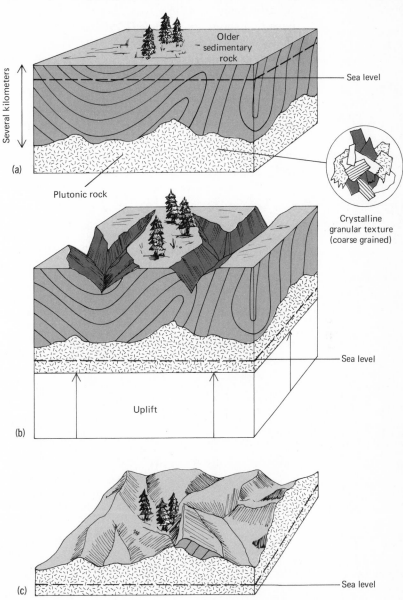

Older sedimentary rock

Several kilometers

Sea level

(a)

Plutonic rock

Crystalline granular texture (coarse grained)

Sea level

Uplift

(b)

Sea level

(c)

system of classification. He must concern himself with the subtleties of rock differences, so it is likely he would have many different names that could serve to distinguish one rock type from another. What should be the goal in making these distinctions? We might design a system of classification that separates large rocks from small rocks, or light-colored rocks from dark-colored ones, or pretty rocks from ugly rocks, or we might apply a host of other criteria. Should we arbitrarily select *any* rock characteristic, or are there logical guidelines that should be followed? We have already discussed the one rock characteristic, texture, that geologists have seized upon as possessing particular significance. Why? Recall the contrasting inferences made concerning igneous rocks of fine-grained porphyritic and coarse-grained granular texture. We concluded that they had solidified—that is, originated—in different environments. Texture was used, in other words, as a criterion for the genesis (mode of origin) of rocks. For this reason, all systems of rock classification begin with a consideration of texture. It serves to provide an important insight into how individual rocks originated. Returning to the specimens shown in Figure 2.6, for

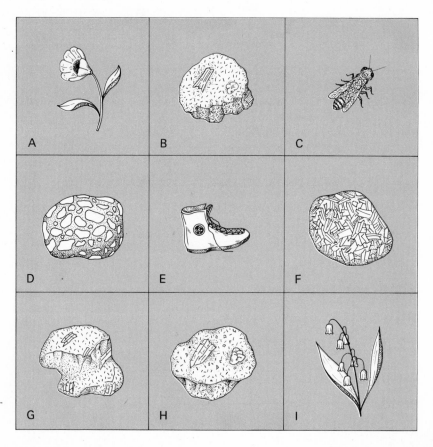

FIGURE 2.6
Objects for classification (see text for discussion).

example, using texture we would conclude that B, G, and H were volcanic, while F was plutonic in origin. This is only the first step. Now we must choose other criteria as a means of making further distinctions. For igneous rocks we have selected mineral content. Specimens G and H are both volcanic, for instance, so we must distinguish between them mineralogically in order to assign different names, such as "basalt," or "andesite," or a variety of others (see Figure 2.7).

For igneous rock classification, then, the properties of texture and mineral composition are the basis for assigning different names. The organization of these rocks is particularly important because igneous rocks are so abundant: they make up about two-thirds of the earth's crust. During our planet's long history most parts of the crust have at some time been heated to form a molten liquid and subsequently cooled to crystallize as igneous rocks (Figure 2.8). As might be predicted from our earlier discussions of the most abundant elements and minerals, the rocks produced by these meltings and coolings are composed almost entirely of a few common silicate minerals—feldspars, quartz, micas, amphiboles, pyroxenes, and olivine. Not all these minerals occur together in a single igneous rock, for when magmas crystallize only certain combinations of minerals are formed. These combinations are the basis for the classification of igneous rocks shown in Figure 2.7. Note that the varying compositions of igneous rocks form a gradational series: from silicon-rich granites, composed mostly of quartz and orthoclase feldspar, at one end of the scale, to magnesium- and iron-rich gabbros and ultramafic rocks, composed largely of pyroxenes and olivines, at the other end. (Recall that pyroxenes and olivines both have iron and magnesium as their dominant positive ions, whereas neither iron nor magnesium occur in quartz or feldspars.) The silicon-rich granites and granodiorites are commonly referred to as silicic igneous rocks, whereas the iron- and magnesium-rich gabbros and ultramafic rocks are called mafic igneous rocks.

Silicic and mafic igneous rocks differ not only in mineral composition, but also in certain gross physical characteristics, such as color and density. Silicic rocks are typically light in color—white, pink, or light gray as in the familiar granites used as building stone (Figure 2.9a). Mafic rocks, on the other hand, are usually very dark in color—dark gray, dark green, or even black—because of the predominance of dark-colored pyroxenes and olivines (Figure 2.9b). Mafic rocks are also denser than silicic rocks. This density difference has important implications for the distribution and origin of igneous rocks, because the light, silicic rocks, particularly granite, make up most of the earth's continents, whereas the ocean basins are primarily basalt (a dense igneous rock), and probably much of the earth's interior is composed of denser mafic rocks.

Figure 2.7 also indicates which silicate structures are dominant in

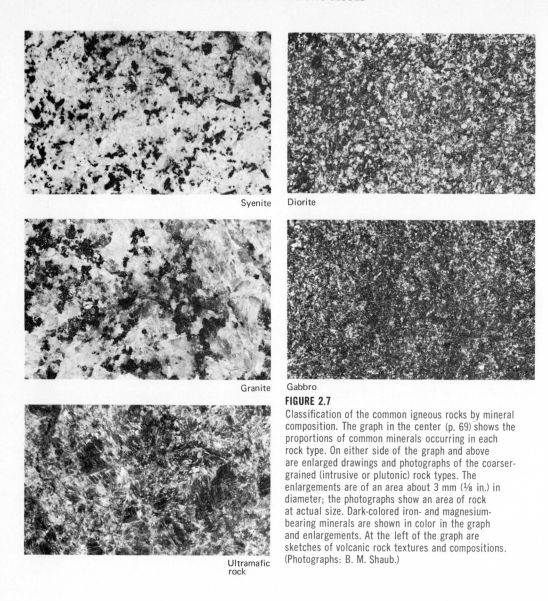

Syenite   Diorite

Granite   Gabbro

FIGURE 2.7

Classification of the common igneous rocks by mineral
composition. The graph in the center (p. 69) shows the
proportions of common minerals occurring in each
rock type. On either side of the graph and above
are enlarged drawings and photographs of the coarser-
grained (intrusive or plutonic) rock types. The
enlargements are of an area about 3 mm (⅛ in.) in
diameter; the photographs show an area of rock
at actual size. Dark-colored iron- and magnesium-
bearing minerals are shown in color in the graph
and enlargements. At the left of the graph are
sketches of volcanic rock textures and compositions.
(Photographs: B. M. Shaub.)

Ultramafic
rock

silicic and mafic rocks. In silicic rocks, the stable framework structures
of quartz and feldspars prevail, with only minor amounts of minerals
with sheet or chain structures (some micas and amphiboles). In contrast, the silicon-poor mafic rocks are dominated by chain structures
(pyroxenes) and isolated tetrahedral structures (olivine). The ratio of
silicon to other elements is much lower in these structures than in the
framework structures, and this accounts for the relative depletion of
silicon in mafic rocks.

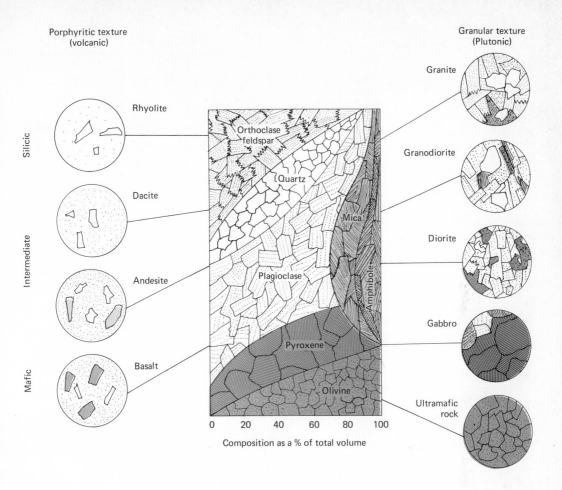

Porphyritic texture (volcanic)

Granular texture (Plutonic)

Silicic

Intermediate

Mafic

Rhyolite

Dacite

Andesite

Basalt

Orthoclase feldspar

Quartz

Mica

Plagioclase

Amphibole

Pyroxene

Olivine

Granite

Granodiorite

Diorite

Gabbro

Ultramafic rock

0   20   40   60   80   100

Composition as a % of total volume

IGNEOUS ROCK ORIGINS

As with all scientific investigations, we shall begin this search for igneous rock origins by describing observable processes and their products—namely, volcanoes and the landforms they create. This will help us in formulating the important questions that need answering.

2.3 Volcanoes and Volcanic Landforms

Much of what is known about volcanoes comes from studies of oceanic islands and continental margins where volcanic rocks can be observed above sea level. Such studies have provided a wealth of data on the composition both of volcanic rocks themselves and of the magnetic liquids and gases from which they form. In general, all volcanoes

FIGURE 2.8
Abundances of the principal rock types in the earth's crust. Igneous basalts and granites are the dominant crustal rocks. (Data from Ronov and Yoroshevsky, American Geophysical Union Monograph 13, 1969.)

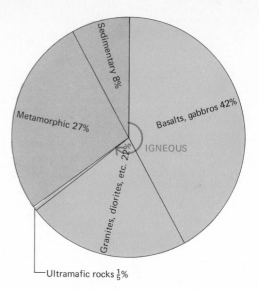

Sedimentary 8%

Metamorphic 27%

Basalts, gabbros 42%

Granites, diorites, etc. 22%

IGNEOUS

Ultramafic rocks $\frac{1}{5}$%

FIGURE 2.9
Exposures of igneous rocks. (a) Granite exposed in a massive dome in the mountains of central California. (b) Basalt exposed along the Snake River in southeastern Washington. (a: Josef Muench; b: U. S. Geological Survey.)

(a)

(b)

associated with oceanic islands as well as some found along the continental margins are composed of basalt just as are the volcanic rocks sampled from the ocean floor. In many regions, however, volcanic outpourings near continental margins are *not* mafic basalt but are, instead, more silicic andesites, the fine-grained volcanic equivalent of diorite (see Figure 2.7). Because of differences in their fluidity, basaltic and andesitic lavas produce somewhat different landscapes at the earth's surface. The fluidity of a magma is largely determined by its silicon content—the more silicon, the thicker and more viscous the magma. This is because the silicon-oxygen tetrahedra are bulky structures, and thus a greater proportion of them in a magma obstructs its movement, causing it to be less fluid.

Basaltic lavas are, as a general rule, the most fluid. Even they, however, exhibit variations of fluidity, depending principally on the amount of gas (mostly H_2O and CO_2) dissolved in the magma as it erupts. As an example of the influence of gas content, contrast the appearance and fluidity differences shown between the two Hawaiian basalt flows in Figure 2.10. The upper photograph depicts the ropy,

(a)

(b)

FIGURE 2.10
Lava flows. (a) Pahoehoe
flow. (b) Aa flow. (a, b:
G A. MacDonald.)

gas-rich, more fluid variety called "pahoehoe." Flows of pahoehoe basalt typically evolve, through loss of their gases, into the blocky, more viscous flows, called Aa, shown in the lower photograph. There are many vesicles (gas-bubble holes) in a hand specimen of pahoehoe.

Very fluid lavas are likely not to form conelike landforms at all. This is the reason why the classic volcanoes of the world such as Mount Fuji in Japan or Mount Rainier in Washington State (Figure 2.11) are not basaltic in composition. For such a volcano to form, eruptions must remain localized at a single position and the outpourings must be sufficiently viscous to build up a conelike feature surrounding the central vent. Basalts, on the other hand, are frequently observed as great sheetlike flows, as shown in Figure 2.9b. A closer look at these "flood" or "plateau" basalts, as they are often called, commonly reveals an internal pattern of vertical fractures called "columnar jointing" (Figure 2.12). How might these remarkable columns have formed? Let us once again apply two important techniques of geologic investigation: allowing the present to serve as the key to the past, and reasoning by analogy. The pattern that most of us have seen developed by mud as it dries out and shrinks (Figure 2.13) is so like the polygonal basalt that a similar origin is suggested. In the case of basalt, contraction results from cooling rather than loss of moisture, but the same polygonal pattern is developed in both cases to accommodate the reduction of volume.

FIGURE 2.11
Composite volcano, Mt. Fuji.
(W. C. Ernst.)

FIGURE 2.12
Devil's Postpile in California. Note the
excellent polygonal columns. (E. A. Hay.)

If fluid basalt flows continue to erupt from the same general region for long periods of time, a volcano having a broad, low, domal form develops. Its form reflects the lava's tendency to flow great distances before solidifying. These large landforms are called shield volcanoes. They are the only volcanoes surfacing from deep ocean basins; they

(b)

FIGURE 2.13
Development of polygonal shrinkage cracks. (a) Modern mudcracks.
(b) Top of Devil's Postpile. (a: N. T. Hall; b: E. A. Hay.)

(a)

are also to be found on continents. The Hawaiian Islands comprise a chain of such shield volcanoes that have achieved truly gigantic dimensions. Compare the sizes of various other volcanic landforms shown in Figure 2.14 to Mauna Loa, the largest of the Hawaiian volcanoes.

In addition to shield volcanoes, there are other volcanic types, each reflecting differences in eruptive styles, which in turn are primarily controlled by the chemical composition and gas content of the erupting lava. **Composite volcanoes** (Figure 2.15a) consist of interlayered flows (typically andesite) and fragmental ejecta. The fragmental material consists of fine ash, called "tuff," and larger angular blocks of andesite, called "breccia." This **pyroclastic** (*pyro* =fire, *clastic* = broken) debris results from periodic eruptions of great explosive violence caused by abnormal build-up of gases. When interlayered with the normally more viscous (compared to basalt) andesite flows, it yields steep-sided volcanoes with concave upward slopes.

Often during the later stages of a series of eruptions the rising magma becomes more viscous. This can be accomplished by either depletion of gas content or by enrichment of silicon content in the magma.* Very sticky magma (typically dacite or rhyolite) pushed up into the throat of a central vent may not have sufficient fluidity to flow at all. Instead, it will rise upward and perhaps bulge outward in the manner of an almost-solid plug. Landforms of this origin are referred to as volcanic domes or **dome volcanoes** (Figure 2.15b).

Rapidly effervescing gas-rich mobile magmas (typically basalt) erupt explosively in a pyroclastic manner. The resultant landform is a **cinder cone** (Figure 2.15c). **Maar volcanoes** (Figure 2.15d) are similar to cinder cones, but differ by containing a large explosion crater. The explosions are caused by steam action generated when large volumes of water come into contact with hot magma at depths of a few hundred feet.

There is one additional landform of volcanic origin deserving of

*A mechanism for the evolution of magmas from silicon-poor to silicon-rich varieties will be discussed in Section 2.6.

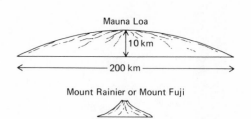

FIGURE 2.14
Comparison of sizes of volcanoes.

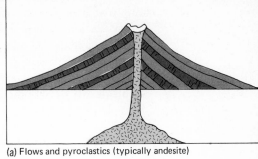

(a) Flows and pyroclastics (typically andesite)

FIGURE 2.15
Types of volcanoes.
(a) Composite cone
(Mayon). (b) Dome vol-
cano (Lassen). (c)
Cinder cone (Sunset
crater). (d) Maar vol-
cano. (a: Gardner
Collection, Harvard
University; b: E. A.
Hay; c: Tad Nichols;
d: NASA.)

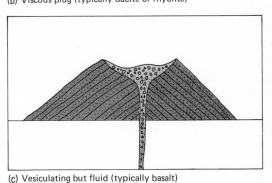

(b) Viscous plug (typically dacite or rhyolite)

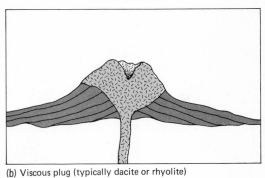

(c) Vesiculating but fluid (typically basalt)

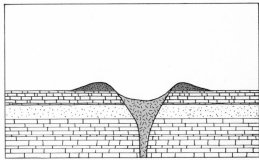

(d) Steam blast by H_2O penetrating the volcanic "plumbing"
     (typically basalt)

mention: the caldera. Calderas are roughly circular collapse depressions that form around the central summit of volcanoes. Collapse is almost certainly caused by the relatively rapid excavation of the magma chamber, thereby leaving the massive summit of a volcano without sufficient support. Crater Lake in Oregon is one of the best known calderas in the world (Figure 2.16). Indian lore of that region establishes its development in very recent, but prehistorical, time.

(a)

(b)

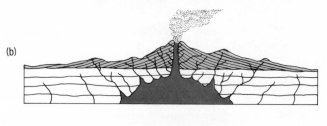

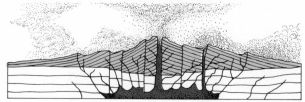

FIGURE 2.16
(a) Crater Lake, Oregon, a recently formed caldera. (b) Sequence of events that formed Crater Lake caldera. (a: John S. Shelton; b: after H. Williams.)

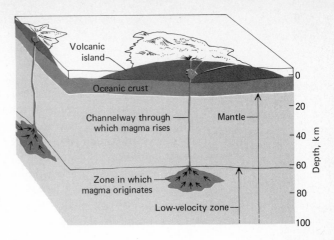

FIGURE 2.17
An idealized diagram of the origin of oceanic basalts, such as those making up the Hawaiian and other oceanic islands. Seismic evidence indicates that the basaltic magmas originate in the low-velocity zone of the mantle.

2.4 The Origin of Magmas

Although studies of volcanoes that are exposed above sea level have provided much information, the most fundamental questions about volcanic rocks concern the enormous volume of ocean-floor basalts which cover about three-quarters of the earth's surface. At what depth within the earth's interior did these lavas form, and how did they make their way to the surface? To answer these questions we must, once again, rely on indirect evidence, principally that provided by earthquake waves.

Most of the evidence upon which we have developed a model for the earth's interior comes from the analysis of seismic (earthquake) waves. Because seismic waves travel at different velocities within earth materials of different rigidity and density, inferences can be made about how the composition and structure of rocks change with depth. A more complete discussion of these techniques will come in Chapter 6, but for our present purposes we shall introduce some of the conclusions. Figure 2.17 depicts a cross-section of the upper earth in an oceanic region. Note the distinction between *crust* and *mantle*. These regions are separated by a surface across which seismic waves abruptly change velocity, called the Moho discontinuity named after Andrija Mohorovičić, a Serbian seismologist.

Earthquake waves penetrating the 8 kilometer (5-mile) thickness of oceanic rocks lying above the Moho discontinuity show rather high velocities comparable to those found in laboratory experiments to be characteristic of basalt. For this reason, it appears most probable that the basaltic rocks exposed on the ocean floor* extend downward to the

---

*The basaltic rocks of the sea floor are actually "exposed" only in rugged submarine cliffs, canyons, and mountains. Over most of the flat areas of the ocean floor they are buried beneath a thin blanket of sedimentary mud and sand. The thickness of this sedimentary cover can be easily determined from seismic evidence, for it transmits seismic waves much more slowly than does the underlying basalt; it has been found normally to be only about a hundred to a thousand meters (330–3300 feet) thick.

Moho discontinuity, where an abrupt increase in wave velocity indicates a change to the denser and more rigid underlying rocks of the mantle. It was long assumed that basaltic lavas originated from melting and upward movement of the lower parts of this basaltic crust, but it is now clear that they form not within the crust itself, but from a melting of the much deeper upper mantle. This is indicated by several lines of evidence.

First, regions of present-day volcanic activity are closely associated with the earth's deep earthquake zones (Figure 2.18). This suggests that volcanoes are related to events in the mantle, for all deep earthquakes occur in the upper mantle far beneath the crust. Ruptures caused by deep earthquakes would provide ideal channels for the escape to the surface of molten magma formed deep within the earth. Second, recent seismological studies of individual volcanic eruptions in Hawaii and elsewhere have shown that the extruded basaltic magmas originate from fluid pockets in the upper mantle formed at depths ranging from 65 to 100 km (40 to 60 mi) beneath the crust (Figure 2.17). This is precisely the depth range at which a plastic, *low-velocity mantle layer* is indicated by other seismic evidence. Local melting within this high-temperature plastic layer is therefore the most prob-

FIGURE 2.18
Comparative positions of the earth's active volcanoes (dark dots) and deep earthquake zones (light shading). The "andesite line" marks the boundary between oceanic volcanoes discharging only basalt, and those near the continents that discharge mostly andesite.

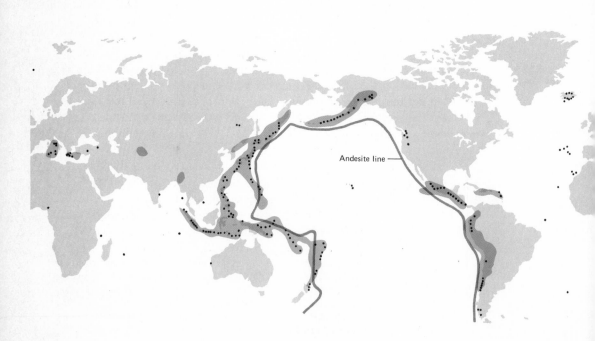

able source for the basaltic lavas which have poured forth to form the oceanic crust.

If the basaltic rocks which make up most of the oceanic crust originate deep within the mantle, then we are faced with a fundamental problem, because the increase in seismic wave velocity at the Moho discontinuity indicates that the mantle is composed of materials that are more rigid and denser than the overlying basaltic crust. Thus the crustal basalts appear to have originated, in some manner, from *heavier* underlying rocks. Many lines of evidence suggest that the uppermost mantle consists largely of olivine and pyroxene and thus is similar in composition to the rare *ultramafic* rocks exposed at the earth's surface. If the upper mantle is composed of heavy, silicon-poor ultramafic materials, then how can their melting give rise to basaltic magmas which are lighter and contain more silicon bound up in the structure of feldspar minerals? The most probable answer relates to simple melting-point differences. Because feldspar melts at a lower temperature than either olivine or pyroxene, heating only slightly beyond this temperature might cause a partial melting of ultramafic mantle rocks which contain only small amounts of feldspar (Figure 2.19). This partial melting, in turn, could form local concentrations of silicon-enriched basaltic magmas surrounded by still-solid olivine and pyroxene. Most geologists now believe that basalts originate in this way even though it would require the partial melting of an enormous volume of silicon-poor mantle materials to account for the huge quantity of basaltic rocks found in the oceanic crust.

We have considered a plausible model to account for the huge volumes of basaltic magma found on earth, but what about the origin of more silicic magmas from which rocks such as andesite (diorite) or rhyolite (granite) are formed? These rocks are unique to continental regions, where they have received extensive study. Reference to Figure 2.7 shows how such studies have led to a system of classification for them as well as for the more mafic rocks that not only dominate oce-

FIGURE 2.19
Partial melting of ultramafic rocks to create basalt. The small amount of feldspar in ultramafics has a lower melting point than either pyroxene or olivine. Heating sufficient only to melt this feldspar plus a small portion of the pyroxene and olivne (dark color) could produce a magma of basaltic composition (light color).

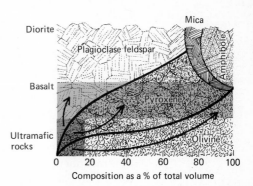

anic regions but also are common on continents. Earth scientists, however, want to do more than just classify these rocks—they want to
explore and seek answers to more fundamental questions of rock
origin such as: Why are some rocks dominated by silicon-rich minerals and others by magnesium- and iron-rich minerals? Why can certain combinations of minerals coexist in the same rock while other combinations never occur? Under what conditions of temperature and
pressure do the common igneous rocks originate? Many clues to such
questions have been provided by careful field and laboratory studies
of natural igneous rocks, but perhaps the most important clues have
come from laboratory studies of silicate melts—artificial, man-made
"magmas" heated and cooled under varying conditions in an attempt
to simulate the natural settings in which igneous rocks form. Such
studies were begun over 50 years ago by pioneering workers, particularly N. L. Bowen, at the Geophysical Laboratory of the Carnegie
Institution in Washington, D.C. This laboratory, still one of the lead-
ers in this work, is today augmented by many university and government laboratories that are also engaged in experimental studies of
igneous rock origins.

Laboratory experiments can seldom *precisely* duplicate natural conditions because most natural magmas have from six to ten or more
chemical constituents, which is too many for accurate, controlled laboratory analysis. Instead, most laboratory experiments are confined to
simpler systems containing only two to four elements or compounds.
These studies have shown that solid minerals do not "freeze" all at
once from the liquid as a magma cools, but instead go through complex series of interactions with the liquid magma; that is, certain minerals crystallize at high temperatures, but become unstable as the temperature drops and react with the magmatic liquid to give still other
minerals.

Among the first minerals to be investigated experimentally were
the feldspars and pyroxenes, two of the most common constituents of
igneous rocks. Figure 2.20 shows the interrelationships of a pyroxene
and a calcium-rich plagioclase feldspar at different temperatures; these
two minerals are the dominant constituents of gabbro and its fine-
grained extrusive equivalent, basalt. The relations are expressed in a
phase diagram which shows the composition of the various liquid
and solid components in the artificial magma at differing temperatures.
In Section 1.5 we discussed a phase diagram for water (Figure 1.10)
which showed its gaseous, liquid, and solid phases at various temperatures and pressures. That phase diagram presented the behavior of a
rather uncomplicated one-component system. Studies of the more
complex phases of multicomponent silicate melts are often conducted
at atmospheric pressure in unsealed containers in which the gaseous
phases are lost. Phase diagrams of such systems show only the chang-

ing proportions of the solid and liquid phases at differing temperatures. The two-component system in Figure 2.20 is of this sort. It was worked out over 40 years ago by Bowen and his associates and is still among the most fundamental because it deals with abundant rock-forming minerals. Recent experimental studies have concentrated on more complex three- and four-component systems and on simpler systems at high pressures such as those to be expected deep within the earth. In addition, the great influence, in the course of magmatic crystallization, of *volatile constituents* such as water and dissolved gases has come to be recognized, and most experiments now examine the gaseous as well as the liquid and solid phases.

From his experimental studies, combined with observations on natural igneous rocks, Bowen in 1922 proposed the following theory to account for the varying compositions of igneous rocks. If a mafic magma (rich in magnesium and iron) crystallizes directly, it will form gabbroic or ultramafic rocks made up of calcium feldspar, pyroxene, and olivine. But if, as crystallization begins, the first-formed crystals of olivine, pyroxene, and calcium feldspar are somehow separated from the remaining liquid (for example, by sinking to the bottom of a chamber filled with the liquid magma), then the remaining magma will

FIGURE 2.20
Pyroxene–plagioclase feldspar phase diagram, showing relations between cooling liquid melt and solid crystals for melts containing various proportions of pyroxene and plagioclase (bottom bars). The circles on the right and arrows on the diagram illustrate the cooling sequence in a melt containing 80 percent plagioclase (gray) and 20 percent pyroxene (color). As the melt cools, plagioclase crystals form while the melt changes composition along the curved line. On reaching the composition and temperature at point E, pyroxene and the small amount of remaining plagioclase crystallize simultaneously to give the same mineral proportions as in the original melt.

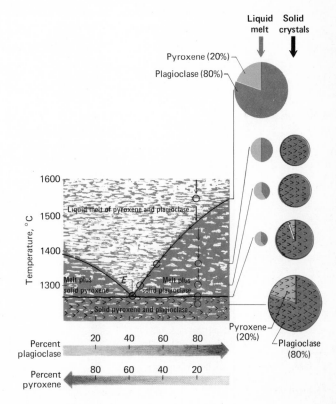

become progressively more silicon-rich. If the separation of iron and magnesium minerals continues long enough, the remaining silicon-rich liquid magma will crystallize as a granite made up largely of quartz and potassium feldspar. Bowen recognized that there is a sequence of different minerals formed in a magma as it cools; he showed that two such sequences of minerals—one for feldspars, the other for olivines, pyroxenes, amphiboles, and micas—coexist in magmas as temperatures are lowered (Figure 2.21). These two series correspond, in a general way, to the changing compositions of igneous rocks shown in Figure 2.7. These findings led Bowen to postulate that all igneous rocks form from "primary" magmas of mafic composition. Silicic rocks originate when these mafic magmas change composition, or "evolve," as early-formed, silicon-poor crystals are separated from the main magmatic mass. Field studies of igneous rocks have shown that such magmatic evolution does indeed occur, and there are other lines of evidence suggesting that many silicic igneous rocks did originate from mafic materials, but in a more complex manner than visualized by Bowen.

2.5   Plutonic Rock Masses

Earlier in this chapter we inferred that coarse-grained silicate rocks have probably crystallized in the plutonic environment deep within the crust. Were it not for profound erosion we would never see such plutons, the general name given to any intrusive body. But we do, in fact, see them abundantly, and they display a variety of sizes, shapes,

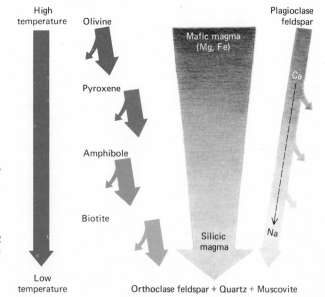

FIGURE 2.21
Changing composition of magma, based on Bowen's work. If crystals of olivine, pyroxene, and calcium-rich plagioclase are removed (small arrows) from a cooling mafic magma, the remaining melt becomes more silicic in composition. Crystallization of amphiboles, micas, and, ultimately, orthoclase feldspars and quartz results from this silicon enrichment. These changes correspond to the changing compositions of the principal igneous rocks (Figure 2.7).

and relationships to the older material they intruded (so-called country rock). Broad categories of plutonic rocks include those that are either **concordant** or **discordant** in their mode of emplacement (Figure 2.22). Igneous bodies emplaced parallel to, or concordant with, the layering fabric of the older intruded rock, are called **sills** or **laccoliths** (*lacco* = cistern, *lith* = rock). **Dikes, stocks,** and **batholiths** (*batho* = deep, *lith* = rock), on the other hand, show crosscutting relationships (discordant emplacement). Choice between the term *stock* and *batholith* depends on the extent to which erosion has exposed the plutonic rock (Figure 2.23 shows this distinction). It seems likely that most stocks will become batholiths with time, after erosion strips away more and more of the country rock. Occasionally fragments or large remnants of country rock are engulfed, or surrounded, by plutonic rock. On a regional scale this gives rise to **roof pendants,** as shown in Figure 2.23. On a much smaller scale, incorporation of old rock material by intruding magma forms **xenoliths** (*xeno* = foreign, *lith* = rock) of the sort shown in Figure 2.24.

So much for the naming of plutonic bodies. What about their ultimate origin? There are several lines of strong geophysical evidence to support the view that continents, sharply in contrast with oceans, consist of thick masses of granitic rock material. Since granite is the dominant rock of the continental crust, geologists have long been concerned with trying to understand its origin. The most fundamental question is: How did the lighter elements and minerals characteristic of granite become concentrated into thick, separated continental masses lying adjacent to the heavier rocks of the ocean floor? No complete answer to this puzzle yet exists, but strong clues are provided

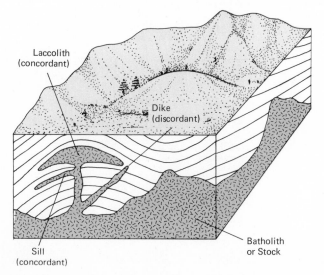

Laccolith (concordant)

Dike (discordant)

Sill (concordant)

Batholith or Stock

FIGURE 2.22
Reconstruction of the appearance of plutonic rock bodies prior to their exposure by erosion. Note both concordant and discordant bodies.

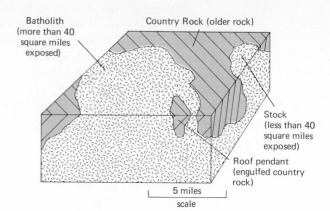

Batholith
(more than 40
square miles
exposed)

Country Rock (older rock)

Stock
(less than 40
square miles
exposed)

Roof pendant
(engulfed country
rock)

5 miles
scale

FIGURE 2.23
Distinction between batholiths and
stocks; batholiths are plutonic masses
exposed over an area greater than 40
square miles. (Note roof pendants are
remnants of engulfed "country rock.")

by studies of *movements of the crust*, which will be considered later
in Chapter 7. A somewhat more restricted problem concerns the local
processes that form granitic rocks.

Detailed mapping of granitic rocks provides evidence of two different origins (Figure 2.25). In some areas the boundaries between large
granite masses and the surrounding rocks are very sharp and distinct,
suggesting that liquid granitic magma was squeezed into the surrounding rocks to make the granitic mass. In contrast, many areas show a

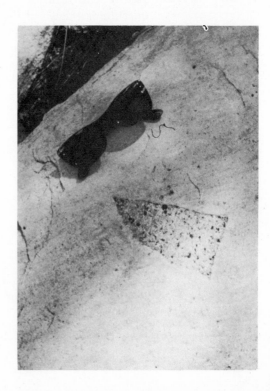

FIGURE 2.24
Xenoliths (foreign rocks) are fragments of older rock that have
been incorporated within younger
igneous material. (E. A. Hay.)

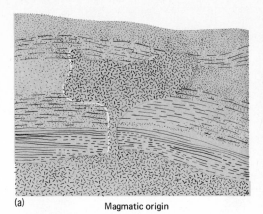

(a)   Magmatic origin

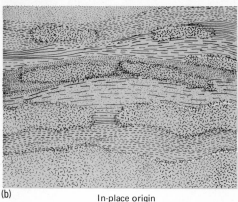

(b)   In-place origin

FIGURE 2.25

Cross sections of shield area rocks suggesting different origins of granite. In (a) the granite apparently crystallized from a molten magma that moved upward from deeper levels. In (b) the granite appears to have formed by in-place melting of pre-existing rocks without any appreciable movement of liquid magma.

complete gradation between granitic masses and surrounding meta-morphic rocks, which suggests that the granite formed from a kind of "in-place melting" of preexisting rocks without any appreciable movement of liquid magma.* Typically such regions of gradation between metamorphics and granites show a zone of rocks composed of thin bands of granite alternating with thin bands of metamorphic rock which occurs between the areas of "pure" granite and "pure" metamorphic rock (Figure 2.26). Such mixed granitic rocks, called migmatites, indicate that the granites formed where the melting was most intense but that in the surrounding areas temperatures did not exceed the melting range of preexisting minerals and only metamorphic rocks were produced.

Many geologists have favorably noted the advantages of the concepts of *in-place melting* and *granitization* in regard to the so-called room problem. The problem is simply this: Where did the material (country rock) go that occupied the pluton's space prior to its intrusion? In many cases field mapping provides evidence for shouldering aside of country rock and forceful emplacement of molten material, but not always. For those regions where migmatites are common and evidence for large-scale magmatic movement is lacking, the room problem is obviated by a model calling for granitization or in-place melting.

Because of the abundance of migmatites and transitional granitic masses in the oldest cores of continents, most geologists now believe

*The question of the actual degree of "melting" that takes place in the formation of granites by this means has been much debated by geologists. Some believe that complete fusion takes place, but others feel that true granites may form from rearrangement of crystal structures in a solid or partially solid state. This suggests a *metamorphic* origin for some granites, and is referred to as "granitization." This problem has not yet been resolved, but it is less important for our purposes than the still more fundamental question of the *place* of origin of the granitic materials.

FIGURE 2.26
Migmatite showing a small mass of granite (light color) surrounded by granitic gneiss (dark color). The hammer resting on the granite provides the scale. (Geological Survey of Great Britain.)

that much of the granite seen in the continents today formed by local melting of preexisting continental rocks, rather than by upward movement of liquid magma from deep levels of the earth's interior. Note, however, that if most of the continental crust has originated from melting of earlier crustal rocks, we are still faced with the fundamental question of *how* the silicon, sodium, potassium, and other elements characteristic of granite became concentrated in continental masses in the first place. This problem remains for future research to unravel.

METAMORPHIC ROCKS

2.6   Metamorphism

At the beginning of this chapter we concluded that some rock types could not be directly observed in the process of formation. Metamorphic (*meta* = change, *morphic* = form) rocks fall into this category. Let us again use the approach of evaluating a rock's texture under the assumption that it will provide a clue about the rock's origin. Figure 2.27 shows an example of a rock with foliated texture, wherein the minerals are oriented in nearly parallel sheetlike planes. Nowhere has anyone observed such a texture in the making. Furthermore, many of the minerals common to rocks having a foliated texture—such as garnet, muscovite, and various aluminum silicates—are comparatively rare among volcanic, plutonic, and sedimentary rocks. Two rather striking differences are thus apparent between the rock in Figure 2.27 and those with which we are more familiar: (1) the texture shows preferred (not random) directions of crystal growth, and (2) the mineral assemblage is different from well-known igneous associations such as quartz, feldspar, pyroxene, etc., or sedimentary suites comprising sandstones, shales, and the like.

It appears that these metamorphic rocks must have originated at depth within the earth. Perhaps they are really nothing more than a special group of igneous (grown-from-a-melt) rocks and do not deserve special classification. But how can we explain the remarkable foliations by growth from a melt? The parallelism of crystal growth suggests the existence of directed pressures in the environment of growth, but since liquids have no rigidity, they flow. It is essentially impossible, therefore, to attain great pressure differences of a directional nature within them (Figure 2.28). Solids, on the other hand, *can* have directions of greater shear stress developed within them. The unique mineral assemblages of these rocks provide an additional basis for rejecting their igneous origin. Recall the work of N. L. Bowen (Section 2.4) regarding the products of crystallization from a silicate melt. Minerals such as garnet and the aluminum silicates, andalusite, kyanite, and sillimanite are never formed in the laboratory from normal silicate melts and are extremely rare in naturally occurring igneous rocks.

FIGURE 2.27
Hand specimen of schist showing foliated texture. (Yale Peabody Museum.)

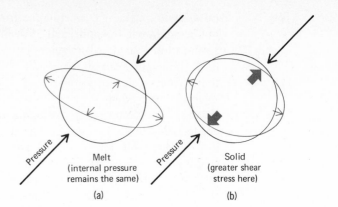

FIGURE 2.28
Application of directed pressure on liquids and solids. (a) Directed pressure on a liquid causes the liquid mass to change shape, thereby maintaining an approximately equal internal pressure throughout. (b) Solids do not readily flow, so greater shear stresses can be generated in some directions as compared to others.

Thus the circumstantial evidence against the igneous origin of what we call metamorphic rocks seems compelling. But what other alternative is there? Difficult though it may be to visualize, consider this: Some preexisting rock (igneous, sedimentary, or even metamorphic) is subjected to significantly increased conditions of *temperature* and *pressure*, and perhaps new *chemically active fluids*, for a long period of time. This new environment, no doubt at considerable depth, imposes nonequilibrium conditions on the minerals and texture of the preexisting rock. That is to say, the minerals are no longer comfortable in their new environment, so they slowly reorganize their lattice structures in a manner to achieve a more stable configuration (Figure 2.29). The process of doing this is called **ionic diffusion**. It is greatly facilitated by elevated temperatures. Metamorphic rocks are thus seen as rocks having formed in response to new environments of increased temperature, pressure, and sometimes the availability of new ions.

As you might expect, there are a great many variations in the intensity of metamorphism to which a rock might be subjected. The upper extreme is melting, at which point metamorphism gives way to igneous processes. Laboratory investigations of the effects of elevated temperatures and pressures on mineral change have provided instructive tools with which to evaluate the intensity of natural metamorphic environments. Figure 2.30 shows, for example, the experimentally determined stability fields for the aluminum silicate ($A1_2SiO_5$) polymorphs (*poly* = many, *morph* = form). Andalusite, kyanite, and sillimanite have the same composition but different lattice structures: each mineral is a different form of $A1_2SiO_5$, and thus is a different polymorph. The lattice structure for each is determined by the conditions of temperature and pressure at which it is most stable. When one or another of these minerals is in a metamorphic rock, we are thus afforded some specific insight into the conditions of its formation.

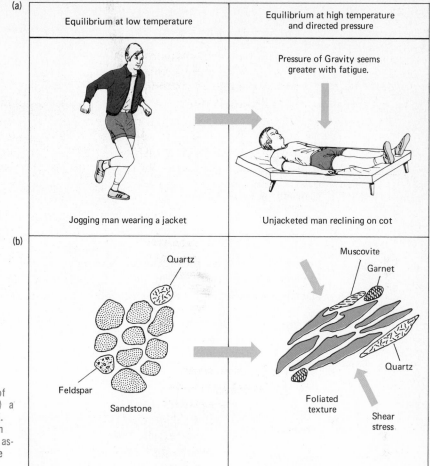

(a)

Equilibrium at low temperature

Equilibrium at high temperature
and directed pressure

Pressure of Gravity seems
greater with fatigue.

Jogging man wearing a jacket

Unjacketed man reclining on cot

(b)

Quartz

Feldspar

Sandstone

Muscovite

Garnet

Quartz

Foliated
texture

Shear
stress

FIGURE 2.29
The achievement of equilibrium for (a) a man and (b) a rock. Rock metamorphism results as mineral assemblages achieve new equilibria.

2.7   Classification of Metamorphic Rocks

It is often possible to distinguish between three broad genetic categories of metamorphic rocks. Contact metamorphism results from increased temperature effects in older intruded rock near contacts with plutonic masses. Regional metamorphism occurs over broad areas in response to strong deformational pressures as well as significant increases in temperature. Dynamic metamorphism is a variety of mineral change wherein strong shearing deformation is the dominant factor. Regional and contact metamorphism are by far the most significant of these categories, accounting for the great majority of metamorphic rocks.

Metamorphic rocks are more varied and complex than normal igne-

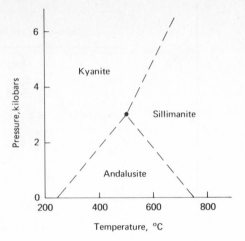

FIGURE 2.30
Aluminum silicates derived from different metamorphic conditions. (After W. S. Fyfe, "Chemical Geology," vol. 2, p. 74, 1967.)

ous and sedimentary rocks, and many names and classification schemes have been proposed for them. Figure 2.31 shows a simple, widely used classification that is based not on mineral composition, as were the

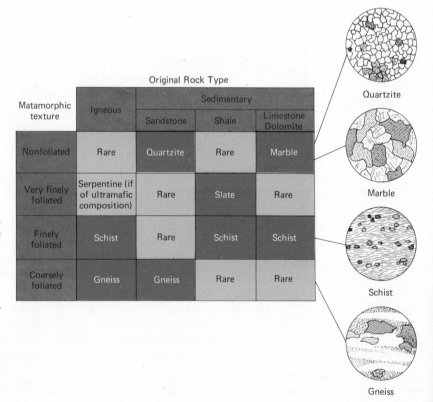

FIGURE 2.31
Classification of the common metamorphic rocks by texture **and** original rock type. The enlarged drawings are of an area about 3 mm (⅛ in.) in diameter; the photographs (on page 91) show an area of rock at actual size. (Quartzite, marble, gneiss: B. M. Shaub; schist: Yale Peabody Museum.)

Matamorphic texture	Original Rock Type			
	Igneous	Sedimentary		
		Sandstone	Shale	Limestone Dolomite
Nonfoliated	Rare	Quartzite	Rare	Marble
Very finely foliated	Serpentine (if of ultramafic composition)	Rare	Slate	Rare
Finely foliated	Schist	Rare	Schist	Schist
Coarsely foliated	Gneiss	Gneiss	Rare	Rare

Quartzite

Marble

Schist

Gneiss

previous classifications of igneous rocks, but on the different textures of the resulting metamorphic rock. As the minerals of preexisting rocks recrystallize during metamorphism, they commonly became oriented in parallel, sheetlike planes yielding **foliated texture.** The basis of classification for these rocks depends on the coarseness of the foliation. **Slates** are so fine-grained that only a distinct luster on some surfaces serves as evidence of parallel grains. **Schists** are coarser-grained and are frequently rich in micas and amphiboles or other sheetlike or needlelike crystals that make the foliation obvious. More coarsely foliated varieties, called **gneisses,** are banded by segregated regions of different minerals. These types of foliation can result from the metamorphism of many kinds of igneous and sedimentary rocks (see Figure 2.32). In addition, a few metamorphic rocks lack a foliated arrangement of the constituent minerals. The most common such rocks are **quartzites** and **marbles** which sometimes form from the metamorphosis of sedimentary sandstones and carbonate rocks, respectively (Figure 2.31).

By far the most abundant metamorphic rocks are gneisses, most of which are similar to ordinary granite in composition and are thus called **granitic gneisses.** It has been estimated that such gneisses account for about 80 percent of all metamorphic rocks, schists for about

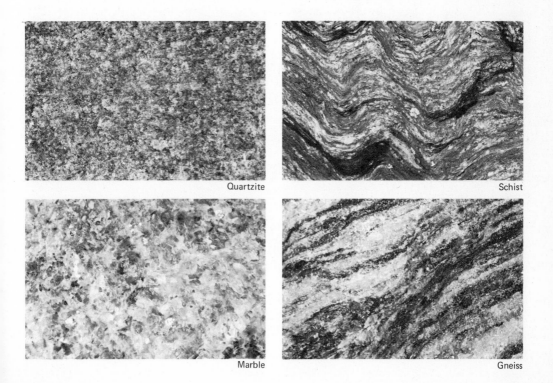

Quartzite

Schist

Marble

Gneiss

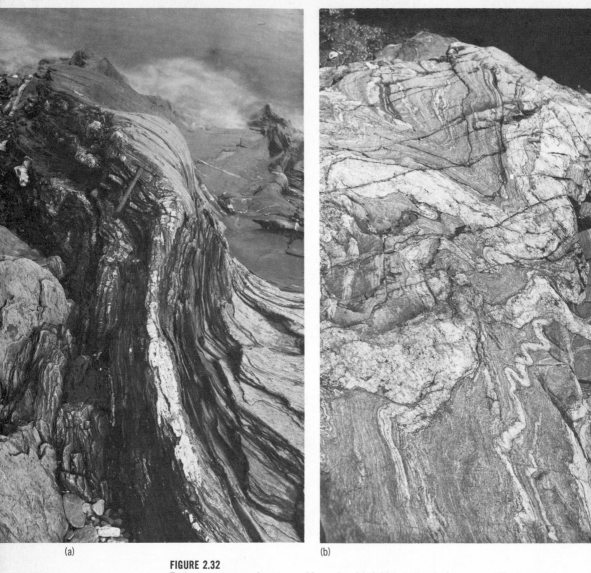

(a)   (b)

FIGURE 2.32
Typical exposures of metamorphic rocks. (a) Schist exposed along sea cliffs in western Scotland. (b) Gneiss exposed along a stream in northern Arizona. (a: Geological Survey of Great Britain; b: John S. Shelton.)

15 percent, and quartzites and marbles for only about 5 percent.

Classifications of metamorphic rocks based on mineral composition are more complex than the textural classification of Figure 2.31 be-

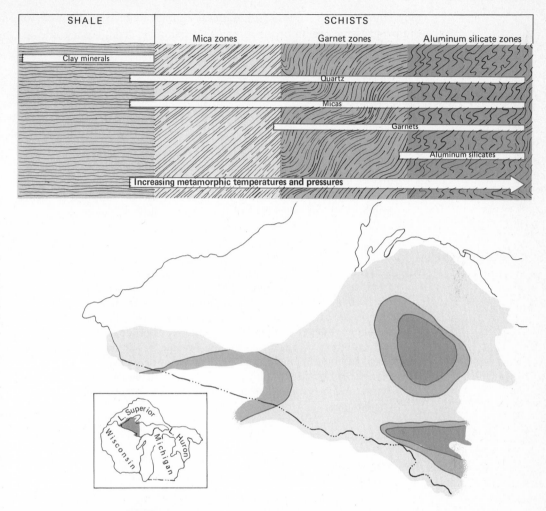

FIGURE 2.33
Changes in the mineral components of shale when metamorphosed to schist at increasing temperatures and pressures. The clay minerals are first altered to quartz and mica, some of which then converts to garnet and, finally, to aluminum silicate minerals. The presence of these key minerals can be used to identify "zones" of increasing metamorphism in the schists. The map shows the progressive metamorphic zones as exposed in schists of upper Michigan. (After Ernst, "Earth Materials," 1969; generalized from James.)

cause different minerals may be formed from the same original rock by differing metamorphic temperatures and pressures. Figure 2.33 shows how increasing temperatures and pressures lead to progressive changes in mineral components during metamorphosis of a sedimentary shale. Similar "zones" of mineral change have been defined for the progressive metamorphism of other common sedimentary and igneous rocks.

SUMMARY OUTLINE

Igneous Rock Diversity

2.1 *Igneous textures:* volcanic rocks are typically fine-grained and porphyritic in texture, whereas plutonic rocks have coarse-grained, granular textures.

2.2 *Classification of igneous rocks:* igneous rock classification is based primarily on the properties of texture and mineral composition.

Igneous Rock Origins

2.3 *Volcanoes and volcanic landforms:* different volcanic landforms result from variations in eruptive style, which are strongly influenced by differences in composition and gas content of the erupting lava.

2.4 *The origin of magmas:* laboratory experiments with artificial silicate melts have suggested that some granite rocks originate from the chemical evolution of basaltic magmas; oceanic basalt magmas appear to result from partial melting of ultramafic mantle rocks.

2.5 *Plutonic rock masses:* plutonic rock masses have crystallized at great depth, and may be either concordant or discordant in their mode of emplacement.

Metamorphic Rocks

2.6 *Metamorphism:* when preexisting rocks are subjected to high temperature and pressure in the presence of chemically active fluids, metamorphism is likely to occur.

2.7 *Classification of metamorphic rocks:* gneisses and schists are the most abundant foliated metamorphic rocks; quartzites and marbles are the most common nonfoliated ones.

ADDITIONAL READINGS

*Ernst, W. G.: *Earth Materials*, Prentice-Hall, Inc., Englewood Cliffs, N.J., 1969. A comprehensive introduction to minerals and rocks.

*Available in paperback.

Huang, W. T.: *Petrology*, McGraw-Hill Book Company, New York, 1962. A standard intermediate text on rocks.

Macdonald, G. A.: *Volcanoes*, Prentice-Hall, Inc., Englewood Cliffs, N.J., 1972. A clearly-written, nontechnical intermediate text.

*Zim, H. S., and P. R. Shaffer: *Rocks and Minerals*, Golden Press, New York, 1957. An excellent elementary guide to mineral and rock identification.

*Available in paperback.

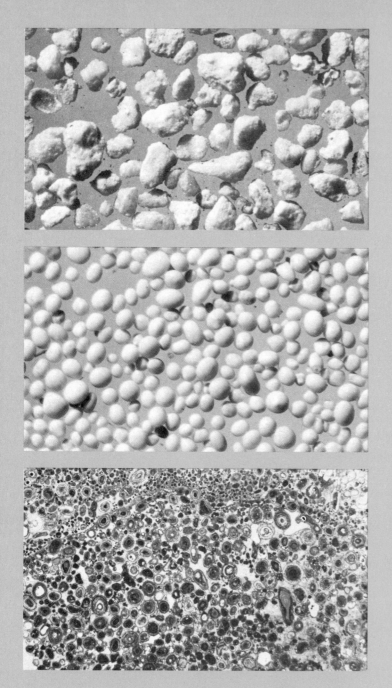

chapter 3

Sedimentary
Processes

The silicate minerals of igneous and metamorphic rocks normally form at high temperatures and pressures and are generally unstable when exposed to the lower temperatures and pressures of the earth's surface. There the action of air and water tends to break down the common igneous minerals and convert them into other minerals that are stable under surface conditions. This process, known as **weathering**, provides the raw materials for sedimentary rocks, which are merely the products of rock weathering transported and deposited by the action of water, wind, or ice. Sedimentary rocks thus originate from the interaction of the minerals of the solid earth with the waters and atmosphere of the fluid earth. Sedimentary rocks constitute only about 8 percent of the earth's crust, yet they are of great importance as the primary documents for understanding the earth's history because they lend themselves so well to comparisons with modern processes of sedimentation.

SEDIMENTS AND SEDIMENTARY ROCKS

Particles accumulating as sediments today have two principal sources: some originate and are transported as *solid* particles derived from weathering of the land—accumulations of such particles are called **detrital sediments;** others originate when the *dissolved* materials derived from weathering are precipitated from the waters of streams, lakes, or the ocean—accumulations of these precipitated particles are known as **chemical sediments** (Figure 3.1). Let us briefly contrast the rock textures resulting from these two processes (Figure 3.2). Clastic texture refers to fragmental accumulations. Whenever weathered bits of mineral or rock are carried in transport from a source area to an environment of deposition, they undergo abrasion. The effect of this wearing-away process is both to round the individual grains and to reduce them in size. A necessary consequence of rounding is that the grains, when deposited, cannot pack together in a manner that fills up the total rock volume, hence detrital sediments always display a texture, having rounded grains and original pore spaces between them.

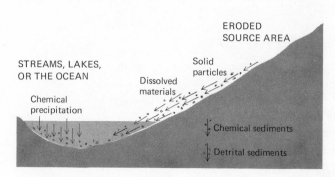

FIGURE 3.1
Detrital and chemical sediments.

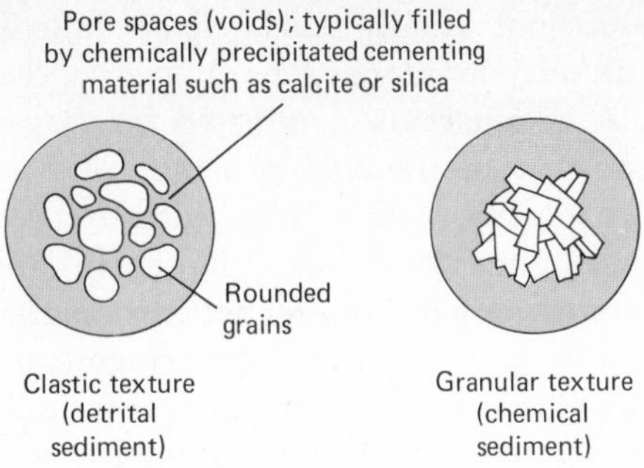

Pore spaces (voids); typically filled by chemically precipitated cementing material such as calcite or silica

Rounded grains

Clastic texture (detrital sediment)

Granular texture (chemical sediment)

FIGURE 3.2
Sedimentary rock textures.

Whenever we observe clastic texture in a rock, we are well assured that it originated by a sedimentary process.

Quite in contrast to detrital deposits, chemical sediments display a **granular** texture, not unlike that seen in silicate-rich plutonic rocks. Confusion between the two is not likely, however, because the sedimentary rocks having this texture are composed of low-temperature nonsilicate materials of the sort that can precipitate from surface waters.

3.1   Detrital Sediments

Complete weathering of igneous rocks normally leads to two kinds of solid particles: *clay minerals*, extremely fine-grained sheet silicates produced by the alteration of feldspars and other igneous minerals; and *quartz particles*, which resist weathering and accumulate relatively unchanged from the original igneous rock. Because clay minerals and quartz grains are the dominant particles produced by weathering, it is not surprising that they are the principal constituents of most accumulations of detrital sediments.

Less commonly, detrital sediments have other compositions. In regions where mechanical weathering is more rapid than chemical weathering, the feldspar minerals which make up a large part of most igneous rocks may *not* be converted to fine clay minerals. Instead they may accumulate, along with quartz, as sand or pebble-sized particles. Such accumulations of both quartz *and* feldspar fragments are called *arkose*. At the opposite extreme, in regions where chemical weathering is intense, sedimentary accumulations often contain fine-grained particles of various silicon-poor iron and aluminum oxide minerals mixed with clay minerals and quartz.

Arkose and oxide-rich deposits are examples of relatively rare sediments which differ in *composition* from the quartz-clay mineral accumulations that are the dominant detrital sediments. Because many detrital sediments are made up primarily of quartz grains and clay minerals, they are not usually classified by mineral composition, but instead by their relative proportions of clay and quartz particles of various sizes. Accumulations dominated by clay-sized particles are called muds, and those dominated by larger quartz particles are called sands. Muds are usually further subdivided by color, which normally ranges from very light gray to black; in general, the darker the color, the more fine-grained organic matter the mud contains. Sands, in turn, are classified by the size of their quartz fragments which may range from tiny grains of *silt* less than one-sixteenth millimeter in diameter to large *pebbles* many centimeters in diameter (Figure 3.3). When such large fragments predominate, the accumulation is usually given the name gravel rather than sand.

Sediments composed largely of either clay particles or sand particles of a single size are known as well-sorted sediments. Mixtures of clay and sand particles of various size are called poorly sorted sediments (Figure 3.4). The degree of sorting of detrital sediments provides a useful index to the conditions under which they were originally deposited. Because moving fluids of differing velocities erode, transport, and deposit particles of differing sizes (Figure 3.5), well-sorted sediments normally result from slow accumulation and long exposure to varying velocities of flow. Poorly sorted sediments, on the other hand, indicate rapid accumulation and little exposure to such velocity variations.

3.2   Chemical Sediments

FIGURE 3.3
Grain sizes of detrital silt, sand, and pebble particles. Still larger pebbles, to many centimeters in diameter, are not uncommon. True scale photographs. (Fundamental Photographs.)

In contrast to the solid products of weathering that accumulate as detrital sediments, chemical sediments are derived from the more soluble elements—particularly calcium, magnesium, sodium, potassium, and silicon—that are transported to lakes and the ocean in solution. This material does not remain indefinitely in solution; instead, much of it is *precipitated* as solid particles that accumulate to form chemical sediments. Today, much of this precipitation results from the life

Silt   **Sand** fine   medium   coarse   very coarse   **Pebbles** fine   medium   coarse

FIGURE 3.4
A poorly sorted sediment, made up of sand and pebble particles of varying size. Exaggerated scale. (Fundamental Photographs.)

processes of water-dwelling plants and animals, many of which can concentrate even those dissolved elements that are present in very small quantities. Such organism-produced sediments are known as biogenic chemical sediments. Less commonly, the dissolved materials may become so concentrated by evaporation—for example in desert lakes or tropical lagoons—that they spontaneously precipitate without

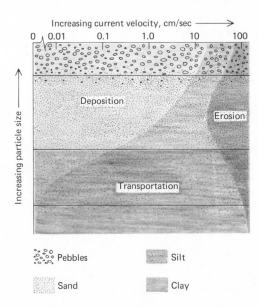

FIGURE 3.5
Approximate velocities of water flow for erosion, transportation, and deposition of detrital particles of different sizes (100 cm/sec=about 2 mi/hr). (Modified from Turekian, "Oceans," 1968, after Heezen and Hollister.)

the intervention of living organisms. Accumulations of such materials are called **nonbiogenic chemical sediments.**

By far the most common chemical sediments forming today are biogenic accumulations of calcium carbonate deposited by marine animals and plants (Figure 3.6a). Many animals living in the ocean secrete hard, mineralized shells composed largely of the two calcium carbonate minerals calcite and aragonite. Corals, snails, clams, sea urchins, starfish, and crabs are a few familiar examples, but there are numerous others. In addition, many simple seaweedlike marine plants also precipitate calcium carbonate from ocean water. Over large areas of the ocean floor, areas where there is little influx of detrital sand and clay to dilute the carbonate minerals, relatively pure concentrations of calcium carbonate shells and shell fragments are accumulating today.

(a)

Calcium is the least soluble of the principal dissolved elements found in streams, lakes, and ocean water, and is readily precipitated not only by animals and plants, but also when the dissolved materials are concentrated by evaporation. Such nonbiogenic accumulations of calcium, usually combined in either calcium carbonate or calcium sulfate minerals, are far less abundant today than biogenic calcium carbonate, but are found in a wider range of environments. Desert soils and the floors of desert lakes commonly contain such calcium sediments concentrated by evaporation; they accumulate in even greater abundance in shallow ocean bays and lagoons in regions where evaporation is intense (Figure 3.6b and c).

(b)

Occasionally the evaporation of water from lakes, bays, and lagoons proceeds so far that even the most soluble dissolved elements—sodium, magnesium, and potassium—are deposited as unusual sulfate minerals and as common salt, sodium chloride (Figure 3.7). The deposition of these very soluble elements requires an almost complete evaporation of the water in which they are dissolved and leads to the accumulation of **evaporite sediments.**

Biogenic calcium carbonate deposits and nonbiogenic evaporites are the most common chemical sediments forming today, but two less common types also deserve mention. The first are deposits of *silica,* a compound of silicon and oxygen with the same composition as quartz, but often lacking its stable, regular crystal structure. Quartz, as we have seen, is extremely resistant to weathering; in contrast, the silicon in more complex silicate minerals is far less stable, and these minerals contribute large amounts of dissolved silicon to streams and the ocean as they are weathered. Most water-dwelling organisms that

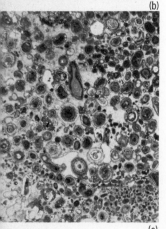

(c)

FIGURE 3.6
Biogenic and nonbiogenic chemical sediments. (a) Biogenic calcium carbonate sand made up largely of shell fragments. (b, c) Nonbiogenic calcium carbonate sand. The surface of the sand particles is shown in b; c shows the internal structure of concentric precipitated spheres. (a-c: Norman D. Newell, The American Museum of Natural History.)

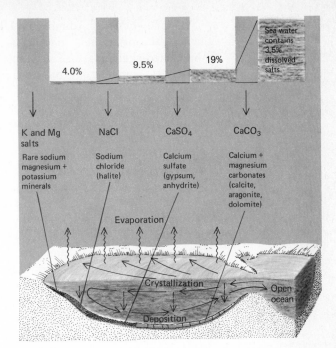

FIGURE 3.7
Sequence of mineral precipitation in evaporating ocean water. Evaporation of lake water sometimes produces different minerals and sequences because of differing proportions of dissolved elements.

secrete shells or other hard parts construct them of calcium carbonate, but a few use this dissolved silicon instead. By far the most important silicon extractors today are *diatoms*, microscopic plants that are abundant in both the ocean and in lakes, and *radiolarians*, microscopic animals that live primarily in the ocean. Both secrete tiny, glasslike skeletons of silica that are accumulating on some parts of the ocean floor to form sedimentary silica deposits (Figure 3.8).

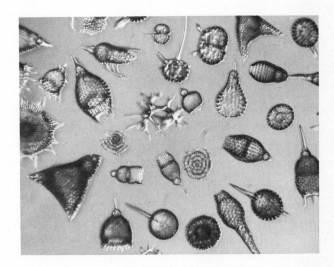

FIGURE 3.8
Skeletons of radiolaria magnified 140 times. The remains of these microscopic animals accumulate in enormous quantities on some parts of the ocean floor to form biogenic silica deposits. (Eric V. Grave.)

The final chemical sediment that we will consider also results from the activities of animals and plants, but in this case it is not the hard, mineralized skeletons that accumulate, but the soft organic materials themselves. Some organic matter from the decay of animals and plants is found in almost all present-day sediments, but such materials usually make up only a small fraction of the total volume. Less commonly, local conditions permit the accumulation of sediments with much higher amounts of organic matter. On land, deposits of *peat* may form when plant remains accumulate in acid waters which act as a preservative to prevent decay. Peat deposits often contain as much as 90 percent organic materials; the remaining 10 percent is usually detrital sand or clay. Likewise, on portions of the ocean floor organic materials produced by animals and plants in the sunlit surface waters may sink and accumulate, along with clay and sand, to make up a significant fraction of the bottom sediments. Both of these types of organic-rich chemical sediment have a significance far out of proportion to their small volume for, when buried and altered in sedimentary rocks, peat deposits become coal, and the organic matter in marine muds becomes petroleum.

3.3   Classification of Sedimentary Rocks

About 99 percent of all sedimentary rocks fall into one of four integrating groups: sandstones, shales, carbonate rocks, or evaporites (Figures 3.9, 3.10, 3.11). Sandstones, as the name implies, are rocks made up of particles of sand compressed and cemented to form rock. Usually the sand particles are composed of quartz, because of all the common igneous minerals only quartz is stable and resists further breakdown to far smaller particles when exposed to weathering at the earth's surface. Under special circumstances, however, other minerals, particularly feldspars, may also be important constituents of sandstones. When particles significantly larger than sand, such as pebbles or cobbles, are abundant in a sedimentary rock, the term conglomerate is applied to it.

Shales are rocks made up of clay mineral particles that are too small to be distinguished except under the extreme magnification of an electron microscope. Clay minerals are the stable silicate phases formed at the earth's surface by weathering of most kinds of igneous silicate minerals, especially feldspars. Because feldspars are the most common minerals in igneous rocks, clay minerals resulting from the weathering of feldspars are also the most common sedimentary minerals. It has been estimated that about half of the earth's sedimentary rocks are clay-rich shales, almost one-quarter sandstones and conglomerates, and about one-quarter carbonate rocks and evaporites.

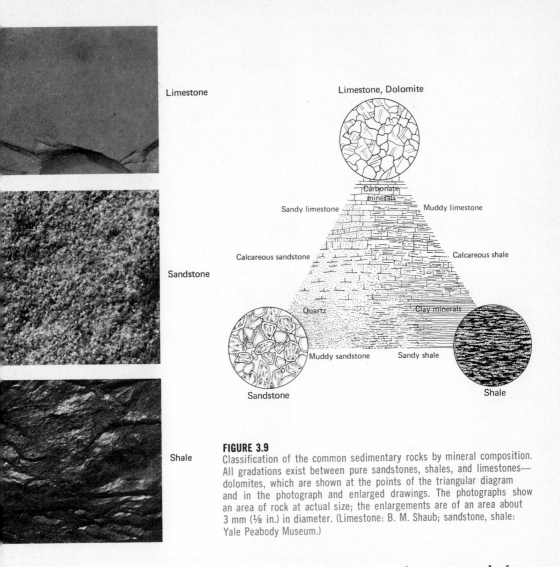

Limestone

Sandstone

Shale

FIGURE 3.9
Classification of the common sedimentary rocks by mineral composition.
All gradations exist between pure sandstones, shales, and limestones—
dolomites, which are shown at the points of the triangular diagram
and in the photograph and enlarged drawings. The photographs show
an area of rock at actual size; the enlargements are of an area about
3 mm (⅛ in.) in diameter. (Limestone: B. M. Shaub; sandstone, shale:
Yale Peabody Museum.)

Carbonate rocks, as the name implies, are those composed of carbonate minerals, principally calcite and dolomite. The two common carbonate rocks are limestone, composed predominantly of calcite, and dolomite, composed principally of the mineral dolomite (this is one of the few cases where the same name applies to a rock and a specific mineral). These carbonate rocks form when dissolved calcium and magnesium ions, derived from the weathering of preexisting rocks, are precipitated either directly from the waters of lakes and oceans or indirectly by the action of organisms in making shells and skeletons, many of which are composed of calcium carbonate.

Evaporites are relatively uncommon chemical sedimentary rocks

(a)

(b)

FIGURE 3.10
Typical exposures of sedimentary rocks. (a) Sandstone exposed in a road cut in western Texas. (b) Shale exposed along a stream bank in eastern New York. (c) Evaporite exposed in a road cut in southern Utah. (b: New York State Geological Survey; c: U.S. Geological Survey.)

(c)

that form when the dissolved materials of lake or ocean waters become highly concentrated through evaporation. Under such circumstances dolomite, gypsum, anhydrite, and halite are the principal minerals deposited. Evaporite rocks usually contain two or more of these minerals.

Sedimentary rocks commonly occur as pure, unmixed sandstones, shales, carbonates, or evaporites, but all gradations exist among these major types. Sandstones, for example, may contain clay or carbonate minerals, giving rise, respectively, to "muddy" or "calcareous" sandstones. Some of these intermediate compositions can be conveniently described by the terms shown in Figure 3.9.

At least one other chemical sediment deserves inclusion in our system of classification: chert. Chert consists of layers of exceedingly fine-grained quartz (Figure 3.12), which typically have a porcelainlike surface. It seems likely that tiny silica-secreting organisms (Figure 3.8) are the principal contributors to these deposits.

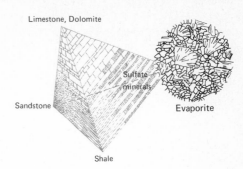

FIGURE 3.11
Gradational relationship of evaporite to the three more common sedimentary rock types shown in Figure 3.9. Areas of photograph and enlargement same as Figure 3.9. (Photograph: Yale Peabody Museum.)

3.4   Sedimentary Structures

In addition to the sedimentary properties of particle size and composition, special features of layering can also be instructive about a rock's origin. Like detrital sand and clay, the precipitated grains of chemical sediments are usually transported by moving fluids before they finally come to rest in sedimentary accumulations. For this reason, both chemical and detrital sedimentary deposits normally show patterns of superimposed layers formed as they are sorted and deposited by fluid motions. These layers, and related features formed at the time of sediment deposition, are called **physical sedimentary structures**. An additional structural pattern is found in sediments that accumulate in aquatic environments where bottom-dwelling animals and plants are abundant, as they are on the floors of most lakes and shallow seas. Under these conditions the sediments are often reworked, burrowed, and otherwise modified by the organisms to produce **biogenic sedimentary structures**.

FIGURE 3.12
Bedded chert from California. (E. A. Hay.)

The most important physical sedimentary structure is the characteristic layering, called **bedding** or **stratification**, found in almost all sediment accumulations (Figure 3.13). Bedding results from the fact that thick sediment accumulations rarely form from a single massive deposition of particles; instead, many small depositional increments are required. A single stream flood or moving ocean current normally transports and deposits only a relatively thin sheet of particles. These sheets seldom exceed a few centimeters in thickness, and it thus requires many of them to form thick accumulations of sediment. Usually each thin increment, or *bed*, shows slight differences in particle size and arrangement which distinguish it from adjacent increments; these slight differences give sediment accumulations their characteristic bedded structure.

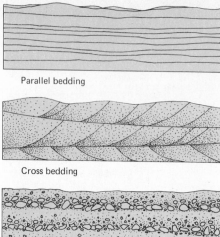

Parallel bedding

Cross bedding

FIGURE 3.13
The principal kinds of sedimentary
bedding.

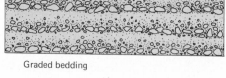

Graded bedding

In the simplest case the boundaries of individual beds in a sediment accumulation are horizontal, and adjacent beds are distinguished by slight differences in color or particle size. Such **parallel bedding** is common in sedimentary accumulations, but since it can form under a variety of conditions, it gives relatively few clues to the original fluid motions that deposited the sediment (Figure 3.14). More useful in such interpretations are two equally common but less regular types of bedding—*cross bedding* and *graded bedding*.

Cross bedding is formed when the fluid motions of sediment deposition cause wavelike undulations on the surface of the sediment. Such sedimentary wave forms are known as *ripples* and are common in accumulations of sand-sized sedimentary particles. When several rippled sand beds are superimposed, the individual beds, viewed in cross section, are not parallel, but have various angular relations to each other (Figure 3.15). Such bedding, called **cross bedding**, may provide important clues to the nature of the depositing fluid. For example, cross bedding formed by wind action usually shows individual beds inclined at a somewhat steeper angle than does cross bedding formed by moving water. In addition, laboratory studies of sediment rippling show that different patterns of cross bedding form under differing conditions of fluid flow. Thus it is sometimes possible to infer the approximate velocity, direction, and other features of the depositing fluid motion from patterns of cross bedding (Figure 3.16).

The second bedding type that gives important clues to the depositing fluid is **graded bedding**, in which each individual bed shows a progressive decrease in particle size upward through the bed (Figure

(a)

FIGURE 3.14
Parallel bedding in
modern beach sand of
eastern Scotland (a)
and in ancient lime-
stones of southern
Utah (b). (a: Geologi-
cal Survey of Great
Britain; b: Tad
Nichols.)

(b)

(a)

FIGURE 3.15
Cross bedding in modern muds and silts of eastern England (a) and in ancient sandstone of northern Arizona (b). (a: Geological Survey of Great Britain; b: Tad Nichols.)

(b)

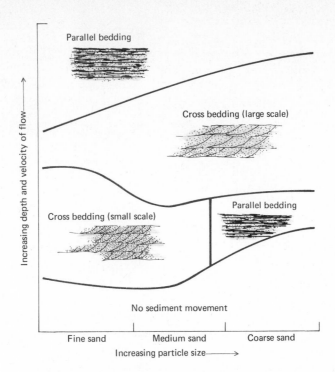

FIGURE 3.16
Relation of bedding patterns to fluid motions and particle size. (From Allen, "Physical Processes of Sedimentation," 1970.)

FIGURE 3.17
Graded bedding (upward decrease in particle size) as seen in a pebble-sized sandstone. (F. J. Pettijohn.)

3.17). In most sedimentary deposits *adjacent* beds may show slight size differences but, normally, the particle size *within* an individual bed, deposited as a single event, remains constant. Graded beds are an exception and require a depositing mechanism which, in a single event, deposits progressively smaller particles. Field observations and laboratory experiments indicate that graded beds form when dense, sediment-laden currents are introduced into relatively still water. When this occurs, the current abruptly loses velocity, depositing first the larger and heavier particles and then the finer and lighter in a gradational sequence. Such debris-laden currents, called turbidity currents, are common where sediment-carrying streams flow into lakes or the ocean; they also occur on the steep slopes of submerged scarps and canyons when large, unstable sediment accumulations break away and slide downward into deeper water (Figure 3.18).

Because of their high concentration of dense sediment particles, turbidity currents normally have a much higher overall density than the surrounding water; they thus move downward under the influence of gravity at high velocities and cover large areas before they are finally dissipated by friction. Their high velocities and densities also commonly cause them to erode or *scour* the previous bottom sediment before dissipating and depositing their own sediment load. This erosion usually forms distinctive channels, scratches, and gouges in the

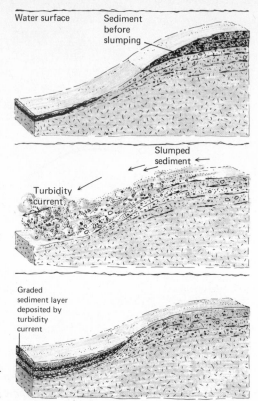

FIGURE 3.18
Origin of graded bedding from sediment slumping and turbidity current flow.

underlying sediment which, along with graded bedding, are characteristic structures of sediments deposited by turbidity currents. Because
they are both formed and buried by a single sedimentary event, these
erosional features are difficult to observe in present-day sediments,
but are a common feature of ancient sedimentary rocks. There, the
scratches, grooves, and channels are most often preserved as impres-
sions on the base of the overlying bed, and for this reason they are
known as **sole markings** (Figure 3.19).

Parallel bedding, cross bedding, graded bedding, and sole markings
are the most important physical sedimentary structures; equally common are various biogenic sedimentary structures formed in the sediments shortly after deposition by living animals and plants. The two
principal kinds are: tracks and trails produced by animals on the *surface* of the sediment, and burrows and borings produced by animals
and plant roots *within* the sediment (Figure 3.20). Both of these types
of biogenic structure can provide important clues to the original depositional environment of the sediment.

In addition to such structural traces of organic activity, animal shells
and other hard parts of animals and plants are also commonly found

FIGURE 3.19
Sole markings, caused by turbidity current scour, exposed on the base of a sandstone. (W. Hiller.)

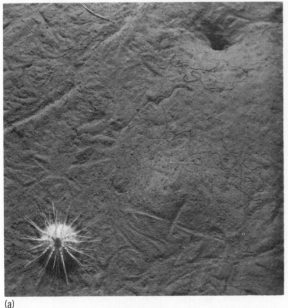

(a)

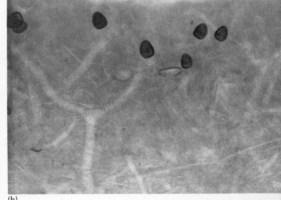

(b)

(c)

FIGURE 3.20
Biogenic structures (a, b). (a) Surface tracks and trails made by bottom-dwelling animals on muds of the deep-ocean floor. (b) X-ray cross section of burrows made within lagoonal muds by burrowing animals. Some of the animals' shells can be clearly seen. (c) An accumulation of animal shells becoming buried in beach sands. (a: D. M. Owen; b, c: Donald C. Rhoads.)

in sedimentary accumulations (Figure 3.20). We have already seen that many chemical sediments, particularly calcium carbonate and silica accumulations, are formed almost entirely from the shells and hard parts of animals and plants. In addition, most detrital sediments also contain scattered animal and plant remains. Usually these biologic remains persist, as *fossils*, when the sediment is altered to sedimentary rocks, providing invaluable clues to the original depositional environment of the sediment.

TERRESTRIAL SEDIMENTARY PATTERNS

Both detrital and chemical sediments are accumulating on the earth today under a host of different environmental conditions—near polar glaciers and around tropical coral reefs, in valleys far above sea level, and in the deepest ocean basins. The principal goal in studies of these present-day sediments is to determine the characteristic associations of compositions, grain sizes, and structures that are forming under each environment so that this knowledge can be used to interpret the environments preserved in ancient sedimentary rocks.

The most important environmental distinction in both present-day and ancient sediments is between those that accumulate on land as terrestrial sediments and those that are deposited on the ocean floor as marine sediments. In this section we shall review the characteristic patterns of sediment accumulating today in a variety of terrestrial environments before moving on to consider the much larger volumes of sediment accumulating on the ocean floor.

3.5 Fluvial Environments

The sedimentary environments formed by continuously flowing streams in tropical and temperate regions are known as **fluvial environments**; because of their relative accessibility, these are the most thoroughly studied and best understood of all present-day sediment accumulations. Sand and clay carried by streams during floods are normally deposited downstream as the flood subsides, only to be transported and deposited again during the next flood. The deposition of fluvial sediments, called alluvium, is therefore usually only a temporary stop in a journey that leads ultimately to the ocean. Nevertheless, in certain regions, such as low-lying areas where the land surface is rapidly sinking as a result of earth-moving forces, the alluvium deposited by one flood may be buried *without erosion* by deposits of the next flood, and thus be removed indefinitely from the possibility of further transportation toward the ocean. Under such conditions fluvial sediments hundreds of meters thick may accumulate and, ultimately, be preserved as sedimentary rocks.

Studies of present-day fluvial sediments show that they usually consist of two strikingly different kinds of detrital particles: thick layers of well-sorted sand or gravel alternate with layers made up of poorly sorted mixtures of finer sand and mud. In addition to differences in grain size and sorting, the alternating sand and mud layers usually show a sharp contrast in sedimentary structures. The sands usually show abundant cross bedding, indicating that they were deposited by swift horizontal fluid motions which strongly rippled the sediment surface; in contrast, the muds usually show a dominance of parallel bedding, indicating vertical settling in relatively still water (Figures 3.21 and 3.22).

Extensive studies of modern stream deposition have shown that these alternating sand and mud layers tend to form simultaneously, but in different horizontal and vertical positions. During floods the highest velocities of water movement tend to occur in the deepest parts of the stream channel; there the heaviest, coarsest particles are concentrated and moved by bouncing, sliding, and rolling along the bottom. The finer particles, on the other hand, are widely dispersed in suspension throughout the moving water which covers the entire floodplain (Figure 3.21). As a result, the coarse and fine particles are spatially separated. The coarse particles accumulate as **channel deposits**—thick masses of well-sorted sand and gravel which are rippled and cross-bedded by rapid channel flow. As the river channel slowly shifts its position, these channel deposits become covered by finer-grained **overbank deposits**—clays and fine sands that are spread in suspension over the entire floodplain and deposited there as the flood-waters recede (Figures 3.21 and 3.22). Unlike the channel deposits, these overbank deposits may accumulate in a variety of subenvironments such as swamps, marshes, or temporary lakes. This horizontal and vertical alternation of channel sands and overbank muds is the characteristic pattern of fluvial sediments.

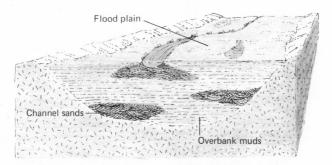

FIGURE 3.21
Characteristic fluvial sediments.

(a)

(c)

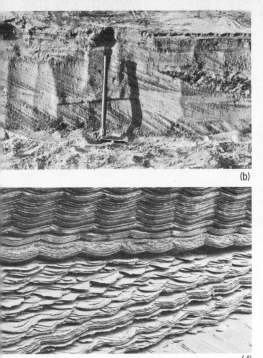

(b)

(d)

(e)

FIGURE 3.22

Fluvial sediments. (a) Channel sands and gravels exposed along a stream in northern Scotland. (b) Close-up view of cross-bedded channel sands exposed in a small trench along a stream in western Indiana. (c) Flooding and deposition of overbank sediments along the Missouri River in western Iowa. The normal river channel can be seen at the right. (d) Close-up view of overbank mud deposits in a 1 m (3 ft) cross section along the Colorado River in western Arizona. (e) Ancient channel sands (light color) and overbank muds (dark color) preserved in sedimentary rocks exposed in southern Connecticut. (a: Geological Survey of Great Britain; b: P. E. Potter; c: U. S. Department of Agriculture; d: Tad Nichols.)

Fluvial sediments are always composed predominantly of detrital sedimentary particles. The only chemical sediments normally associated with fluvial environments are organic materials; these are rare in channel deposits, but sometimes occur in abundance in overbank muds, particularly those deposited in marshes and swamps. In such environments the muds are commonly very dark gray or black because of abundant fine-grained plant debris. In addition, larger plant fragments, such as leaves, stems, or even logs, sometimes are found in such deposits. Occasionally these larger plant fragments may accumulate in such high concentrations that they overshadow the muds to form peat, but peat occurs in only a small fraction of present-day fluvial sediments. Most contain relatively few animal and plant remains because such material quickly decays when exposed to the air between periods of flood.

3.6 Desert Environments

Much of the sediment accumulating today in desert regions is deposited by streams just as are the fluvial sediments of temperate and tropical areas. These desert sediments differ markedly from those of humid regions, however, because temporary desert streams provide little opportunity for separating the sedimentary particles into distinctive channel and overbank deposits. Most desert streams flow only during infrequent periods of heavy rainfall, and even then they flow only over relatively short distances—usually from desert highlands down to the surrounding lowlands, where they quickly disappear by evaporation and absorption into the dry, porous soil. Under such conditions the sedimentary debris moved by the streams tends to be concentrated at the base of the highlands in wedge-shaped **alluvial fans** (Figures 3.23 and 3.24). Because there are no permanent streams to further transport alluvial fan deposits toward the ocean, they can accumulate to great thicknesses as older fan deposits are buried by younger.

Alluvial fan deposits are almost always composed of a poorly sorted mixture of detrital particles ranging in size from the finest clay to extremely coarse gravel. They commonly contain abundant feldspar fragments, along with the more typical quartz and clay particles, because the relative lack of chemical weathering and the rapid burial of the sediments prevent much of the eroded feldspar from weathering to clay minerals. Cross bedding may be present, but in general these poorly sorted detrital mixtures show few distinctive sedimentary structures.

In addition to alluvial fan deposits, two less common but very distinctive kinds of sediment also accumulate in desert regions. The first are chemical evaporites formed in temporary lakes that sometimes occur on the floors of desert basins (Figures 3.23 and 3.24). These lakes are filled during the infrequent periods of heavy rainfall as ephemeral streams flow from desert highlands into the surrounding lowlands. Between rainfalls the waters are partially or completely lost to the atmosphere by evaporation; when this occurs the dissolved elements in the lake water are deposited as chemical evaporites. Evaporites deposited in different desert lakes show a wide range of chemical

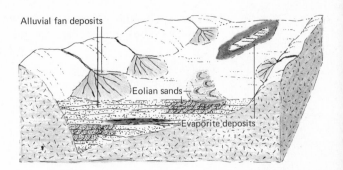

FIGURE 3.23
Characteristic sediments of desert regions.

(d)

FIGURE 3.24
Desert sediments. (a) Cross section of poorly sorted sediments of a
small alluvial fan in southern California. (b) Desert evaporite
deposits, Death Valley, California. (c) Cross bedding in desert dune
sands of southern California, exposed in a small trench. (d) Ancient
alluvial fan deposits (above) and dune sands (below) preserved in
sedimentary rocks exposed in southeastern Utah. (a, c: John S. Shelton;
b, d: U. S. Geological Survey.)

compositions because of differences in the dissolved elements transported to the lakes by local streams. Calcium and sodium evaporite
minerals, such as calcite, gypsum, and halite, are the most common
deposits, but some evaporites have high concentrations of unusual
minerals containing such rare elements as boron and lithium.

Desert basins where evaporation is intense provide the only major
setting where *non*biogenic chemical sediments accumulate above sea
level. Biogenic chemical sediments almost never occur in desert environments, both because plants and animals are scarce and because the
lack of water, and high temperatures, lead to rapid decay of the small
amounts of organic matter present.

In addition to alluvial fan and evaporite deposits, a third type of
sediment may occur in desert environments. These eolian sediments,
as they are called, originate when the sands and clays of alluvial fans
are further transported and deposited by wind action (Figures 3.23
and 3.24). Fan deposits normally lack a protective cover of vegetation
and are therefore continually subject to varying velocities of wind flow
which tend to sort them into different size groups just as does the
flowing water of fluvial environments. During periods of high winds,
fine silt and clay particles are transported in suspension as *dust storms*
and may be carried long distances, often even beyond the desert into
more humid environments, before the wind subsides.

Sand particles, in contrast, are usually too heavy for suspension transport by wind; instead, they are bounced and rolled along the desert floor to be concentrated as dunes on the windward side of desert basins. There they may be covered and preserved by later alluvial fan deposition. Wind-deposited sand accumulations show many of the characteristics of fluvial channel sands in that they are well sorted and show strong cross bedding and surface rippling caused by the rapid horizontal flow of the depositing fluid. Certain details of eolian cross bedding differ, however, from those of water-deposited sands. In particular, the angular bedding is commonly much steeper in wind deposits, and this, combined with the associated alluvial fan and evaporite sediments, normally makes it possible to recognize ancient desert sands.

3.7 Glacial Environments

Sediments accumulating today in glacial environments show many of the same characteristics as desert alluvial fan deposits, and the two types of deposits are often difficult to distinguish in ancient sediments. Like desert alluvium, most glacial deposits are unsorted mixtures of clay, sand, and gravel which lack distinctive sedimentary structures. Because glacial environments, like deserts, normally lack a dense cover of vegetation, wind action may locally sort the surficial glacial deposits into eolian sand dunes similar to those of desert regions. Unlike desert sediments, however, those of glacial regions rarely show evaporite deposits, which normally originate only in warm climates. As in desert environments, organic matter and animal and plant remains are rare in glacial deposits. The most characteristic feature of glacial sediments is the presence of pebbles that have been scratched and abraded by moving against solid bedrock while frozen into the flowing ice; such striated pebbles are the surest criterion for the recognition of ancient glacial deposits (Figure 3.25).

3.8 Coastal Environments

Only a relatively small fraction of the sediment accumulating on the earth today is found above sea level in fluvial, desert, and glacial environments. Instead, most sedimentary debris is ultimately transported to the ocean where it accumulates in various *marine* environments under the influence of ocean waves and currents. The sediments of marine environments will be the subject of the next section, but before turning to them we need to consider those environments along the edges of the continents where sedimentary accumulations are forming under the influence of *both* ocean water *and* the dominant terrestrial depositing agents—streams, wind, and ice.

All such coastal sedimentary environments occur in the narrow zone

FIGURE 3.25
Ancient glacial sediments exposed in southwest Africa, showing striated pebbles scratched by moving ice. (J. W. Hälbich.)

where land, sea, and air come in contact. Coasts, in general, are regions of intense and varied fluid motions: waves stir the bottom and pound the land, tides rise and fall, and both forces act to create strong currents parallel to the shoreline. At the same time, streams flow into the ocean supplying new sediment and creating additional current motions as fresh water mixes with the heavier salt water of the sea. Because of these varied fluid motions, relatively little sediment accumulates in coastal regions. Instead, most is ultimately transported beyond the shore to come to rest in deeper water where there are fewer opportunities for fluid transport. There are, nevertheless, thick sequences of coastal sediments accumulating today in a few regions where the underlying crustal rocks are slowly sinking and thus allowing the older sediments to be buried by younger sediments before they have an opportunity to be transported farther. Studies of such accumulations show that the sediments are normally well sorted because of the varying velocities of flow to which they are subjected.

In general, the larger, heavier sand and gravel particles are concentrated in linear belts parallel to the shore; such belts make up the *beaches* and *barrier islands* that are a familiar feature of so many coasts (Figures 3.26 and 3.27). Both form in zones of strong wave or current action which remove the finer sedimentary particles and concentrate the coarser sand and pebbles. As in the well-sorted sands of fluvial channels and desert dunes, beach and barrier sands tend to be rippled and cross-bedded from the motions of the depositing fluid and contain little organic matter or preserved animal or plant remains. Indeed, ancient beach and barrier deposits are often indistinguishable from channel or dune sands except by their association with other, more distinctive, sediments.

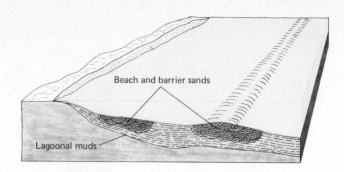

FIGURE 3.26
Characteristic sediments of coastal regions.

Beach and barrier sands

Lagoonal muds

FIGURE 3.27
Coastal sediments.
(a) A beach made up
of large pebbles, New
Zealand. (b) Ancient
lagoonal deposits
preserved in sedimen-
tary rocks exposed in
western Wales. (a:
Tad Nichols; b: Geo-
logical Survey of
Great Britain.)

In contrast to beach and bar sands, finer silt and clay particles in coastal regions accumulate primarily in sheltered *lagoons* where fluid motions are less intense (Figures 3.26 and 3.27). Lagoons develop wherever the configuration of the coastline or the presence of energy-absorbing sand barriers protect shallow bodies of water from strong wave and current motions. Such regions receive water and sediment both from streams on their landward side and from the ocean on their seaward side. Because of their sheltered setting, the clay and silt particles are not normally removed by fluid motions, but instead accumulate to form thick mud or muddy sand deposits.

Lagoons usually support an abundance of plant and animal life; indeed, in very shallow lagoons, bottom-dwelling plants often grow above the surface of the standing water to form *marshes*. Shells, plant fragments, and other larger remains of bottom-dwelling animals and plants are therefore common in lagoon and marsh sediments. The presence of such remains helps distinguish ancient lagoonal muds

(a)　　　　　　　　　　　　　　　　　　　　　　　(b)

from muds formed as fluvial overbank deposits, which lack abundant organic remains. The exact kinds of animal and plant remains found in lagoon sediments are determined by the proportion of fresh water and ocean water that normally occurs in the lagoon. Most ocean-dwelling life cannot tolerate fresh water and vice versa, so that lagoons dominated by ocean water contain characteristic marine organisms and those dominated by streams have characteristic fresh-water animals and plants. Lagoons which fluctuate rapidly between salt and fresh water support relatively little life because neither group can become established.

The presence of bottom-dwelling organisms in lagoons also strongly influences the sedimentary structures of lagoonal deposits. Like fluvial overbank deposits, lagoonal muds are usually deposited as horizontal layers with parallel bedding. In some lagoons that lack abundant bottom life this bedding persists, but more commonly it is partially or wholly destroyed by the burrowing activities of bottom animals. This activity, in time, creates an abundance of biogenic sedimentary structures which are also a characteristic feature of most lagoonal sediments.

Beach and bar sands, and lagoonal muds, are the principal sediments forming today in coastal environments. In polar and temperate regions they are normally composed of land-derived quartz fragments and clay minerals transported to the shore by streams. Such deposits are also common along tropical coasts, but in tropical regions where there is no nearby source of land-derived particles beaches may be made up entirely of sand-sized fragments of biogenic calcium carbonate and lagoons may be filled with calcium carbonate mud. In these settings the sedimentary particles are not derived from rock weathering on land, but from the removal of dissolved elements from ocean water by animals and plants living near the shore.

MARINE SEDIMENTARY PATTERNS

Most sediment accumulation takes place on the ocean floor in various marine environments. In present-day marine sediments a primary distinction can be made between those deposited under relatively shallow water on the submerged continental shelves and those accumulating beyond the continents in the deep-ocean basins (Figure 3.28). Continental shelf sediments tend to accumulate both more rapidly, and by somewhat different processes, than do those of the deep oceans. Shelf and deep-ocean environments resemble each other, however, in that each has some large areas where land-derived sands and muds are accumulating as detrital sediments, and each has other large areas dominated by chemical accumulations which, for brevity, are usually referred to as *carbonate* sediments because they are dominated by biogenic calcium carbonate. We shall consider both detrital and car-

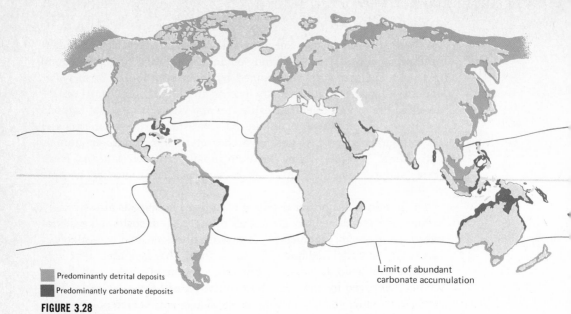

Predominantly detrital deposits

Predominantly carbonate deposits

Limit of abundant
carbonate accumulation

FIGURE 3.28
Present-day continental shelves, the submerged margins of the continents where most sediment accumulation takes place. (Compiled from various sources.)

bonate accumulations of the continental shelves before examining the somewhat different detrital and carbonate accumulations of the deep-ocean basins.

3.9 Detrital Shelf Environments

Most of the sand and mud delivered to the oceans by streams is deposited on the submerged shelves which surround the continents. These shelves today have an average width of about 80 kilometers (50 miles) and slope gradually seaward to an average water depth of about 200 meters (600 feet); beyond them lie the steeper slopes of the continental margins that lead to the deep-ocean basins (Figure 3.28). Because the ocean floor on the shallow, near-shore parts of these shelves is subjected to much more intense wave and current motions than is the deeper shelf, we might predict that large sedimentary particles should be sorted and concentrated on these high-energy near-shore bottoms as accumulations of sand and gravel. Conversely, only finer silt and mud particles should reach the outer shelf region where relatively little wave and current energy is available to transport larger particles (Figure 3.29a).

Studies of sedimentary rocks have shown that this offshore decrease in particle size *was* common in ancient marine sediments, for they normally show a concentration of sand and gravel in near-shore regions that grades seaward into finer silt and mud deposits. This pattern does *not* occur, however, on the continental shelves today. In-

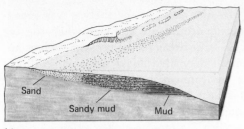 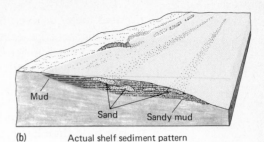

(a)   Predicted shelf sediment **pattern**   (b)   Actual shelf sediment pattern

FIGURE 3.29
Predicted and actual patterns of sediment distribution on the continental shelves. The relatively recent expansions and contractions of continental ice sheets have caused repeated migrations of the shoreline and deposition of shallow-water sediment over much of the shelves. The more regular "predicted" pattern of offshore size decrease occurs in many ancient sediments.

stead, studies of modern shelf sediments (made by sampling or photographing the bottom from ships or, in shallow water, by diving with scuba gear) have shown that well-sorted *sand* is the dominant sediment over much of the present-day shelves (Figure 3.29b). Furthermore, sand is particularly abundant on the outermost shelf, where it should be rare or absent. In addition, mud accumulations are far more common today in shallow, near-shore settings than would be predicted from the distribution of wave and current energy.

We have already anticipated the reasons for this anomalous pattern —the relatively recent changes in sea level caused by the expansions and contractions of large continental ice sheets. These sea-level changes led to many oscillations of the shoreline back and forth across the continental shelves (Figure 3.30). As a result, much of the sediment

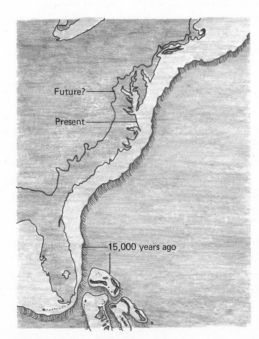

FIGURE 3.30
Fluctuation of the Atlantic shoreline of the United States. Fifteen thousand years ago enough ocean water was locked up in large continental ice sheets to drop sea level about 200 m (660 ft) and shift the shoreline to the edges of the present continental shelf. If the present Greenland and Antarctica ice sheets were to melt, sea level would rise and the shelves would expand to approximately the line marked "future?".

exposed today on even the outermost parts of the shelves was deposited in high-energy near-shore environments at times when the sea level stood much lower. Indeed, large areas of the present shelves appear to be covered with materials deposited above sea level in fluvial or, on polar shelves, glacial environments. Most present-day shelf sediments, therefore, do not reflect their present environments but, instead, are *relict* sediments of earlier environmental patterns.

Too little time has elapsed since the most recent melting of continental ice sheets, and the consequent rise of sea level, to allow sediments now being delivered to the shelves to reach an equilibrium with their environments. Presumably, if sea level were to remain at its present position long enough (an unlikely event, as we shall see in Chapter 10), fine muds now accumulating near shore would be slowly transported seaward to cover the near-shore sands that now dominate the shelf, leading to the expected offshore decrease in grain size. In the meantime, only a few large rivers that carry huge volumes of mud and sand are able to deposit these materials beyond the shore region by building out broad deltas onto the continental shelves. In contrast, the lower reaches of most streams have been "drowned" by rising sea level to create elongate, lagoonlike bodies of water known as estuaries. Most of the sediment carried by these streams is accumulating in these coastal estuaries, which must be filled before delta build-out, and transport of particles onto the shelf, can begin.

Because present-day continental shelves are mostly covered with sediments that were deposited under previous environmental conditions, studies of modern shelf sediments do not provide clear sedimentary patterns that can be applied to interpreting sedimentary rocks. Nevertheless, certain generalizations about shelf-deposited sands and muds are possible.

Perhaps the most characteristic feature of present-day shelf sands and muds is that they contain abundant evidence of the prolific life that is almost everywhere present on the ocean floor. Shells, bones, and other hard remains of bottom-dwelling marine animals are common in shelf sediments, as are biogenic sedimentary structures and fine-grained organic matter (Figure 3.31). About the only exceptions are those sediments, usually fine muds, that accumulate in deep, isolated basins where there is little water movement. Under such conditions there is usually too little oxygen in the bottom waters to support life, and such environments are the "deserts" of the ocean floor. Fine-grained organic matter produced elsewhere, however, commonly accumulates in quantity in such environments because there is no bottom life to utilize the particles as food. The results are black, evenly bedded, organic-rich mud deposits that lack both shells and biogenic sedimentary structures (Figure 3.31).

FIGURE 3.31
Detrital shelf sediments. (a) Accumulation of oyster shells and (b) vertical biogenic
burrowed structures preserved in ancient shelf sediments exposed in western South Dakota.
(c) Ancient organic-rich shelf mud deposits, preserved as shales exposed in central New
York. (d) Alternating bands of parallel-bedded mud and silt preserved in ancient shelf
sediments exposed in western Wales. (a-c: Donald C. Rhoads; d: Geological Survey of Great
Britain.)

Another characteristic feature of offshore detrital shelf sediments, in those few deltaic areas where they are being deposited today, is the frequent occurrence of small-scale alternations of parallel-bedded muddy and sandy layers, each usually from a few millimeters to a meter thick (Figure 3.31). These alternating layers are also very common in ancient shelf sediments, but their exact origin is obscure. Most probably they result from large, infrequent storms which agitate the water to unusual depths and transport near-shore sand particles seaward. Between these events, mud slowly accumulates to give the alternating muddy layers. On the more steeply sloping parts of the shelf, gravity transport of sedimentary particles by turbidity currents can also lead to alternating mud and fine sand sequences which, however, usually show graded bedding and other distinctive structures. Such turbidite sequences, as they are called, are less common in shelf sediments than in sediments accumulating in deeper environments beyond the shelf.

3.10   Carbonate Shelf Environments

Land-derived sand and mud are today the dominant sediments on polar and temperate shelves, but are replaced on many tropical shelves by extensive accumulations of calcium carbonate that have either been precipitated directly from the local ocean water or, more commonly, removed indirectly by the activities of animals and plants (see Figure 3.28). Direct precipitation is favored in tropical regions because the solubility of calcium carbonate in ocean water varies *inversely* with water temperature: the higher the temperature the less dissolved calcium carbonate it can contain, and vice versa. Only in shallow, tropical oceans does the water temperature rise high enough to cause spontaneous precipitation of calcium carbonate. Even then, some reduction in the volume of water by evaporation is probably necessary to further increase the calcium concentration and thus decrease its solubility, although the exact chemistry of the process is still poorly understood. Such conditions are met on many shallow tropical shelves today and result in sedimentary accumulations of directly precipitated calcium carbonate grains (Figures 3.32 and 3.33). Most commonly these grains are rounded, sand-sized particles, called ooliths, which show a characteristic concentric pattern of growth (see Figure 3.6).

Under exceptional circumstances, evaporation of ocean water may proceed far enough to precipitate still more soluble dissolved elements (see Figure 3.7). Such conditions are met today only in a few tropical lagoons that receive infrequent inflows of normal ocean water. Between such influxes, intense evaporation may lead to deposition of *marine evaporite* sediments (Figure 3.32). Usually such sediments are

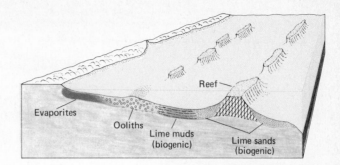

FIGURE 3.32
Characteristic sediments of carbonate shelves.

made up mostly of the minerals halite and gypsum. Less commonly various unusual sodium, potassium, and magnesium minerals are deposited in such settings.

Directly precipitated calcium carbonate and marine evaporites are not forming today in polar or temperate regions and make up only a small fraction of the shelf sediments of tropical regions. Most present-day tropical shelves are covered instead by calcium carbonate particles of *biogenic* origin (Figures 3.32 and 3.33). The exact reasons for the dominance of biogenic calcium carbonate deposits on tropical shelves are still uncertain. Probably the precipitation of calcium carbonate by animals and plants is favored in the tropics because of the lowered solubility of calcium carbonate just as is direct precipitation. Unlike directly precipitated calcium carbonate, however, shell-bearing marine animals are also common on temperate and polar shelves, yet in such regions they generally make up only a small fraction of the shelf sediments, which are predominantly detrital sands and muds. Perhaps the most likely explanation is that streams carrying land-derived sand and mud are less common in the tropics, particularly in the subtropical zones of arid climates. In these regions there is little detrital material supplied to the shelves to dilute the accumulating carbonate sediments.

When shell-bearing animals and carbonate-secreting plants die, their skeletons do not normally remain intact, but are progressively reduced to smaller and smaller particles as the surrounding organic matter decays and by such secondary processes as wave abrasion and reworking by predators (Figure 3.34). As a result, most biogenic carbonate sediments consist not of whole shells, but of sand- and mud-sized particles derived from whole shells. These lime sands and lime muds are the dominant shelf carbonate sediments and generally show the same patterns of size sorting and sedimentary structures as do detrital sands and muds. The lime sands tend to be concentrated in near-shore areas of high-energy water motions and show cross bedding and other structures indicative of such motions. The lime muds, on the other hand, tend to accumulate in sheltered or deeper environments where there

(a)

(b)

(c)

(d)

FIGURE 3.33

Carbonate shelf sediments. (a) Aerial view of a part of
the Bahama Bank, a broad shelf off the Florida coast
dominated by carbonate deposition. (b-d) Close-up views
of nonbiogenic lime mud (b), biogenic lime sand (c),
and organic reef (d), all on the Bahama Bank. (e) Ancient
lime mud preserved as sedimentary rock exposed in
central Alabama. (a-d: Norman D. Newell, The American
Museum of Natural History.)

(e)

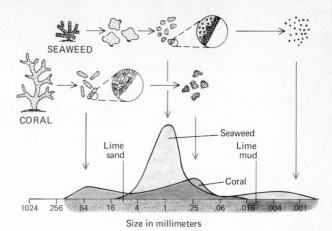

FIGURE 3.34
The breakdown of calcium-carbonate-depositing seaweeds and corals to produce lime sand and lime mud. (From Laporte, "Ancient Environments," 1968, after Folk and Robles.)

is less water motion. As on temperate and polar shelves, anomalous relict accumulations of lime sand, formed during recent intervals of lowered sea level, are probably common on deeper tropical shelves today, but this is not yet certain since tropical shelves have been less extensively explored than have those of temperate regions.

In addition to lime mud and lime sand, a third important type of biogenic carbonate accumulation occurs locally on most tropical shelves. These are organic reefs made up of closely interlocked skeletons of living and dead marine life (Figures 3.32 and 3.33). Many organisms contribute to present-day reefs, but the dominant groups are corals, sponges, and certain seaweedlike plants which thrive only in relatively warm, tropical waters. Because corals are often a conspicuous part of modern reefs they are commonly called "coral reefs," but this term tends to obscure the other equally important reef-building organisms that are always present.

Unlike isolated shells and skeletons, the closely packed and cemented calcium carbonate of reefs does not readily break down into smaller particles when the reef-building organisms die and decay. As a result, massive structures composed of dead carbonate skeletons persist indefinitely to form a foundation upon which the animals and plants of the living reef are but a thin surface film. The largest such reef that is still being actively built today is the Great Barrier Reef, which extends for 1,600 km (1,000 mi) along the coast of northeastern Australia; similar but smaller reefs are common on most tropical shelves.

Because of the presence of scattered organic reefs, tropical shelves usually show less regular profiles and more local relief than do the gently sloping shelves of temperate and polar regions. Basins and depressions around the massive reefs usually trap thick, *interreef* accumulations of lime sand and lime mud derived both from local, non–reef-building organisms and by wave and current erosion of the reef itself. Such reef and interreef accumulations of calcium carbonate are also a common feature in ancient sedimentary rocks.

3.11 Deep-Ocean Environments

Deep-ocean sediments are mostly either detrital sands and muds or biogenic carbonates just as are those of the shallower continental shelves. There are, however, significant differences in the depositional processes of these materials in the deep ocean.

Land-derived sand and clay particles reach the deep-ocean basins by either of two routes (Figure 3.35). Some are *bottom-transported*; that is, they are moved seaward from the continents across the shelves by bottom currents and into the deep basins by gravity slumping down the continental margins. Such slumping results in turbidity currents that convey the sediments into the deep basins, usually through submarine canyons, and distribute them over wide areas in thin, graded beds. Most of the deep basins of the Atlantic Ocean are covered with thick sequences of such graded sand and mud layers transported across the continental shelves (Figures 3.36 and 3.37).

The second process by which land-derived sands and muds reach the deep oceans is by transport not along the bottom, but by motions in either the upper layers of water or the overlying atmosphere. Clay and fine sand particles may be carried long distances by such movements. When the current motions or winds that transport them cease, they slowly sink to accumulate on the ocean floor as *surface-transported deposits*. Limited observations suggest that wind transport is the more important contributor to such deposits, for sand and clay seem to be transported across the shelves in surface waters only during very large storms.

Relatively thin, parallel bedded accumulations of surface-transported sediments are characteristic of many deep basins far from the continental margins, particularly those of the Pacific Ocean (Figures 3.36, 3.37). Unlike the Atlantic, much of the Pacific is bordered by deep trenches that extend downward thousands of meters below the surrounding ocean floor. These trenches tend to trap the bottom-transported materials that move down the continental margins and thus

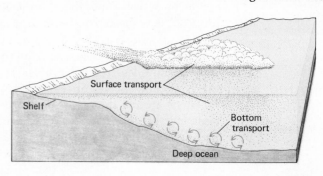

FIGURE 3.35
Transport of land-derived sand and mud to the deep ocean.

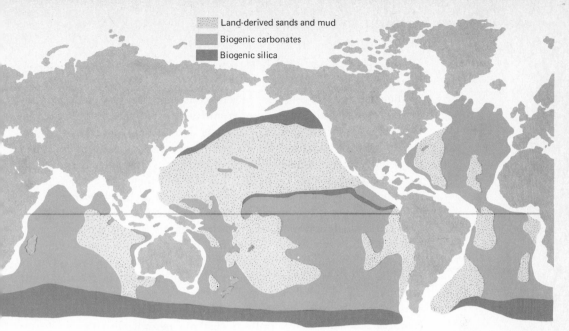

Land-derived sands and mud
Biogenic carbonates
Biogenic silica

FIGURE 3.36
Present-day deep-ocean sediments. (Modified from Hill, "The Sea," vol. 3, 1963.)

FIGURE 3.37
Deep-ocean sediments. (a) Land-derived bottom muds on the floor of the deep Atlantic. Note the abundant animal tracks and burrows. (b) Biogenic calcium carbonate sand on a submarine hill in the deep Atlantic. The smaller particles are skeletons of foraminifers and the larger, light-colored particles are pteropod shells. (a, b: Richard M. Pratt, U.S. Geological Survey.)

prevent them from spreading throughout the deep basins as in the Atlantic. As a result, thin, slowly accumulating deposits of surface-transported muds and silts tend to floor most deep Pacific basins beyond the deep marginal trenches, which contain thick, bottom-transported sequences. In the Atlantic, the principal surface-deposited accumulations occur in equatorial basins east of the Midatlantic Ridge, where winds transport and deposit much debris from the adjacent deserts of northern Africa.

Although the volume of land-derived sand and mud ultimately reaching the deep ocean by either bottom or surface transport is large, it nevertheless represents only a small fraction—probably about one-tenth—of all such particles delivered by streams to the ocean. The great bulk accumulates instead on the continental shelves.

In addition to land-derived sands and clays, the deep ocean, like the

(a)

(b)

continental shelves, also has large areas dominated by biogenic deposits which, however, differ fundamentally from the biogenic sediments of the shelves. The principal difference is that deep-sea biogenic sediments are generally formed not from fragments of the relatively large skeletons of bottom-dwelling animals and plants, but from the tiny skeletons of the floating animals and plants that are collectively called *plankton*. Such organisms thrive in enormous numbers in the sunlit surface waters of the oceans. As they die and their soft tissues decay, their tiny skeletons sink to the ocean floor and accumulate there.

On those parts of the ocean floor that receive large quantities of land-derived silt and mud, such as the Atlantic basins, the steady rain of planktonic skeletons is diluted and makes up only a small fraction of the bottom materials. In regions where the land-derived influx is small, however, planktonic skeletons may accumulate as the dominant sediment. In the Atlantic such biogenic accumulations are found mostly on submarine hills and on the slopes of the Midatlantic Ridge (Figures 3.36 and 3.37). Bottom currents periodically transport any accumulated clay particles from these higher areas into the surrounding basins, leaving a residue of the larger and less easily transported skeletons. In the Atlantic, the principal contributors to such deposits are two groups of floating animals called *foraminifers* and *pteropods* and one group of tiny plants called *coccolithophores*, all of which build skeletons of calcium carbonate (Figure 3.37).

Carbonate deposits similar to those of the Atlantic also occur in some regions of the Pacific; unlike the Atlantic, however, the deep Pacific also has large areas dominated by planktonic skeletons composed not of calcium carbonate but of *silica*. Two groups of planktonic organisms secrete skeletons of silica, the plantlike *diatoms* and the animallike *radiolarians*. Diatoms are particularly common in surface waters of polar regions, and their shells dominate bottom deposits in the North Pacific and in a broad belt under the ocean that surrounds Antarctica. Radiolarians, in contrast, are most abundant along the equator, and accumulations of their shells floor large regions of the equatorial Pacific (Figure 3.36).

SUMMARY OUTLINE

Sediments and Sedimentary Rocks

3.1 *Detrital sediments:* fine-grained clay minerals and quartz and feldspar particles of various sizes accumulate as muds, sands, or poorly sorted mixtures of the two.

3.2 *Chemical sediments:* materials in solution in lakes and the ocean may be precipitated directly as their concentrations are increased by evaporation, or indirectly through removal by animals and plants.

3.3 *Classification of sedimentary rocks:* most sedimentary rocks fall into one of four intergrading groups: sandstones, shales, carbonate rocks, and evaporites.

3.4   *Sedimentary structures:* detrital and chemical particles moved and deposited by fluids show various kinds of layered bedding which give clues to the nature of the depositing fluid; other structures originate from the actions of sediment-dwelling organisms.

Terrestrial Sedimentary Patterns

3.5   *Fluvial environments:* these are dominated by alternations of well-sorted sands deposited in the stream channel and overbank muds deposited during floods.

3.6   *Desert environments:* these show poorly sorted alluvial fan deposits, well-sorted sand dunes, and lake evaporites.

3.7   *Glacial environments:* these have sediments similar to those of desert alluvial fans, but often show striated pebbles abraded by moving ice.

3.8   *Coastal environments:* these are dominated by well-sorted sands and gravels deposited on beaches and bars, and muds that accumulate in sheltered lagoons.

Marine Sedimentary Patterns

3.9   *Detrital shelf environments:* these normally show an offshore decrease in particle size; modern shelves have anomalous deep sands caused by recent sea level fluctuations.

3.10   *Carbonate shelf environments:* these replace detrital sands and muds on many tropical shelves; most are composed of fragments of carbonate-secreting animals and seaweeds.

3.11   *Deep-ocean environments:* these have accumulations of land-derived sands and muds as well as biogenic carbonate and silica derived from planktonic animals and plants.

ADDITIONAL READINGS

*Allen, J. R. L.: *Physical Processes of Sedimentation*, American Elsevier Publishing Co., Inc., New York, 1970. An authoritative intermediate text on detrital sediments and their environments of deposition.

Berner, R. A.: *Principles of Chemical Sedimentology*, McGraw-Hill Book Company, New York, 1971. An advanced text on chemical sediment.

Blatt, H., G. Middleton, and R. Murray: *Origin of Sedimentary Rocks*, Prentice-Hall, Inc., Englewood Cliffs, N.J., 1972. A comprehensive and up-to-date intermediate text.

Degens, E. T.: *Geochemistry of Sediments*, Prentice-Hall, Inc., Englewood Cliffs, N.J., 1965. An intermediate text on chemical sediments.

*Laporte, L. F.: *Ancient Environments*, Prentice-Hall, Inc., Englewood Cliffs, N.J., 1968. A brief, readable introduction to sedimentary environments.

Pettijohn, F. J., and P. E. Potter: *Atlas and Glossary of Primary Sedimentary Structures*, Springer-Verlag New York, Inc., Berlin, 1964. Outstanding photographs of sedimentary structures.

*Available in paperback.

chapter **4**

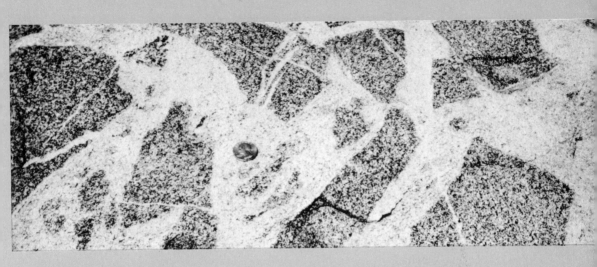

Earth
Chronology

Our understanding of events in the earth's past rests almost entirely on studies of crustal rocks exposed today at or near the earth's surface. The volatile fluids of past oceans and atmospheres are never preserved to leave a historical record, whereas crustal rocks, particularly those making up the continents, have been slowly accumulating throughout most of the earth's long history. Many of these rocks have survived with little change for millions or even billions of years and may thus provide revealing insights into the earth's past. For example, plutonic igneous and metamorphic regions indicate intervals of mountain-building deformation deep within the crust, while volcanic and sedimentary rocks reflect changing landscapes at the earth's surface at the time they were formed. Sedimentary rocks are particularly useful because they may suggest past climates, and thus provide *indirect* clues about earlier states of the ocean and atmosphere. In order to decipher the long sequence of events recorded in the many kinds of crustal rocks, however, we must have some means of determining their relative ages so that a worldwide chronology of events can be established.

There are two complementary, but basically independent, methods for establishing the age relations of ancient rocks. The first is applicable primarily to sedimentary rocks and relates to the sequential nature of sediment deposition and to progressive changes in the animal and plant remains preserved in ancient sediments. The second method applies primarily to igneous and metamorphic rocks and relates to the steady rates of decay of certain radioactive nuclides incorporated into the rocks as they were formed. We shall look briefly at both of these techniques for determining earth chronology before considering the universal time scale of earth history that they have provided.

RELATIVE AGES AND EARTH CHRONOLOGY

4.1 Physical Relationships

As long ago as 1669, Steno recognized that the physical process of sediment deposition provides a means of establishing relative chronology that is not available for most rocks that solidify from molten magmas. Sediments, we have seen, normally accumulate under the influence of gravity as relatively thin horizontal sheets called *beds* or *strata*. Each such stratum is usually deposited on top of older beds and is, in turn, buried by younger ones. This simple relationship means that when thick sequences of sediment are preserved without deformation, the *underlying beds are always older than those overlying them*. (Note that this relationship is not necessarily true of igneous rocks, which are normally melted from below and may crystallize

(a)

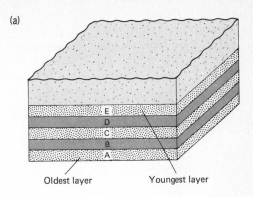

Oldest layer   Youngest layer

FIGURE 4.1
The law of superposition. (a)
In sequences composed entirely
of sedimentary rocks, younger
beds always overlie older. The
sequence shown thus becomes
progressively younger from A
to E. (b) Igneous rocks, on the
other hand, are not necessarily
older than overlying rocks. In
the sequence shown the igneous
unit (G) was emplaced into the
sedimentary layers (A-F) and
thus is younger than sediments
which overlie it.

(b)

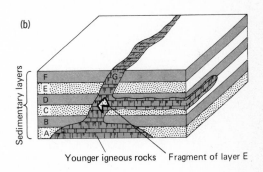

Younger igneous rocks   Fragment of layer E

while overlain by older rocks; see Figure 4.1.) Recall (from the Introduction) that this seemingly obvious principle—that younger sediments always overlie older—is known as the law of superposition and
provided the first key to deciphering crustal history.

Sequences of ancient sediments thousands of meters thick, representing many different ages and sedimentary environments, are found
on every continent. Although about 95 percent of the total *volume* of
rocks making up the continental crust is of igneous or metamorphic
origin, the remaining 5 percent of sedimentary rock is spread as a
veneer that covers much of the *surface area* of the continents (Figure
4.2). As a result, igneous or metamorphic rocks are directly exposed
on only about 25 percent of the continental surfaces; the remaining
75 percent is covered by a layer, averaging about 2 km (kilometers) in
thickness, of sedimentary rocks that lie on top of a much greater thickness of igneous and metamorphic rocks. Once the law of superposition
was understood, it became possible to infer sequences of ancient sedimentary events by merely observing the progressive upward changes
in sediment types preserved in this continental veneer.

Igneous events could also be related to the sedimentary stories told
about a region by the application of two additional relative age-dating

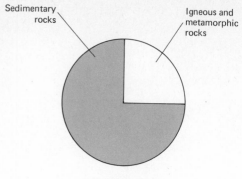

SURFACE AREA

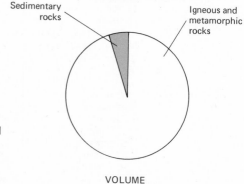

FIGURE 4.2
Surface area versus volume of
sedimentary rocks of the continental
crust. Although they make up only
5 percent of the volume of crustal
rocks, sedimentary rocks cover 75
percent of the continental surfaces.

VOLUME

principles: (1) intrusives are younger than the rock they intrude, and
(2) rock fragments are older than the rock that contains them. Figures
4.1b and 4.3 depict these two relationships. Clearly the igneous rocks
must be younger than the sediments they intrude. Further, the sedimentary layer E *must have* existed prior to igneous intrusion in order
for a fragment of it to be incorporated in the solidified igneous rock
(Figure 4.1b).

Using these three relative age-dating tools—superposition, intrusive
relationships, and fragment relationships—eighteenth-century geologists began attempts to organize the rock record of the earth's crust.
Some of the earliest efforts were conducted in Italy and Germany,
where investigators independently arrived at similar conclusions. As
a general rule, it was concluded that throughout the earth's crust plutonic and matamorphic rocks are the oldest and that they are uncon-
formably overlain by highly deformed sedimentary rocks, which in
turn are overlain by highly fossiliferous, gently folded sedimentary
layers.

It was within the framework of these observations that A. G.
Werner promoted his once-ruling concept of Neptunism (discussed in

(a)

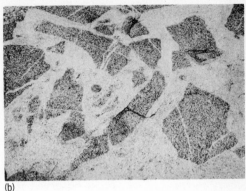

(b)

FIGURE 4.3
Relative age-dating tools. (a) Intrusive relationship. (b) Fragment of older rock incorporated in younger rock. (a, b: E. A. Hay.)

the Introduction). Werner decided that granular-textured plutonic rocks everywhere must have been chemically precipitated from a very deep primordial ocean. Following this the ocean lowered (according to his theory) and eventually yielded mechanically derived sediments (sand, clay, etc.) that were deposited at lower elevation, but were nevertheless younger than the crystalline rocks because of superposition (Figure 4.4). Although such a model is consistent with rock relationships that are frequently the case, it soon became apparent there are many exceptions. For example, granite or other primary rocks were seen to intrude highly fossiliferous strata, and hard, unfossiliferous sedimentary layers were found overlying highly folded, yet richly fossiliferous, formations. Werner's attempt to *impose* simplicity and order on the earth simply failed to account for reality. Once geologists began to make close observations of rock types and careful determinations of age sequences using the three techniques we have discussed, Neptunism was sunk. More thorough studies showed the rock histories of different regions to be complex and unique—there was no "universal system" such as Werner had proposed.

A major difficulty in satisfactorily conducting such careful and thorough studies, however, has continued to plague geologists even today. The problem stems from lack of adequate *exposures* of rocks. On most parts of the land surface the sedimentary rock veneer is itself overlain

FIGURE 4.4
Idealized cross section showing rock relationships as visualized by the late eighteenth-century Neptunists. Younger rocks are (1) less deformed, (2) more fossiliferous, (3) at lower elevation, and (4) less well consolidated.

Tertiary

Quaternary

Secondary

Primary

by a much thinner veneer of soil and vegetation, which largely hides the underlying rock. In such regions the sedimentary sequence is exposed only where streams cut into "bedrock" along hillsides or where deep, man-made excavations, such as road cuts, quarries, or mines, extend into the underlying rock. Even where such local exposures of bedrock are abundant, they commonly do not reveal more than a surficial fraction of the underlying sedimentary rock, which has an average thickness of about 2,000 m (meters). Only in mountainous regions of high relief is the complete sedimentary sequence normally exposed for study. As a result of these limitations, it is always necessary to correlate—that is, determine age equivalencies between the sedimentary rocks observed at widely scattered exposures.

The simplest kinds of sedimentary rock correlations are those based on the physical nature of the rocks themselves. Where exposures are continuous or very closely spaced, as sometimes occurs in desert regions, it may be possible to follow single beds or groups of beds over long distances. More commonly, physical correlations are made by recognizing similar rock types, or sequences of rock types, in discontinuous exposures (Figure 4.5).

The difficulty with physical correlations is that they are useful only for rocks originally deposited over relatively small areas and in the

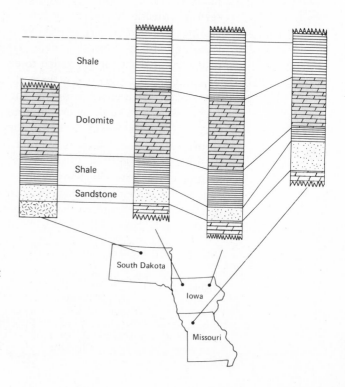

FIGURE 4.5
Physical correlations of sedimentary rocks. Similarities in rock sequence often permit interrelation of the continent's sedimentary veneer over wide areas. The illustration shows a distinctive sequence of sandstone (oldest), shale, dolomite, and shale (youngest) that can be recognized over much of the north-central United States.

same sedimentary environment or sequence of environments. Physical techniques are useless between continents, or over distances of thousands of kilometers on a single continent, where sedimentary environments and resulting sediment accumulations must have varied at any time in the past just as they do today. Physical correlations based only on similarities in rock type, in short, can never provide a worldwide chronology of sedimentary events, even though they sometimes permit local events to be worked out in great detail (Figure 4.6). What is needed to establish a worldwide chronology is an independent means of determining the relative ages of sedimentary rocks wherever, and in whatever environment, they were deposited. A close approach to such a chronology is provided by the fossil remains preserved in many sedimentary rocks.

4.2   Fossil Contents

In Chapter 3 we saw that animal and plant remains commonly become buried in sediment accumulations where they may be preserved indefinitely as **fossils.** Such fossil remains have long provided the most useful and widely applicable means of establishing the relative ages of sedimentary rocks. Late in the eighteenth century, about a hundred years after the concept of superposition was first appreciated, an English surveyor, William Smith, noticed that each unit of sedimentary rock that he encountered in excavations for a canal west of London was characterized by a distinctive assemblage of fossil shells. These characteristic fossils allowed each unit to be easily recognized wherever it occurred. Using this knowledge, and the law of superposition,

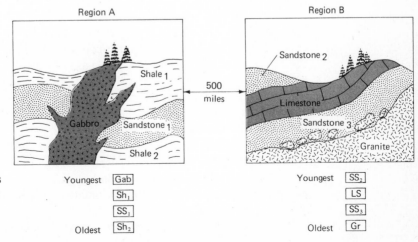

FIGURE 4.6
The exact sequence of relative rock ages can be determined by physical criteria for Region A and Region B, but there is no sure way to interrelate them to each other by physical criteria alone.

Smith over many years worked out the complete sequence of rocks and fossils exposed in England and Wales, and in 1815 he published a geological map illustrating his findings that is a landmark in our understanding of earth chronology (Figure 4.7).

Smith's map clearly demonstrates what later came to be known as the "law of faunal succession" and paved the way for the construction of a worldwide geological time scale, for it was soon discovered that sequences of fossil assemblages very much like those in England occurred in the sedimentary veneer of the European continent and even as far away as North America. This discovery showed that widely separated sedimentary rocks *can be correlated if their fossil remains are essentially the same* as those found in other regions where the fossil sequence has been carefully established. Smith's example, and the law of faunal succession he had so successfully elucidated, served as springboards for the compilation of the relative ages of rock units mapped throughout Europe. By the mid-1800s a geologic column of essentially modern aspect had been established depicting rock units of the world that had been chronologically ordered (Figure 4.8). Such an ordering process required widespread recognition and description of sedimentary rock units called **systems.** Systems were defined in different localities on the basis of distinctive characteristics such as degree of folding, types of rock, and fossil content. Then by application of the "law of faunal succession," different systems throughout Europe were placed in their correct chronology. In Figure 4.7b, for example, the Chalk unit was used to establish the Cretaceous System (*creta* = chalk), and the Brickearth and London Clay, above it, were assigned as lower formations of the Tertiary System, seen resting upon the Cretaceous. (Note that while the modern term *Tertiary* is a holdover from the eighteenth century, it no longer has the same meaning.)

Smith and his followers, who established the sedimentary column, based their work on progressive changes in the kinds of fossil animals and plants, but had little understanding of *why* such changes occurred. We now recognize that the underlying cause of faunal succession is **organic evolution:** the expansions, contractions, and modifications of the living world that have taken place since life originated early in earth history. Because of this continuous change of living organisms, the fossil record of past life provides a dynamic framework for judging relative time, a framework that is not deducible from examining only the physical properties of the sedimentary particles themselves. So successful has been the "fossil-dating" technique that it is still the principal means of unraveling crustal history in mines and wells that penetrate the sedimentary veneer and in surface rock exposures in relatively unexplored areas of the continents. (In any event, the general age relations of the sedimentary veneer for much of the earth's continental surface have long been established by such studies.)

A

DELINEATION
OF THE

STRATA
OF

ENGLAND AND WALES,
WITH PART OF

SCOTLAND:
EXHIBITING

THE COLLIERIES AND MINES.

THE MARSHES AND FEN LANDS *ORIGINALLY OVERFLOWED* BY THE SEA.
AND THE

VARIETIES OF SOIL

ACCORDING TO THE VARIATIONS IN THE SUBSTRATA.

ILLUSTRATED by the MOST DESCRIPTIVE NAMES

BY W. SMITH.

(a)

FIGURE 4.7
William Smith's landmark geological map, published in 1815. (a) Title sheet. (b) Rock units of the London region. (c, d) Some characteristic fossil shells of the London Clay (c) and Chalk (d). (e) A cross section across line X-X'. (Fossils from Smith, "Strata Identified by Organized Fossils," 1816.)

(b)

London Clay
Brickearth
Chalk

Tertiary System

London Clay

Brickearth

Chalk

Cretaceous System

(e)

(c)

(d)

	SYSTEMS	WHERE FIRST DESCRIBED
	Quaternary	Europe
	Tertiary	Europe
	Cretaceons	England
	Jurassic	Jura Mts. Europe
Youngest	Triassic	Germany
	Permian	Russia
	Carboniferous	England
	Devonian	England (Devonshire)
	Silurian	England (Wales)
	Ordovician	England (Wales)
	Cambrian	England (Wales)
Oldest	PRECAMBRIAN ERA	Non-fossiliferous, therefore not well subdivided

FIGURE 4.8
The Geologic Column, a composite from many different places around the world.

In spite of its enormous value, a chronology of earth history based only on fossils also has serious limitations. In the first place, not all sedimentary rocks contain fossils. In general, they are most common in marine sediments, particularly those that accumulate on the shallow, submerged margins of continents rather than in deep-ocean basins. Bottom-dwelling organisms, particularly shell-bearing animals, are abundant today, and were for much of the geologic past, almost everywhere on the shallow ocean floor. In contrast, most terrestrial sediments contain very few fossils because even hard skeletal remains rather quickly decay, weather, and disappear when exposed to the atmosphere. Only in relatively rare terrestrial sediments deposited beneath the standing waters of lakes or swamps are animal and plant remains common.

Because fossils are most abundant in marine sedimentary rocks, it follows that it is usually far easier to establish the age relations of such rocks than it is for those deposited in terrestrial environments. If sea level had always stood as low in relation to the continents as it does today, there would be few ancient marine sediments preserved above sea level, and correlations of the continents' sedimentary veneer would be correspondingly difficult. Fortunately, however, shallow seas covered much of the surface of the continents for long intervals during the geologic past (Figure 4.9). For this reason, thin layers of fossil-rich marine sedimentary rock cover a large fraction of the continental surfaces even in regions that today lie far above sea level. The age relations of this sedimentary veneer can thus be far more easily established than would be the case if terrestrial sediments made up a larger proportion of it. In those regions where thick sequences of fossil-poor terrestrial sediments *do* occur, it is often difficult or impossible to determine their relative ages.

Still other difficulties arise in applying fossil dating. Fossils, of course, are never found in igneous rocks, and only rarely can they be

| 500 million years ago | 380 million years ago | 100 million years ago |

FIGURE 4.9

The distribution of land and sea on the North American continent at three intervals of the geologic past. Note the occurrence of shallow seas over much of the continental surface. Fossil-bearing marine sediments that accumulated in such seas permit long-range correlation of the continent's sedimentary veneer.

recognized in slightly metamorphosed sedimentary rocks. More commonly, igneous or metamorphic rocks may be interbedded with, or otherwise related to, fossil-bearing sedimentary rocks in such a way that their relative ages are apparent (Figure 4.10). Usually, however, it is difficult or impossible to relate the bulk of the crust's igneous and metamorphic rocks to the ages established for the sedimentary veneer.

A still greater problem arises from the fact that abundant shell-bearing life arose relatively late in earth history. For this reason, even marine sedimentary rocks representing much of the earth's past cannot be dated by the fossil technique because they contain few fossils.

A final, and most important, difficulty with the fossil time scale is that it provides only *relative* ages, not absolute ones. Fossil-bearing rocks can be readily placed in a sequence based on the continuous change of the living world, yet fossils provide no means of knowing exactly how long these changes required *in years* (or some other absolute time unit). A particular evolutionary sequence might have taken place over thousands, millions, or even hundreds of millions of years, but the fossils themselves provide no means for determining this.

Because of these difficulties in the fossil time scale, there is a clear need for an independent means of determining the age relations of rocks. The most useful would be a technique that would establish *absolute* ages of rocks and also be applicable to the great volume of crustal igneous and metamorphic rocks that cannot be dated by fossils. We shall see in Section 4.4 that a chronologic method filling both of these requirements is made possible by the slow decay of certain

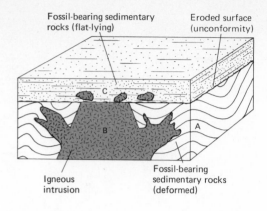

Fossil-bearing sedimentary rocks (flat-lying)

Eroded surface (unconformity)

C

B

A

Igneous intrusion

Fossil-bearing sedimentary rocks (deformed)

E

D

C

B

A

Igneous lava flows

Fossil-bearing sedimentary rocks

FIGURE 4.10
Typical relationships of igneous and fossil-bearing sedimentary rocks. (a) The age of the igneous intrusion (B) is "bracketed" by the ages of sedimentary unit A, which is clearly older than the intrusion, and sedimentary unit C, which is clearly younger. (b) The lava flows, having formed at the earth's surface, are clearly younger than underlying sedimentary rocks and older than those which overlie them.

radioactive nuclides incorporated into rock minerals when they were formed. Radioactivity has given geologists a powerful tool for converting the *geologic column* into a *geologic time scale*. At first glance the time scale (Figure 4.11) looks no different from the column. Familiar names such as Cambrian, Cretaceous, and Tertiary are seen in both, but the time scale contains new terms and is calibrated by numerical age estimates that have been made for the rock units composing the column.

Today we speak of the Paleozoic Era as having begun about 570 million years ago, the Cretaceous Period as having ended approximately 65 million years ago, and the Pleistocene Epoch as having commenced 2 to 3 million years ago. The terms *era*, *period*, and *epoch* refer to different subdivisions of this geologic time scale. To speak of the Cretaceous Period is roughly analogous to speaking of the Middle Ages or the Victorian Period—or perhaps "February." As *time* units, none of these possesses any tangible physical entity; one cannot be hit in the head by Monday, March, the Victorian Period, or the Cretaceous Period. On the other hand, we might sustain a severe injury if struck by some of the *rock* to which the name "Cretaceous System" was originally assigned in England (Figure 4.12). The point is this:

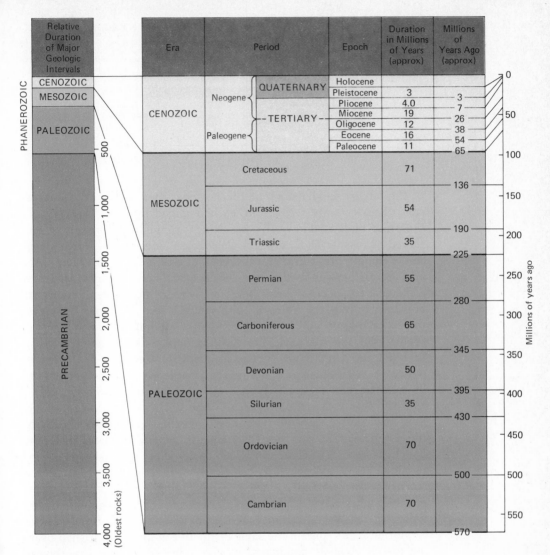

FIGURE 4.11
The time scale of
earth history.

the geologic time scale was originally abstracted from a geologic column of primarily sedimentary rocks that had been synthesized through
the application of *relative* age-dating techniques. It was many years
later before geologists were able to assign to the subdivisions of the
geologic time scale reasonably accurate ages in "years b.p." (years before the present). Even when the techniques of radioactive age-dating
were fairly well established, it was not a trivial task to determine that
the Cretaceous Period ended approximately 65 millions years ago. We
shall return to this important problem following the discussion of
"absolute" age-dating in Section 4.6.

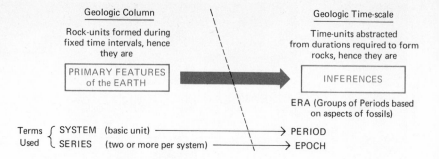

FIGURE 4.12
Relationship between
the geologic column
and the time scale.

The largest subdivisions of the geologic time scale are known as
eras; three eras are recognized, each bounded by times of *relatively*
profound, sudden, and worldwide changes in the living organisms preserved as fossils. These are: the Paleozoic ("ancient life") Era, Mesozoic ("middle life") Era, and Cenozoic ("recent life") Era. The three
eras are further subdivided into periods, each bounded by episodes of
somewhat less profound change in the living world. Eleven periods are
now recognized: the Paleozoic Era includes six of them—Cambrian
(oldest), Ordovician, Silurian, Devonian, Carboniferous, and Permian
(youngest); the Mesozoic Era has three periods—Triassic (oldest),
Jurassic, and Cretaceous (youngest); and the Cenozoic Era includes
only two—Paleogene and Neogene.* The eleven geologic periods are
in turn further subdivided into still smaller units of time, called
epochs. Most of the many names of these smaller divisions of geologic
time need not concern us. The most important are the six epochs of
the Cenozoic Era, the next-to-youngest of which, the Pleistocene
Epoch, includes the recent intervals of ice-sheet expansion and contraction.

The oldest sedimentary rocks bearing abundant fossils are those
of the Cambrian system, which were deposited at the beginning of the
Paleozoic Era. Wherever they occur around the world, fossil-bearing
Cambrian rocks contain the same distinctive association of shell-bearing marine animals dominated by the long-extinct trilobites (Figure
4.13), which were distant relatives of modern crabs and shrimps.
These earliest Cambrian sediments always overlie older rocks that *lack*
abundant fossils.

Commonly, early Cambrian sediments rest upon a "basement complex" of older igneous and metamorphic rocks which, we have seen,
make up the bulk of the continental crust. In many regions, however,
the earliest Cambrian fossils occur above thick sequences of sedimentary rocks that are identical to the overlying Cambrian rocks *except*

*Many geologists divide the Cenozoic Era into the Tertiary and Quaternary Pe-
riods, as shown in Figure 4.11.

FIGURE 4.13
Cambrian trilobites. (Vladius J. Okulitch.)

that they contain no fossils. This abrupt* appearance of trilobites and other relatively advanced kinds of animal life in the sedimentary record is perhaps the most significant milestone in the earth's long history, for it serves to divide the earth's past into two great divisions. Before this event lies "Precambrian time," whose rocks contain few fossils and therefore provide no worldwide time scale based on changing life. With the great early Cambrian expansion of animal life comes "Phanerozoic ('exposed life') time"—and the worldwide applicability of the sedimentary time scale that we have been considering.

Because Precambrian igneous, metamorphic, and sedimentary rocks make up a large percentage of all the rocks of the continental crust, it has long been assumed that they represent a very long span of earth history. Before the widespread application of radiometric dating, however, no worldwide chronology was available for this great volume of Precambrian rocks. As with younger, fossil-bearing rocks before the discovery of the "law of faunal succession," local sequences of Precambrian rocks, and the events they record, had been worked out by purely physical means, but there was still no reliable way to interrelate these events over long distances. This situation has been profoundly changed by radiometric dating.

4.3 Sedimentary Rocks and Past Environments

So far we have emphasized only how the physical relationships and fossil contents of the continents' sedimentary veneer permit a deter-

*Abrupt, that is, in a geological sense—occurring perhaps within 10 to 20 million years.

mination of the relative ages of ancient sediments. As yet, however, we have said almost nothing about the *kinds of events* in the earth's past that may be inferred from the sedimentary rocks once their ages are understood.

As might be predicted from our discussion of present-day sediments in Chapter 3, sedimentary rocks provide much information about ancient environments on the earth's surface at the time and place of their deposition. Thus they may answer such questions as: Was the area submerged by the sea, or did terrestrial sediments accumulate above sea level? If below sea level, was the region receiving only carbonate sediments, as do many tropical shelves today, or were sands and muds dominant as on present-day polar and temperate shelves? If the sediments were deposited above sea level, do they reflect fluvial, desert, or glacial environments?

In Chapter 3 we saw that the composition and structure of the sediments themselves usually provide criteria for answering such environmental questions. When these answers are pieced together for sedimentary rocks of many different ages distributed over much of the present-day surface of the continents, they provide an invaluable records of past changes in the geography of the earth's surface.

The principal problem in reconstructing earth history from sedimentary rocks relates to areas where no sediment accumulates. Much of the surface area of the continents today is not covered by sites of sediment deposition, but is instead being actively eroded; such areas can obviously leave no future sedimentary record. Similar conditions existed through much of the earth's past, and thus the continents' veneer of sediments never reveals the *complete* sequence of surface events, but only those times and places where sediment *accumulation* was taking place.

As a further complication, even areas that receive sediment at one time may, because of mountain-building deformations, lowered sea level, or other causes, be subject to later erosion which destroys their record of earlier events. For these reasons, sedimentary rocks seldom reflect a *continuous* sequence of local sediment accumulation through long intervals of earth history; instead, the sedimentary veneer at any one place always shows discontinuities representing long time intervals when no sediment was accumulating in that area. We have already mentioned such discontinuities, called unconformities, in the Introduction (see Figure I.11). Even though unconformities reflect intervals of no local sediment accumulation, however, it is often possible to make inferences about the events they represent from the nature of the contact between the rocks above and below the surface of unconformity.

Three significantly different kinds of unconformities exist, but in each case the surface separates rocks of distinctly different age and the younger of the two formations lies upon, and essentially parallel to, a buried surface of erosion. Angular Unconformities separate layered rocks where those beneath the contact have been *deformed* and eroded so as to lie at distinctly different angles from the younger strata (Figure 4.14a). Disconformities are the most subtle of all unconformities, because the strata above and below the surface of contact are essentially parallel to each other (Figure 4.14b); sometimes only by careful fossil age determinations can the existence of a disconformity be discovered. Nonconformities result from deposition of sediments upon eroded surfaces of nonstratified rocks such as granites or metamorphics (Figure 4.14c). Almost without exception nonconformities represent *large gaps* in the rock record due to the large volumes of erosion required to expose plutonic and metamorphic rocks at the surface prior to the deposition of younger strata upon them.

In spite of the problem of an incomplete sedimentary rock record, the continents' sedimentary veneer, when viewed on a worldwide scale, preserves a remarkably complete record of events taking place through much of the earth's long history. Undoubtedly, the reason the record is so good is that through much of the earth's past the continents have stood much lower in relation to sea level than they do today. During long intervals shallow seas repeatedly covered the continents, thus both minimizing erosion and allowing the accumulation of the thin but widespread layers of marine sediment that make up much of the continents' cover of sedimentary rock.

NUCLEAR CLOCKS AND EARTH CHRONOLOGY

The time scale of earth history based on fossil remains has been established for well over a hundred years; for most of that time, fossils provided the *only* worldwide scheme for working out earth chronology. Within about the last 25 years, however, fossil-dating techniques have been supplemented by a powerful new chronologic tool based on the decay of certain radioactive atomic nuclei.

4.4 Radioactive Decay

In discussing our model for the structure of matter in Chapter 1, we noted that the small nucleus at the center of every atom is composed of two principal kinds of particles: uncharged *neutrons* and positively charged *protons*, the latter being the same in number as the negative electrons which orbit around the nucleus. All atoms of a particular element have the same number of protons in the nucleus, but they may have different numbers of uncharged neutrons, giving rise to different *nuclides* of the element.

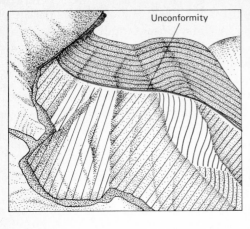

(a)

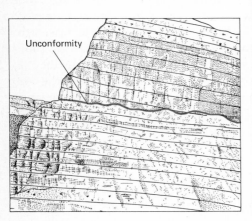

(b)

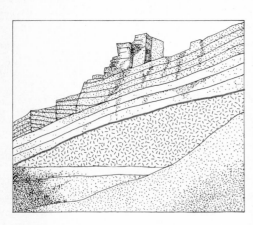

(c)

FIGURE 4.14
Typical unconformities. (a) The steep tilt and irregular topography of the rocks beneath the surface of unconformity suggest that an interval of mountain-building deformation and erosion occurred between the time of deposition of the sedimentary units separated by the unconformity. The rocks are exposed in southeastern California. (b) The parallel relationship of the beds below and above the surface of unconformity suggests relatively minor erosion of the underlying unit during the time interval represented by the unconformity. The rocks are exposed in northern Arizona. (c) Sedimentary rocks deposited on eroded plutonic rocks, representing a long time gap. (a, b: John S. Shelton; c: E. A. Hay.)

Nuclides are distinguished by two important numbers. The first is the *atomic number*, which is simply the number of protons in the nucleus (or of orbital electrons, since the numbers of each are always the same); the atomic number is constant for all nuclides of a particular element. The second is the *mass number*, which is the total of both protons and neutrons in the nucleus and which differs for different nuclides of the same element. Calcium atoms, for example, always have 20 nuclear protons, whose positive charges are balanced by 20 orbital electrons; thus the atomic number for calcium atoms is always 20. Naturally occurring calcium, however, is a mixture of six different calcium nuclides which have either 20, 22, 23, 24, 26, or 28 neutrons in the nucleus. These neutrons, added to the constant number of 20 nuclear protons, give these six nuclides mass numbers of 40, 42, 43, 44, 46, and 48 respectively. For brevity, nuclides are normally referred to by the name of the element and their mass numbers—for example, "calcium-40" or "carbon-14."

Many different nuclides of almost every element can be prepared artificially in nuclear reactors or particle accelerators by adding protons and neutrons to, or subtracting them from, atomic nuclei. Most such artificial nuclides, however, are very unstable or "radioactive," which merely means that they rapidly decay to more stable nuclides of other elements by one of the three processes summarized in Figure 4.15. During decay, the original radioactive nuclide is called the *parent* nuclide and the nuclide or nuclides produced by the decay are called *daughter* nuclides. Each different radioactive nuclide decays into daughter nuclides at a constant rate which is unaffected by such physical conditions as temperature or pressure. Most artificial nuclides decay very rapidly (usually in a few hours or days) to more stable nuclides, but some may persist for many years. The average lifetime of a radioactive parent nuclide is defined as the time required for half the atoms in any given mass of the nuclide to decay, and is termed the nuclide's **half-life** (Figure 4.16). Because some few atoms of even the most rapidly decaying nuclides persist indefinitely, the time required for *complete* decay is infinite.

In contrast to the many radioactive nuclides produced in reactors and accelerators, most natural nuclides occurring in rocks, the ocean, and the atmosphere are stable—that is, they have no tendency to decay to other nuclides. Most natural elements are mixtures of several (usually from two to eight) such stable nuclides of that element. If all naturally occurring nuclides were stable, however, they would be of no value for earth chronology. Fortunately, there are several *natural radioactive nuclides* that make possible the use of radioactive decay as a tool of earth chronology.

All radioactive nuclides found in rocks, the ocean, and the atmosphere come from one of two sources. The first group, called **primary**

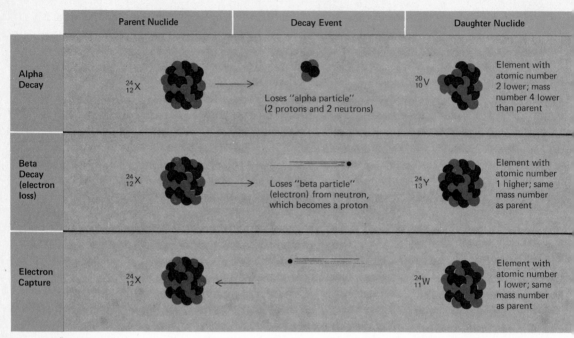

Parent Nuclide		Decay Event	Daughter Nuclide	
Alpha Decay	$^{24}_{12}X$	Loses "alpha particle" (2 protons and 2 neutrons)	$^{20}_{10}V$	Element with atomic number 2 lower; mass number 4 lower than parent
Beta Decay (electron loss)	$^{24}_{12}X$	Loses "beta particle" (electron) from neutron, which becomes a proton	$^{24}_{13}Y$	Element with atomic number 1 higher; same mass number as parent
Electron Capture	$^{24}_{12}X$		$^{24}_{11}W$	Element with atomic number 1 lower; same mass number as parent

FIGURE 4.15
The three principal processes of radioactive decay.

nuclides, have such extremely long half-lives that they have persisted since the earth first formed (Figure 4.17). About twenty such nuclides have been detected, but only four are sufficiently widespread and abundant to be generally useful as chronologic tools: potassium-40, which decays to argon-40; rubidium-87, which decays to strontium-87; uranium-235, which decays, through a series of intermediate radioactive nuclides, to lead-207; and, finally, uranium-238, which decays, also through an intermediate series of nuclides, to lead-206 (Figure 4.17).

In addition to these long-lived radioactive nuclides left over from the earth's formation, there is a second group, of much shorter-lived radioactive nuclides, that are continually being produced in the earth's upper atmosphere by cosmic rays. These "rays" are really extremely high-energy nuclear particles moving in space from unknown sources. When such particles enter the earth's atmosphere they collide with atmospheric gas particles and produce nuclear reactions similar to those of man-made particle accelerators. Some of these reactions form short-lived radioactive nuclides. At least eight such cosmic-ray-induced nuclides have been identified, but only one, carbon-14, has so far proved to be a widely useful chronologic tool.

Carbon-14 is produced by cosmic rays acting on atoms of nitrogen, the most abundant atmospheric gas (Figure 4.18). Atmospheric nitro-

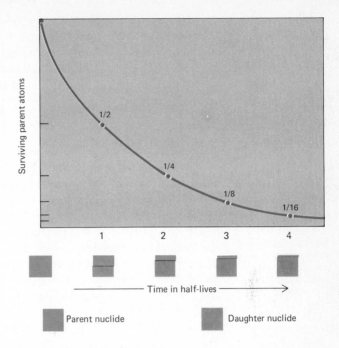

FIGURE 4.16

Half-life of a radioactive nuclide. The half-life is the time required for half of any initial mass to decay to a daughter nuclide. With each half-life interval, the remaining mass is reduced by 50 percent The initial mass never decays completely because some small proportion of the initial atoms persists indefinitely.

gen is composed of only stable nuclides, principally nitrogen-14. Through a complex series of nuclear reactions, high-energy cosmic rays may subtract a proton and add a neutron to some of these nitrogen atoms to produce radioactive carbon-14, which then decays back to stable nitrogen-14 with a half-life of only about 5,600 years.

Once produced, the carbon-14 atoms quickly combine with atmospheric oxygen to become carbon dioxide. Most atmospheric carbon dioxide is made up of stable carbon-12 or carbon-13 nuclides; the relatively small quantity of cosmic-ray produced "carbon-14 dioxide" quickly mixes with this more abundant and stable atmosphere carbon dioxide and, in this manner, enters into the earth's general carbon

FIGURE 4.17

The principal primary nuclides used as geologic clocks. All are capable of dating the earth's oldest rocks. (See also Figure 1.9.)

Radioactive Parent Nuclide	Type of Decay (Figure 10.8)	Stable Daughter Nuclide	Half-life (years)	Minimum ages that can be determined
Potassium-40 ($^{40}_{19}$K)	Electron capture	Argon-40 ($^{40}_{18}$Ar)	1.3 billion	100,000 years
Rubidium-87 ($^{87}_{37}$Rb)	Beta	Strontium-87 ($^{87}_{38}$Sr)	47 billion	50 million years
Uranium-235 ($^{235}_{92}$U)	7 alpha and 4 beta	Lead-207 ($^{207}_{82}$Pb)	0.7 billion	10 million years
Uranium-238 ($^{238}_{92}$U)	8 alpha and 6 beta	Lead-206 ($^{206}_{82}$Pb)	4.5 billion	10 million years

cycle (Figure 4.18). There it may be dissolved in the oceans, precipitated as carbonate minerals, or utilized in the structures of animals and plants. The small amounts of radioactive carbon-14 which utimately enter animals, plants, and minerals in this fashion provide an extremely useful tool for dating the very latest interval of earth history.

4.5 Radiometric Ages

FIGURE 4.18
The carbon-14 cycle. Radioactive carbon-14 is produced from atmospheric nitrogen by cosmic rays (inset). It then enters into CO_2 molecules and becomes incorporated into carbon-bearing sediments and organic remains. The amount of remaining carbon-14 is used to date such materials.

All natural nuclides of the same element, whether radioactive or not, have the same general chemical behavior and thus tend to occur mixed together wherever the element is found. When the mixture contains a radioactive nuclide, however, small quantities of a different element —the daughter element produced by decay of the radioactive nuclide —are also present in the mixture. In theory, all that is necessary to use radioactive nuclides as geologic clocks is to measure the amounts of both the radioactive parent nuclide (normally either potassium-40, rubidium-87, uranium-235, uranium-238, or carbon-14) and the stable daughter nuclide (normally either argon-40, strontium-87, lead-207, lead-206, or nitrogen-14, correspondingly) that are present today in the rock or mineral to be dated (Figure 4.19). Since the decay rates of the radioactive nuclides are known with considerable precision from

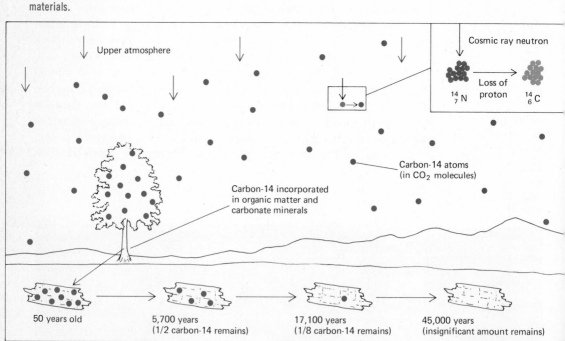

Upper atmosphere

Cosmic ray neutron

Loss of proton

$^{14}_{7}N$ $^{14}_{6}C$

Carbon-14 atoms
(in CO_2 molecules)

Carbon-14 incorporated
in organic matter and
carbonate minerals

50 years old

5,700 years
(1/2 carbon-14 remains)

17,100 years
(1/8 carbon-14 remains)

45,000 years
(insignificant amount remains)

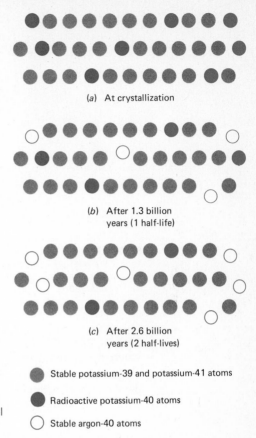

(a) At crystallization

(b) After 1.3 billion
years (1 half-life)

(c) After 2.6 billion
years (2 half-lives)

⬤ Stable potassium-39 and potassium-41 atoms

● Radioactive potassium-40 atoms

◯ Stable argon-40 atoms

FIGURE 4.19
The progressive decay of potassium-40 atoms incorporated into a mica mineral at the time of crystallization. The ratio of remaining potassium-40 atoms to atoms of the daughter nuclide argon-40 is used to determine the age of the mineral (see also Figure 1.27).

laboratory measurements, a simple calculation (the ratio of parent to daughter nuclides multiplied by the decay rate) gives the *length of time* since the radioactive parent nuclide was first isolated in the mineral and thus provides a radiometric age for the mineral. Precise techniques have been developed for measuring very small quantities of parent and daughter nuclides, but there still remain several complications in obtaining ages by this means.

In order for radiometric ages to be accurate, the radioactive parent nuclide and the stable daughter nuclide must have been a *closed system* since their formation; that is, there must have been no later additions or subtractions of either parent or daughter. Unfortunately, these conditions are seldom met. Loss of some of the daughter nuclide is the most usual problem for, being a different element with different chemical properties, the daughter nuclide is not in equilibrium with its surroundings once formed (Figure 4.20a). Daughter nuclide escape is

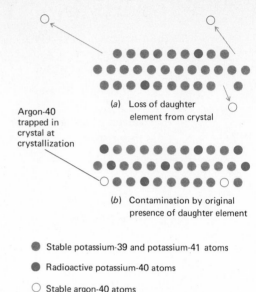

Argon-40
trapped in
crystal at
crystallization

(a) Loss of daughter
     element from crystal

(b) Contamination by original
     presence of daughter element

FIGURE 4.20
Two of the principal complicating factors in radiometric age determinations (see text).

⬤ Stable potassium-39 and potassium-41 atoms

⬤ Radioactive potassium-40 atoms

◯ Stable argon-40 atoms

particularly common when the nuclide is a gas, such as argon-40, or when the mineral to be dated has been secondarily heated by earth movement or deep burial. Although ingenious techniques have been developed to estimate the amounts of daughter elements lost from various minerals, this remains one of the principal difficulties in radiometric dating.

Because of daughter element loss, the most accurate radiometric ages are always considered to be those where calculations on two or more parent-daughter pairs give the same result. In such cases it is unlikely that the different daughter elements, each having different chemical properties, would be lost at the same rate and thus provide the same erroneous ages. Fortunately, many rocks and minerals contain more than one of the principal radioactive nuclides, so that cross-checking is frequently possible. Uranium-bearing minerals are particularly useful in this regard, since they always contain both uranium-235 and uranium-238, each of which provides an independent test of the age.

A second major difficulty in radiometric dating is that the parent nuclide must have been free of contamination by the daughter nuclide when it was originally incorporated into the mineral to be dated. That is, *all* the daughter nuclide subsequently measured must result from radioactive decay rather than original contamination (Figure 4.20b). Seldom is this condition fulfilled, but here again ingenious methods have been developed for estimating the amount of daughter element originally present, so that this difficulty can frequently be overcome.

Various techniques for getting around the daughter element loss and contamination problems have been developed for the long-lived primary nuclides, but these problems are insoluble for carbon-14, the principal short-lived nuclide that is useful in radiometric dating. Carbon-14, we have seen, decays to nitrogen-14, which is the principal constituent of the atmosphere. The daughter element is thus present in large, contaminating quantities both when the carbon-14 is produced in the atmosphere and, commonly, when it decays as well. For this reason, carbon-14 dating is not based on measurements of both parent and daughter nuclides; instead, it requires the assumption that carbon-14 has been produced at a constant rate in the atmosphere and is quickly mixed, in constant proportions, with nonradioactive carbon to enter the earth's overall carbon cycle. If these assumptions are correct, then carbon at the time it is incorporated into carbonate minerals and living organisms always has a constant proportion of carbon-14, which steadily decreases by decay after the materials are formed. It is therefore necessary only to measure the present ratio of carbon-14 to nonradioactive carbon in such materials to determine their ages by this method (Figure 4.18).

So far we have said very little about the specific kinds of rocks, minerals, or other materials that contain the five principal radioactive nuclides in measurable amounts and can therefore be dated by radiometric techniques. Both the potassium-40 and rubidium-87 methods have been applied most successfully to mica minerals found in a variety of igneous and metamorphic rocks. In addition, some other igneous and metamorphic minerals—for example, certain types of feldspar and amphibole—are commonly dated by these techniques. Normally, the individual mineral grains are separated from the rest of the rock to minimize contamination, but it is sometimes possible to obtain accurate dates for very fine-grained igneous and metamorphic rocks simply by analyzing crushed samples of the whole rock rather than individual mineral grains.

The uranium dating technique was the first to be established, but for many years it could be applied only to certain rare minerals that contain relatively large quantities of the element. More recently, the development of precise analytical techniques has permitted accurate measurements of very small traces of uranium that occur in the silicate mineral zircon. Zircon occurs as a minor accessory mineral in many igneous rocks, and thus permits a wide application of the uranium dating technique.

Radiometric dating by means of potassium, rubidium, and uranium nuclides is applicable mostly to minerals found in igneous and metamorphic rocks. In contrast, the carbon-14 technique is principally useful for sedimentary materials: carbonate sediments and fossil shells, bones, and wood fragments are the materials most commonly dated by this technique.

4.6   Absolute Radiometric Dating

The application of radiometric dating techniques has, over the past two decades, made three principal contributions to our understanding of earth chronology. First, these techniques have led to an absolute calibration, in years, of the relative geologic time scale. In addition, they have revolutionized our understanding of the enormous volume of Precambrian rocks that cannot be dated by fossils. Finally, they have greatly clarified the events of the past 50,000 years, a span of time too short to include significant changes in living organisms and thus also beyond the application of fossil-dating techniques. We shall briefly consider each of these contributions before turning to the more complex question of the overall age of the earth.

Perhaps the most significant achievement of radiometric dating has been to provide absolute dates for the relative geologic time scale. Since the early nineteenth century, attempts have been made to calibrate the sedimentary time scale. Most of these involved estimates of the rates at which sediments are currently accumulating and extrapolation of these rates into the past to account for the total thicknesses of sedimentary rocks of different ages. Such estimates all suggested that many millions of years were required to account for the crust's sedimentary veneer, but only with the advent of radiometric dating could the precise number of years be determined.

Radiometric dating techniques are usually applicable only to the minerals of igneous or metamorphic rocks, and for this reason radiometric dating of the sedimentary time scale requires rather unusual associations of sedimentary and igneous rocks (see Figure 4.10). Such associations are far from common, but fortunately enough are known to give a reasonably complete calibration.

Let us more specifically consider the method used to achieve such numerical calibrations. Figure 4.10 suggests that the Cretaceous Period ended approximately 65 million years ago, yet the rocks used to establish the close of Cretaceous time are sedimentary limestones from Dover, England, and thus cannot be directly dated by radioactivity. Radiometric dating in combination with an analysis of certain geologic relationships could, however, yield such an "absolute" age for the Cretaceous Period. With reference to Figure 4.21, the following considerations can lead to a solution to this problem:

1. The *"law of faunal succession"* makes possible a correlation between the limestone formation shown in the western United States and the youngest of the rocks from Dover, England, where the Cretaceous system was defined.

2. Uranium-bearing minerals (usually zircon) from the granite yield an "absolute" age of 70 million years b.p. (before the present).

3. We know the limestone (hence the late Cretaceous) to be

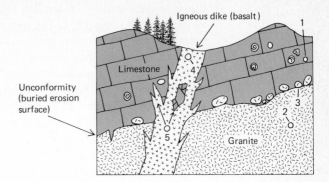

FIGURE 4.21
Cross section of relationships that might be observed somewhere in the western United States. See text for discussion.

younger than 70 million years by applying the *law of superposition*, and also by noting the fragments of older granite contained within the younger limestone.

4. The igneous dike can be dated by the potassium-argon radioactive-decay method. Suppose an age of 60 million years b.p. is determined.

5. We know the limestone (and late Cretaceous) to be *older* than the igneous dike by observing fragments of limestone in the dike and also by recognizing that any intrusive rock must be younger than what it is intruding into.

6. Given the fortuitous observations described above, we can conclude that the age for the youngest part of the Cretaceous Period is less than 70 million yeras old and greater than 60 million years old— 65 million years b.p. is a reasonable *estimate*, therefore, for the end of the Cretaceous Period.

What has been described is a very special set of geologic circumstances. Such ideal relationships are rare. What if, for example, the granite had been 370 million years old? Or suppose the limestone had not contained fossils suitable for correlation. Nevertheless, this is the basic approach that geologists have used in calibrating the geologic time scale. Note again that these conceptions of *time* are derived from *rocks*. Sometimes, datable lava flows and volcanic ash falls are layered among sediments in a manner that simplifies the process described above, but ultimately the problem of making bridges between dates determined for igneous rocks and "ordered" *sedimentary* rocks remains as a major task for the geologist.

The reasonably complete calibration shown in Figure 4.11 shows that the early Cambrian expansion of animal life took place about 570 million years ago. The oldest *Precambrian* rocks so far dated are about 4,000 million (4.0 billion) years old, and thus Phanerozoic (exposed life) time represents only the most recent 15 percent of the time since the formation of the oldest known crustal rocks. Within this 570 million years of Phanerozoic time, the Paleozoic Era accounts for about the first 345 million years, the Mesozoic Era the next 160 million years, and the Cenozoic Era only the last 65 million years. The average length of the 11 Phanerozoic periods is about 55 million years, but radiometric dating has shown that the changes in the living world which bound the periods were not regularly spaced. Instead, the periods range in age from a minimum of 35 million years (Silurian Period) to a maximum of 71 million years (Cretaceous Period). Radiometric dating has also shown that the great ice sheet expansions and contractions of Pleistocene time—events that have had such a profound effect on the earth's present climates, landscapes, and sediments —have been concentrated in about the last 2 or 3 million years, less than 1 percent of the total of Phanerozoic time.

In addition to furnishing an absolute calibration for the Phanerozoic time scale, radiometric dating has provided the *only* chronologic tool applicable to the great mass of Precambrian rocks which represent over 3,400 million (3.4 billion) years of earth history. As a result, a general worldwide chronology and sequence of events are beginning to be established for this long and significant interval. Finally, the carbon-14 dating technique has permitted a refined understanding of events over the most recent 50,000-year span of Pleistocene time.

4.7   The Age of the Earth

The oldest crustal rocks so far dated, granites from southwestern Greenland, have radiometric ages of about 4 billion years. In other regions metamorphosed sedimentary rocks surround similar ancient granites that were injected into, and thus are younger than, the deformed sediments that surround them. It is therefore clear that crustal erosion, sediment deposition, and the formation of sedimentary rocks were all taking place very early in earth history. Apparently these processes have completely recycled, and thus obliterated, any original rocks formed as the crust first consolidated. Direct dating of crustal rocks can therefore only indicate that the earth is older than 4.0 billion years; to answer the question "how much older?" we must turn to less direct evidence of two sorts.

The first piece of evidence comes from radiometric dating of meteorites and rocks recovered from the surface of the moon. All the solid material of the solar system, including the earth, moon, and the small particles that fall on the earth as meteorites, are believed to have had a common origin in the materials that formed the sun. Neither the rocks of the moon nor, presumably, meteorites have been subjected to the intense, continuous erosion and weathering imposed by an atmosphere on rocks of the earth's crust. For this reason, some of them might be expected to give ages indicating the time that they—and by extrapolation the earth as well—first consolidated from solar matter (sun material). Most significantly, almost all meteorites have been found to have radiometric ages between 4.5 and 5.0 billion years, suggesting that the earth has about the same overall age. In addition, the oldest rocks so far recovered from the surface of the moon have ages of about 4.5 billion years, suggesting that the earth is at least that old.

The second kind of indirect evidence for the overall age of the earth is based on the present-day abundance of the various nuclides of lead that occur in minerals of the earth's crust. Natural lead is a mixture of four stable nuclides: lead-204, -206, -207, and -208. Three of these (206, 207, and 208) are produced by the radioactive decay of uranium and other less common radioactive elements. The fourth, lead-204, is not produced by radioactive decay: *all* of it present on the earth today originated when the earth was formed, whereas only a *part* of present-day lead-206, -207 and -208 originated at that time—the rest has been slowly added through the course of earth history by radioactive decay of other elements (Figure 4.22). Now if there was some means of de-

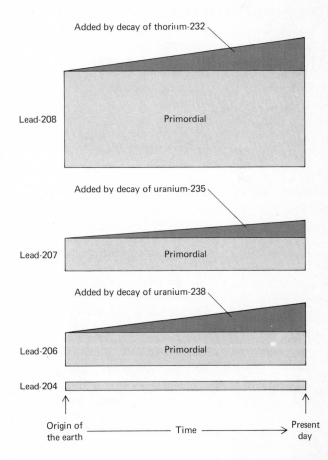

FIGURE 4.22
Determination of the age of the earth from its initial and present-day abundances of lead nuclides. The initial abundance is estimated from meteorites and the present abundance from the average of many rock measurements. The difference represents additions from the decay of radioactive uranium and thorium through earth history.

termining the *original* relative abundance of the four nuclides (at the time the earth was formed), then the earth's age could be estimated by first measuring their *present* relative abundance and then calculating the time required for the additional lead-206, -207 and -208 to have been added by radioactive decay (the decay rates producing each are constant and are precisely known from laboratory measurements). Although there is no direct way of estimating the original lead abundances, certain meteorites which contain no uranium or other radioactive elements that decay into lead are thought to provide a reasonable approximation of the earth's original lead. Using this information, about 4.5 billion years of radioactive decay would be necessary to produce lead having the average nuclide abundances found on earth today. This estimate further confirms the suggestion that the earth originated at least 4.5 billion years ago.

SUMMARY OUTLINE

Relative Ages and Earth Chronology

4.1 *Physical relationships:* the law of superposition and other relative age-dating principles provide a means of establishing local sequences of events in rocks.

4.2 *Fossil contents:* progressive evolutionary changes in ancient life provide a worldwide chronology based on fossil-bearing sedimentary rocks.

4.3 *Sedimentary rocks and past environments:* sedimentary rocks reveal much about ancient geography at their time of deposition; areas of nondeposition and erosion leave unconformities in the sedimentary record.

Nuclear Clocks and Earth Chronology

4.4 *Radioactive decay:* radioactive parent nuclides gradually decay into stable daughter nuclides of other elements.

4.5 *Radiometric ages:* the decay rate of radioactive nuclides is constant and can be determined by laboratory measurements; by measuring amounts of parent and daughter nuclides and applying the decay rate, the time of origin of many igneous and metamorphic minerals can be determined.

4.6 *Absolute radiometric dating:* radioactive nuclides provide absolute ages for the 600 million years of Phanerozoic history, and also permit establishing a chronology for the preceding 3.4 billion years of Precambrian rocks, which lack abundant fossils.

4.7 *The age of the earth:* the oldest known crustal rock formed 4.0 billion years ago and indicates a still earlier, unrecorded phase of earth history; various lines of evidence suggest that the original consolidation of the earth took place at least 4.5 billion years ago.

ADDITIONAL READINGS

*Berry, W. B. N.: *Growth of a Prehistoric Time Scale*, W. H. Freeman and Co., San Francisco, 1968. A brief, readable history of the development of the sedimentary time scale in the eighteenth and early nineteenth centuries.

Donovan, D. T.: *Stratigraphy*, Thomas Murby and Co., London, 1966. A concise intermediate text on the techniques of earth chronology.

*Eichler, D. L.:*Geologic Time*, Prentice-Hall, Inc., Englewood Cliffs, N.J., 1968. An introduction to all aspects of earth chronology.

*Faul, H.: *Ages of Rocks, Planets, and Stars*, McGraw-Hill Book Company, New York, 1966. A good introduction to radiometric dating.

*Harbaugh, J. W.: *Stratigraphy and Geologic Time*, William C. Brown Company, Dubuque, Iowa, 1968. A brief introduction to earth chronology.

Matthews, R. K. :*Dynamic Stratigraphy*, Prentice-Hall, Inc., Englewood Cliffs, N.J., 1974. An up-to-date advanced text on sedimentary chronology.

*York, D., and R. M. Farquhar: *The Earth's Age and Geochronology*, Pergamon Press, Oxford, 1972. A brief intermediate-level survey emphasizing radiometric techniques.

*Available in paperback.

chapter **5**

Deformation
of
Crustal Rocks

5.1 Continents and Ocean Basins

Imagine for a moment that the oceans could somehow be drained to leave the entire surface of the earth exposed as dry land. An explorer traveling in this strange world would discover that the earth's surface consists of two fundamentally different landscapes. The most obvious would be broad, flat plains covering millions of square miles, interrupted only by long, narrow chains of high mountains. These plains would contrast sharply with the second major feature, plateau areas— also relatively flat with linear mountain ranges—that would tower several miles above the surrounding plains and would be separated from them by the highest and most continuous scarps (steep slopes) on earth. The lower plains would, of course, be the floors of today's oceans, and the higher plateaus the surface of today's continents. This division into low *ocean basins* and much higher *continents*, the most obvious feature of the earth's crust, is caused by fundamental differences in the composition and structure of continental versus oceanic rocks.

In general, we know far more about continental rocks than about those making up the ocean floor. Continental rocks are more accessible at the earth's surface and in mines and bore holes which reach maximum depths of 8 km (5 mi), or about one-quarter of the depth to the Moho discontinuity at the base of the continental crust. The Moho lies only about 6 km (4 mi) beneath the deep ocean floor, but because of the difficulties of drilling from floating platforms through several miles of water, only a relatively few ocean-bottom holes have so far been drilled, and the deepest extend only about a thousand meters into oceanic rocks. Rocks from the *surface* of the ocean floor, however, have been recovered by dredging along submarine cliffs and canyons in many parts of the oceans, and additional information about oceanic rocks is provided by islands that are far from the nearest continental land masses. These data on oceanic rock composition, combined with information from hundreds of thousands of surface rock exposures and tens of thousands of mines and wells on the continents, have shown that *continents are largely composed of silicic igneous rocks, particularly granite,* whereas *ocean basins are largely underlain by heavier, mafic igneous rocks, particularly basalt.* These direct observations of rock composition are further confirmed by seismic wave velocities which show values in the range expected for granitic rocks throughout the continental crust, but are generally higher in the oceanic crust, thus suggesting a predominance there of heavier, basaltic materials.

In summary, the differences between continents and ocean basins are primarily related to the concentration, in the continents, of the relatively light elements that are abundant in the minerals of granite —particularly silicon, sodium, and potassium; the rocks under the ocean floors, in contrast, have a higher concentration of heavier iron-magnesium— and calcium-rich minerals and are therefore closer in composition to the inferred rocks of the underlying mantle than are continental rocks (Figure 5.1).

5.2 Continental Rocks

Although most of the surface rock of the continents is covered by a thin layer of soil, it is usually easy to determine the nature of the un-derlying "bedrock," either by digging through the soil or, more com-monly, by observing the rock where it is exposed in stream banks and cliffs or in man-made quarries, excavations, and road cuts. In addition, data from surface rock exposures may be supplemented by informa-tion on the "subsurface" rock distribution in areas that have been mined or explored for petroleum. In studying the rocks of a particular continental region, geologists normally seek out all of this evidence and use it to prepare a **geologic map** showing the distribution of dif-ferent rocks as they would look if all the soil were removed from the surface (Figure 5.2). The degree of detail shown in geologic maps varies with the purpose of the map and the amount of field study that has gone into its preparation. At one extreme, areas with valuable min-

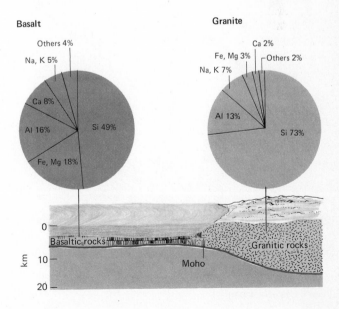

FIGURE 5.1
Generalized chemical compositions of the basaltic rocks that underlie the ocean floor and the granitic rocks that underlie the continents.

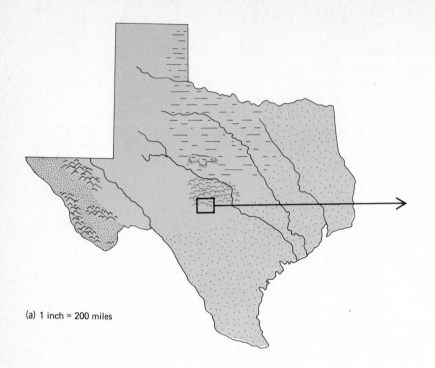

(a) 1 inch = 200 miles

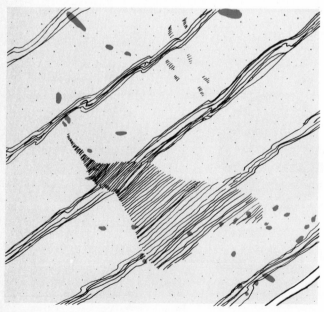

(d) 1 inch = 1/13 mile (400 ft)

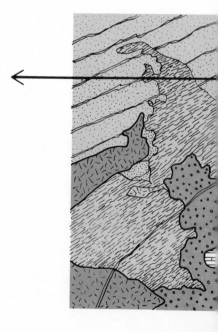

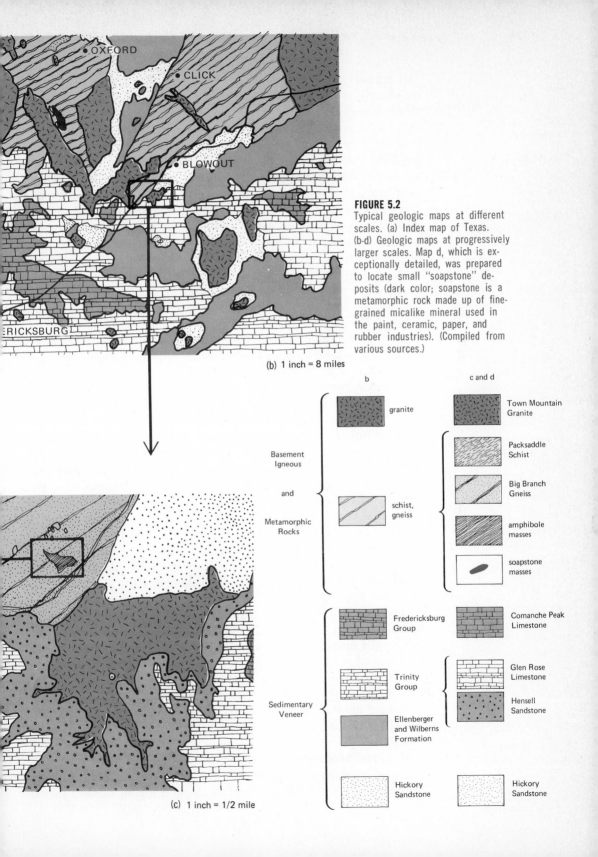

(b) 1 inch = 8 miles

FIGURE 5.2
Typical geologic maps at different scales. (a) Index map of Texas. (b-d) Geologic maps at progressively larger scales. Map d, which is exceptionally detailed, was prepared to locate small "soapstone" deposits (dark color; soapstone is a metamorphic rock made up of fine-grained micalike mineral used in the paint, ceramic, paper, and rubber industries). (Compiled from various sources.)

OXFORD

CLICK

BLOWOUT

RICKSBURG

b

c and d

granite — Town Mountain Granite

Basement Igneous

and

Metamorphic Rocks

schist, gneiss — Packsaddle Schist

Big Branch Gneiss

amphibole masses

soapstone masses

Fredericksburg Group — Comanche Peak Limestone

Trinity Group — Glen Rose Limestone

Hensell Sandstone

Sedimentary Veneer

Ellenberger and Wilberns Formation

Hickory Sandstone — Hickory Sandstone

(c) 1 inch = 1/2 mile

eral deposits may be mapped in such detail that 1 inch on the map equals only a few hundred feet on the ground (Figure 5.2d). At the other extreme, large areas are commonly mapped by rapid reconnaissance methods, usually with the help of aerial photographs, at scales in which 1 inch on the map equals 8 miles or more on the ground (Figure 5.2b). Most geologic mapping falls between these extremes (Figure 5.2c). Most of the United States, for example, has been geologically mapped at scales of 1 in. = 1 mi, and many local regions at scales as great as 1 in. = ¼ mi. In preparing a geologic map each distinctive unit of rock, called a formation, is given its own two-part name. The first part of the name is some geographic locality where the rock is well exposed; the second part describes the general rock type. Thus Manhattan schist, St. Louis limestone, and Columbia River basalt are typical formation names used in the preparation of local geologic maps. Other examples, from a single region, are shown in Figure 5.2.

Geologic mapping not only shows the distribution of different formations, but also the rock's three-dimensional appearance, particularly as revealed by sedimentary layers (the technique for showing this is described in Section 5.6). An extremely significant conclusion is forced upon us by our geologic maps—or, for that matter, by an observant stroll through the countryside: *most rocks have been appreciably changed from the positions they occupied at their time of origin.* Sedimentary rocks that were once flat-lying are found to be intensely folded and broken; plutonic and metamorphic rocks are observed at the earth's surface, a site quite different from environments of their genesis. The present state of all these rocks provides abundant evidence that crustal rocks are continually being moved and altered by dynamic forces originating beneath the crust. Flat-lying sedimentary rocks can be seen to have been crumpled into huge mountain ranges by enormous lateral pressures, while even larger blocks of crustal rocks have moved hundreds of miles along lengthy ruptures. These and other kinds of evidence indicate that most of the topographic features of the earth's surface—features such as mountain ranges, basins, and plateaus on the continents, and island chains, deep trenches, and submerged mountains on the ocean floor—have their origin in complex *crustal movements.* For the remainder of this chapter we shall consider the nature of these fundamental movements, particularly as they apply to the more easily observed rocks of the continents. In Chapter 7 we shall seek a more complete understanding of the mechanism for this deformation when the topic of plate tectonics is considered.

Until the 1960s most evidence for crustal movements came from study of continental rocks, which are of course far more accessible than are those of the ocean floor. Geologic mapping showed that intense movements had not taken place randomly over the surface of the continents, but instead tended to be concentrated in long, narrow belts that usually parallel continental margins. Continental **mountain ranges**, formed in these belts by the compression and distortion of previously flat-lying and undeformed rocks or by the uplift of plutonic masses, provide the principal evidence for movements of the continental crust.

MODES OF ROCK DEFORMATION

When rocks of the crust are subjected to mountain-building forces they can respond in two ways: either they bend or they break. These two responses define the two major structural features seen in deformed crustal rocks: **folds**, the result of rock deformation of a primarily plastic nature (Figures 5.3 and 5.4); and two kinds of *fractures*,

(a)

FIGURE 5.3
Folds. (a) Small-scale, in limestones of northern Scotland; the camera case at top left provides scale. (b) Intermediate-scale, in limestones of northern England. (a, b: Geological Survey of Great Britain.)

(b)

FIGURE 5.4
Large scale folds. (R. H. Chapman, U.S. Geological Survey.)

called **joints** and **faults.** Joints are fractures along which no appreciable movement has occurred (Figure 5.5), while faults show displacements, or offset, of one side of the break with respect to the other (Figures

FIGURE 5.5
Geometrically regular jointing. (Geological Survey of Great Britain.)

5.6 and 5.7). Jointing is so commonplace as to have perhaps gone un-noticed by many, but these fractures, which typically display some geometrically regular pattern, are noteworthy examples of small-scale rock deformation. Folds and faults are much more significant features and occur in limitless variations of size and shape. Entire mountain ranges may be formed by a single huge fold or fault while, at the other extreme, pieces of rock small enough to be held in one's hand may be broken by tiny folds and faults that can be seen only under a micro-scope.

Careful descriptions and geologic maps of the folds and faults seen in continental mountain ranges have been made since the nineteenth century, and these have provided a wealth of information about the geometry and distribution of various kinds of deformed rocks. We shall first introduce the terminology used to describe folds and faults and then deal with the problem of graphically depicting such three-dimensional features on flat, two-dimensional pieces of paper in the form of geologic maps and structure sections.

FIGURE 5.6
An offset center line: California Highway 46,
San Andreas fault (1966). (E. A. Hay.)

(a)

FIGURE 5.7
Faults. (a) Small dip-slip fault in shales of northern England. (b) Reconstruction of (a) as it appeared prior to erosion. (a: Geological Survey of Great Britain.)

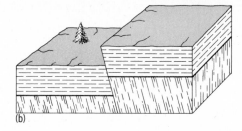

(b)

5.3 Faults

When crustal rocks are dislocated there are three varieties of movement possible: dip-slip, where the movement is primarily in an up-and-down sense (Figure 5.7a and b; strike-slip, commonly called transcurrent, having essentially horizontal movement (Figure 5.6); and oblique-slip, with elements of both directions of movement (Figure 5.8).

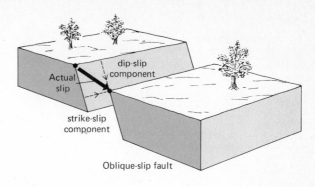

FIGURE 5.8
Oblique-slip faulting.

The complex models that geologists have developed to account for these different types of movement are beyond the scope of this book; nevertheless, we can point out some general relationships existing between directions of stress within the crust and the variety of faulting that results. Dip-slip movements manifest either *compressional* forces, which give rise to *shortening* of the crust in the case of **reverse** (*thrust*)-**slip faulting,** or tensional forces, causing crustal *lengthening,* for **normal-slip faulting** (Figure 5.9). Transcurrent faults, on the other hand, appear to have resulted from crustal movements of a more nearly shearing nature. These faults are described as either right-lateral-slip or left-lateral-slip, depending on the direction each side of the fault has moved with respect to the other (Figure 5.10).

Figure 5.10a shows the results of movement that occurred a few miles north of San Francisco on the famous San Andreas Fault during the great 1906 earthquake. Figure 5.6 also depicts right-lateral-slip on the San Andreas Fault that occurred in 1966. In addition to these recent offsets, we have excellent evidence based on the displacements

FIGURE 5.9
Distinction between normal-slip and reverse-slip faults. (a) Reverse-slip. Note overlapping due to crustal shortening. (b) Normal-slip. Note extension due to crustal lengthening. (a, b: E. A. Hay.)

(a)

(b)

(a)

FIGURE 5.10
Right-slip and left-slip faults. (a) Right-slip, offset road near Olema, California (1906). (b) Block diagram showing left-lateral movement by faulting. A′ and B′ were at the same point prior to faulting. A person walking toward A′ in Block A would have to turn left at the fault to reach B′; therefore the offset is left-lateral. (A person walking toward B′ in Block B would also have to turn left to reach A′.) (a: G. K. Gilbert.)

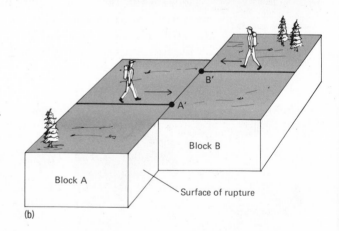

(b)

of different-aged rock formations that this major fault has moved hundreds of kilometers in a right-lateral-slip manner during the last 30 million years or more (Figure 5.11). Recently the San Andreas Fault

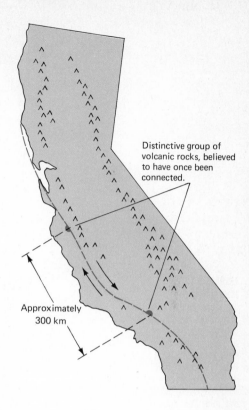

Distinctive group of volcanic rocks, believed to have once been connected.

Approximately 300 km

FIGURE 5.11
Example of a right-lateral slip that has accumulated on the San Andreas fault during the last 25 million years.

has been referred to as a "transform" fault, a judgment based on models to explain its movement using the new concepts of plate tectonics. Transform faults are a special variety of transcurrent faults. They were recently defined by the prominent geologist J. Tuzo Wilson as a means of explaining the relationships between ocean-floor fracture zones and mid-ocean ridges. We shall discuss this entire topic in some detail in Chapter 7.

Because we usually observe the faults in the three-dimensional earth on a surface that is, for all practical purposes, two-dimensional—such as in a roadcut or on flat terrain—it is possible to be misled or uncertain about the actual sense of *movement* that has taken place on a fault. We must make a distinction between the fault *slip*, which reflects its *actual* movement, and the fault *offset*, which reflects only *apparent* movement resulting from erosional effects (Figure 5.12). Sometimes such distinctions are not possible, but when they are, terminology should be applied in a manner to differentiate between slip and offset. In Figure 5.12, for example, we would describe the fault as a

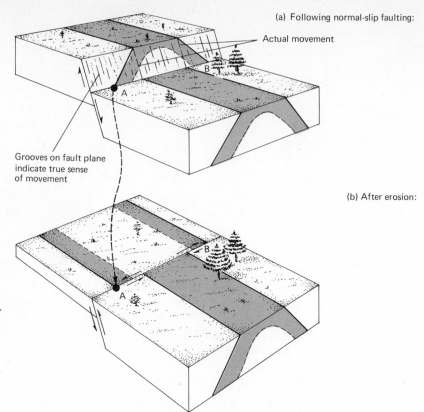

(a) Following normal-slip faulting:

Actual movement

Grooves on fault plane
indicate true sense
of movement

(b) After erosion:

FIGURE 5.12
Distinction between
apparent and actual
movement. In the fig-
ure the arrows at A
and B indicate **appar-
ent** movement: at A
it is left-lateral and
at B it is right-lateral.
They describe **offset,
not slip.** The actual
slip is **normal.**

normal-slip fault showing left-lateral offset at A and right-lateral offset at B.

5.4   Folds

The classification of folded rocks basically employs two categories:
archlike folds are called **anticlines** (*anti* = opposed, *cline* = inclination), where the *limbs* (opposite sides) of the fold are inclined downward away from each other; and troughlike folds are called **synclines**
(*syn* = together, *cline* = inclination), having their limbs dipping toward each other. Figure 5.13 shows useful terminology for describing
the elements of folds. It is no surprise that folded rocks have curved
archlike and troughlike appearances when viewed in cross section
(shown on the front surface of Figure 5.13a), but an explanation for

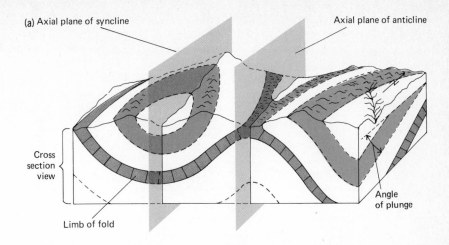

(a) Axial plane of syncline

Axial plane of anticline

Cross section view

Limb of fold

Angle of plunge

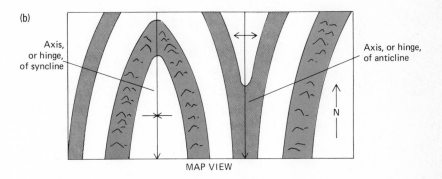

(b)

Axis, or hinge, of syncline

Axis, or hinge, of anticline

N

MAP VIEW

(c) Limb: the sides of a fold

Axial plane: imaginary surface dividing the opposite limbs.

Axis: imaginary line on the surface that divides the opposite limbs.

Plunge: the condition where the axis is inclined downward.

FIGURE 5.13
Fold terminology. (a) Block diagram. (b) Map view. (c) Definitions of fold terminology.

the fact that most folds display similar curved patterns when viewed on the earth's surface is not so obvious (Figure 5.13b). If the axes (hinges) of all folds were parallel to the earth's surface of erosion, such map patterns would not be common; as a general rule, however, not only are the limbs folded, but also the axis of these limbs undergoes tilting or folding (Figure 5.14). The result is an eroded pattern of rock exposures that bend around in a "hairpinlike" manner. Figure 5.15 (a and b) shows that the map pattern of a plunging fold is the

(a)   Originally flat-lying sedimentary layer

Horizontal
reference plane

(b)   Folded, with fold axes remaining horizontal

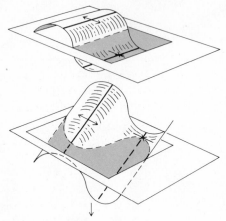

FIGURE 5.14
Origin of plunging folds. In (c) note
that a horizontal plane (such as a
typical eroded surface) cuts the deformed layer to expose a curved outcrop pattern.

(c)   Axes are tilted downward (plunging)
toward the viewer

FIGURE 5.15
Plunging folds. (a) Plunging anticline in Wyoming. (b) Plunging syncline in California.
(a: John S. Shelton; b: E. A. Hay.)

(a)   (b)

FIGURE 5.16
Age relationships of
folded sedimentary
rocks. Imagine walk-
ing across the sur-
face of this terrain.
At the anticlinal axis
you would be on the
oldest exposed rocks,
and at the synclinal
axis you would be on
the youngest. The
arrows on the cross
section show the
direction toward
which the tops of the
beds are facing:
away from the axial
plane of the anticline
and toward the axial
plane of the syncline.

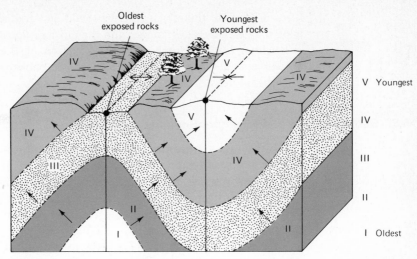

same for both antcilines and synclines. The distinction is that in anticlines the limbs dip away from each other and in synclines they dip
toward each other.

Because of the law of superposition, two important generalizations
can be made regarding the relationships between sedimentary strata
and the folds in which they are found: (1) Rocks become progressively
older toward the axes of *anticlines* and *younger* toward synclinal axes;
and (2) the *tops* of beds face *away from anticlinal* and *toward synclinal* axial planes (Figure 5.16). In regions of intense deformation,
folds are frequently overturned, sometimes becoming isoclinal (*iso* =
equal, *clinal* = inclination), where both limbs are dipping at the same
angle in the same direction (Figure 5.17). Methods for distinguishing
between the top (youngest) and bottom (oldest) side of a sedimentary
layer become exceedingly important as tools for analyzing such folds.
Recall from Chapter 3 that there are a variety of *sedimentary struc-*

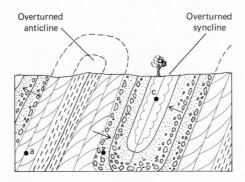

FIGURE 5.17
Isoclinal folds. Arrows indicate
the facing of tops of sedimentary
layers, as indicated by (a) cross-
bedding, (b) graded bedding, and
(c) ripple marks. (Review Section
3.4 regarding sedimentary structures.)

tures useful in this regard: cross bedding, graded bedding, ripple marks, and mud cracks, to mention the most obvious.

Isoclinal folds are not the only variation of the basic pattern of anticlines and synclines. Figure 5.18 shows patterns and terminology for other styles of folding.

5.5 Geologic Maps and Structure Sections

Folds and faults are commonly very large features—large enough so as to be not readily apparent to a person standing at a single location. This makes it necessary to make observations of many different places in order to fully recognize the manner in which a particular area of the earth has been deformed (see Figure 5.2). This activity involves the making of a geologic map.

Geologic maps are products of strenuous physical and mental labor. Although aerial photography often facilitates coverage of inaccessible areas, most geologic mapping is done by a geologist walking over the earth's surface in order to trace contacts that delineate formational boundaries and to plot other symbols, called "attitudes," used to indicate the three-dimensional orientations of rock units. An attitude consists of two elements, a direction of strike and an amount and direction of dip (Figure 5.19). A geologist must carefully observe and measure rock strata in the field in order to plot these symbols on a map. Figure 5.20 depicts a variety of attitudes and their field and map appearances. In many cases, large-scale folding is made apparent only

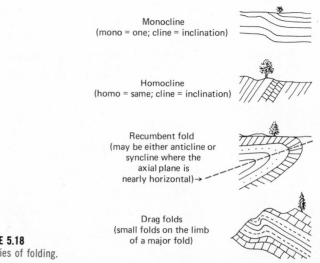

Monocline
(mono = one; cline = inclination)

Homocline
(homo = same; cline = inclination)

Recumbent fold
(may be either anticline or
syncline where the
axial plane is
nearly horizontal)→

Drag folds
(small folds on the limb
of a major fold)

FIGURE 5.18
Varieties of folding.

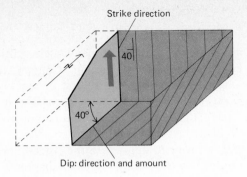

Strike direction

40

40°

Dip: direction and amount

FIGURE 5.19
Determination of an attitude.
Strike is the directional trend of
the bed as seen on a horizontal
surface. **Dip** is the direction and
amount (in degrees) of downward
inclination measured from the
horizontal.

FIGURE 5.20
Map appearance of an eroded stratum. Note that in the bottom map-drawings the direction
of dip can be determined from the appearance of the stratum as it crosses the valley.

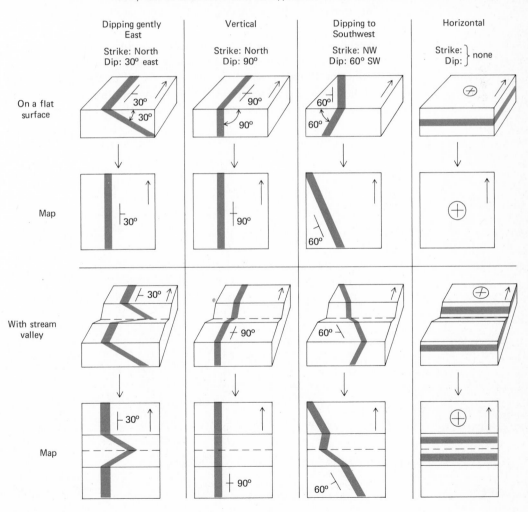

	Dipping gently East	Vertical	Dipping to Southwest	Horizontal
	Strike: North Dip: 30° east	Strike: North Dip: 90°	Strike: NW Dip: 60° SW	Strike: none Dip: none

On a flat
surface

Map

With stream
valley

Map

by the accumulation and plotting of many such attitudes on a map. In addition to being concerned with folding geometry, the field geologist must impose some order on what at first commonly seems to be a bewildering array of different kinds of rock. The immediate goal of all this is to make a geologic map; the ultimate goal is usually to understand earth history.

Obviously it is easier to appreciate the three-dimensional relationships in Figure 5.20 by consulting the block diagrams with their cross-sectional views on the front faces than by studying the map views. It is for this reason that side views called **geologic structure sections** are almost always prepared from geologic maps. Following are examples indicating how such geologic structure sections are prepared. First, the reader should look again at the map view in Figure 5.21 and review the terminology found on it.

Now let us redraw this map, this time showing only the contour lines (Figure 5.22a): this illustrates the technique of constructing a **topographic profile** along the line of section X–X' in the space immediately below the map (Figure 5.22b). A topographic profile, like the profile of a person's face, shows a side view, with its highs and lows,

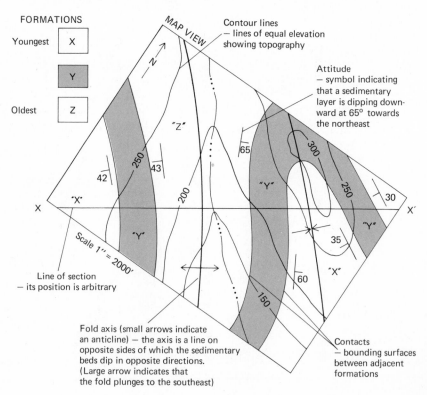

FIGURE 5.21
Geologic map. (See text for discussion.)

MAP VIEW

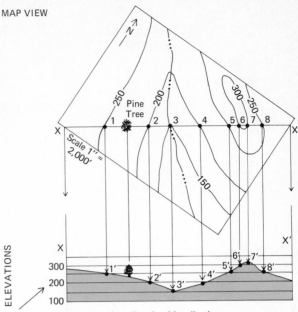

FIGURE 5.22
Contour map and topographic profile
derived from Figure 5.21. (See text.)

The distance between these lines is arbitrarily chosen.
Note that the vertical scale is exaggerated with respect to
the horizontal scale.

of how the earth's surface appears along a given line. By way of analogy, the X–X′ map and topographic profile views respectively correspond to the A–A′ front and profile views of the man's face shown in Figure 5.23.

Every time the line X–X′ crosses a contour line on the map, we have precise information about the elevation above sea level for that point as shown on the profile. It is for this reason that the points 1, 2, 3, etc. on the map can be transferred to the points 1′, 2′, 3′, etc. below as a means of constructing the topographic profile.

Finally we return to our original geologic map (Figure 5.24), add

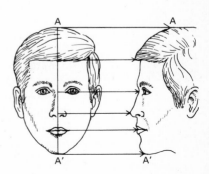

FIGURE 5.23
Relationship between plan and profile
views. (See text.)

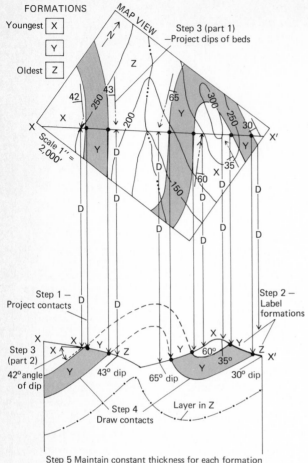

Step 5 Maintain constant thickness for each formation
(e.g. Formation Y is the same everywhere.)

FIGURE 5.24

Making a geologic structure section from a geologic map.
Note sequence of steps (1 through 5) shown in the figure.

The stages in the construction of a geologic structure section can best be done by proceeding in a step-by-step fashion. Note that each of these steps is indicated in the figure; study it carefully.

Step 1   Using a straight-edge, carefully project the points where contacts cross the line of section on the map to the corresponding points on the topographic profile.

Step 2   These projected contacts represent the bounding surfaces between the rock formations they divide. Label (or color) each formation's location on the profile so as not to lose track of the location of each.

Step 3   Project the attitudes (⌐ symbols) on the map into the line of section and then down to the topographic profile. Use these symbols to determine the direction and amount of dip for each of the formations to be used in the structure section. Remember that these dips are only valid right at the surface of the profile and should not be extended indefinitely into the subsurface.

Step 4   Using the surface dips as guidelines, and contacts as starting points, extend the formational contacts into the subsurface. Endeavor to connect like-surfaces in the subsurface, as in the case of a syncline, and dash their projections above the surface, as in the case of a partially eroded anticline.

Step 5   Always maintain the thickness of a formation throughout the geologic structure section.

the topographic profile below it, and proceed with the construction of a geologic structure section to show the appearance at depth of the rock formations along the line X–X′ as they would appear below the surface (topographic profile) of the earth. This time we shall transfer contacts between formations, rather than contour elevations, from the map view to the topographic profile, again recognizing that any point on the map X–X′ line has a precisely corresponding position on the

X–X' topographic profile. Recall that if these sedimentary formations were originally deposited in essentially horizontal, blanketlike layers, with "Z" on the bottom and "X" on the top, the contacts between them have become exposed at the earth's surface because erosion has cut into this pile of layered rocks after the folding occurred. If there had been no erosion, we would see only the top layer formation "X."

5.6   Experimental Rock Deformation

It is readily apparent in Figure 5.24 that rocks of the earth's crust have responded to deforming forces in a variety of ways, sometimes breaking and other times folding in a more plastic manner. Descriptive mapping studies have raised questions about rock deformation that can best be approached by theoretical or experimental studies.

Most of the rocks exposed at the earth's surface are very brittle. If forces are applied to them—for example, by striking with a hammer or squeezing with a vise—they yield only by breaking. Because familiar rocks respond in this way, it is easy to visualize how rocks can deform by breaking along faults, but more difficult to understand how they can deform by *bending* to form folds. Obviously, folded rocks now exposed at the earth's surface must have originated under other conditions, conditions which made the rocks less brittle and more *ductile*—that is, able to flow and deform without fracture. Theoretical and laboratory studies have shown that ductility is not constant for a particular kind of rock, but varies with changing environmental conditions. Three environmental factors are particularly important in affecting rock ductility: pressure, temperature, and the rate at which deforming forces act on the rock.

The effect of variations in pressure and temperature upon rock deformation can be investigated by subjecting rock specimens—usually cut into the shape of cylinders about 3 to 15 centimeters (1 to 6 inches) long—to deforming forces at varying temperatures while they are held in special high-pressure presses (Figure 5.25). These presses simulate the increased pressures and temperatures to which rocks are subjected as they are buried deeper and deeper in the earth. Such presses can achieve pressures in excess of those occurring anywhere in the crust and can be operated at temperatures high enough to melt most rocks. Studies using these presses show that most rocks tend to become more ductile as pressures and temperatures increase. Thus the brittle behavior we associate with rocks at the earth's surface may be replaced by ductile deformation and by folding when the same rocks are subjected to deforming forces while deeply buried. By observing the patterns of deformation seen in rocks now exposed at the surface, it is sometimes possible to infer, from laboratory experiments on sim-

Increasing pressure

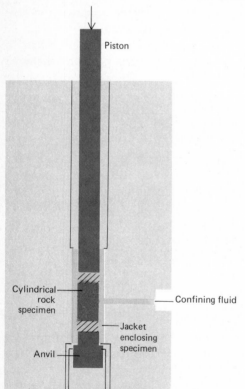

Piston

Cylindrical rock specimen

Confining fluid

Jacket enclosing specimen

Anvil

FIGURE 5.25
Experimental rock deformation. (a) Schematic drawing of a rock deformation press. (b) Cylinders of limestone shortened 15 percent at increasing pressures (left to right) in a deformation press. At low pressures the specimens break as in faulting; at higher pressures they deform without breaking as in folding. (Photos courtesy F. A. Donath; from "American Scientist," vol. 58, 1970.)

ilar materials, the approximate depths, temperatures, and pressures at which the rocks were deformed.

The effects of temperature and pressure on rock ductility are relatively easy to study in laboratory experiments and are now reasonably well understood. In contrast, the third major factor controlling rock ductility—the rate at which the deforming forces act—is an area of great uncertainty, because very small forces, when applied over very long times, can lead to ductile deformation in otherwise brittle materials. A familiar example is a wax candle, which may over a long period bend under its own weight but which can also be easily broken by a sharp blow. Similarly, gravestones made of brittle natural rocks such as marble have been observed to bend under their own weight over many years. These observations suggest that time is an extraordinarily important factor in rock ductility; over long periods even brittle rock masses exposed at or near the earth's surface might deform by ductile flow under the action of relatively small forces, such as the force of gravity. For larger forces of shorter duration, the same rocks would break like brittle solids. Thus either folding or faulting might occur depending on the rate at which the deforming forces acted. Unfortunately, the effect of time on rock deformation has been little studied because laboratory experiments cannot duplicate the effects of small forces acting on rocks for thousands or millions of years.

SUMMARY OUTLINE

Rocks of the Crust

5.1 *Continents and ocean basins:* crustal rocks show a basic division into high-standing continents and low-lying ocean basins.

5.2 *Continental rocks:* continents are composed mostly of granite and granitic gneiss covered by a thin veneer of sedimentary rocks; most of these rocks now occupy positions quite different from their sites of origin.

Modes of Rock Deformation

5.3 *Faults:* crustal rocks may be dislocated to produce strike-slip, dip-slip, or oblique-slip movements.

5.4 *Folds:* anticlines are archlike folds, and synclines the troughlike folds.

5.5 *Geologic maps and structure sections:* these depict the distribution, relative age, and geometry of different rock units (formations).

5.6 *Experimental rock deformation:* folding occurs only when rocks are made ductile by burial at high temperatures and pressures.

ADDITIONAL READINGS

Billings, M. P.: *Structural Geology*, Prentice-Hall, Inc., Englewood Cliffs, N.J., 1972. A standard intermediate text on crustal folds, faults, and mountain building.

*Cailleaux, A.: *Anatomy of the Earth*, McGraw-Hill Book Company, New York, 1968. A popular introduction to the solid earth, emphasizing crustal rocks and structures.

*Clark, S.P., Jr.: *Structure of the Earth*, Prentice-Hall, Inc., Englewood Cliffs, N.J., 1971. Chapter 2 provides a good summary of crustal structures.

*Available in paperback.

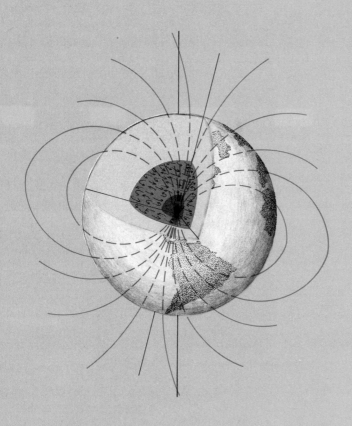

chapter 6

Inside
the Earth

The only rocks and minerals that earth scientists can study *directly* are those exposed on the earth's surface and those which can be sampled in mines and bore holes. The deepest mines extend downward only about 3 kilometers (2 miles), and the deepest bore hole drilled to date (an oil well in Texas) reached a depth of about 8 km (5 mi). Since the earth's radius is almost 6,400 km (4,000 mi), clearly only its most superficial skin can be directly sampled (Figure 6.1). For this reason, our understanding of the earth's interior depends on various indirect clues, most of which are provided by four physical properties: its *wave transmission, heat transmission, magnetism,* and *gravity.* Each of these can be measured at the surface and then used to infer characteristics of the materials below. By combining evidence from these properties, geophysicists have developed a model for the structure and composition of the earth's vast interior; the nature of this model and the supporting evidence for it will be the subject of this chapter.

EARTHQUAKES AND THE EARTH'S INTERIOR

The most important clues to the earth's interior are provided by the manner in which various shock waves are transmitted through the body of the earth. Such shock waves have two sources: some result from man-made explosions of dynamite or nuclear devices; others are caused by earthquakes, which are natural movements along ruptures bounding large blocks of the solid earth. Large earthquakes, when they occur in densely populated regions, are among the most devastating natural phenomena on earth. In Chapter 11 we shall introduce techniques for measuring the sizes of earthquakes and also discuss the variety of hazards they pose for man.

In general, shock waves from explosions are most useful for determining relatively local properties of the earth's crust, whereas earthquakes may reveal both small- and large-scale patterns of the earth's internal structure. Indeed, the study of earthquake wave transmission has provided more information about the earth's interior than have clues from the additional properties of gravity, magnetism, and heat transmission combined.

6.1 Earthquake Waves

The rock movements responsible for earthquakes are not randomly distributed throughout the solid earth, but occur only in the outermost 700 km (400 mi), a little over 10 percent of the distance to the center of the earth (Figure 6.2). Only a few occur at depths as great as 700 km; most take place in the outermost 70 km representing only about

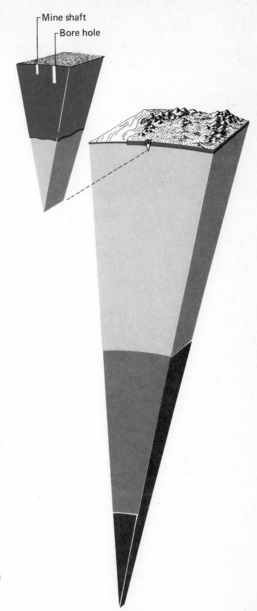

FIGURE 6.1
Scale drawing of a deep mine, bore
hole, and the earth's interior.
Crustal thickness is exaggerated.

1 percent of the radius of the earth. In addition to these depth restric-
tions, earthquakes show a geographic concentration along three great
belts. Most large earthquakes, and virtually all of those occurring at
depths below 300 km (180 mi), take place in a zone around the edges
of the Pacific Ocean; a second zone of smaller earthquakes occurs

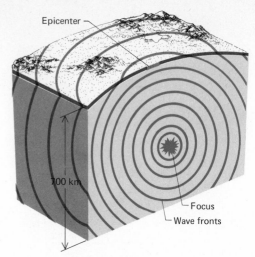

FIGURE 6.2
Relative locations of focus and epicenter (point on earth's surface directly above focus) of an earthquake. Earthquake foci do not occur below 700 km.

along chains of submerged mountains that extend through the ocean basins; a third zone borders the Mediterranean and southern Asia (Figure 6.3).

Earthquakes arising in these regions cause various kinds of shock waves to move upon and through the earth in much the same way that waves and ripples are caused by dropping a pebble into quiet water. These waves, known collectively as seismic waves, are of three principal kinds: body waves, which travel from the source or focus of the earthquake through the main mass of the earth to emerge at the surface; surface waves, which travel only in the outermost layers of the earth; and free oscillations, vibrations of the earth as a whole, caused by only the largest earthquakes, which can be likened to the vibrations of a bell when struck by a hammer (Figure 6.4). Most of our knowledge of the earth's interior has come from study of body waves, but in recent years analyses of surface waves have provided additional information, particularly about the outermost layers of the earth. Free oscillations were first recorded by special instruments after the disastrous Chilean earthquake of 1960, although their existence had been predicted earlier. Like surface waves, they are useful in interpreting the structure of the earth's outermost layers, but they also provide some independent clues to the deep interior. All three wave types are measured by sensitive instruments called seismographs which, when anchored to solid rock, record even the slightest earth shocks (Figure 6.5). Universities and government research institutes have established around the world about 150 major *seismic observa-*

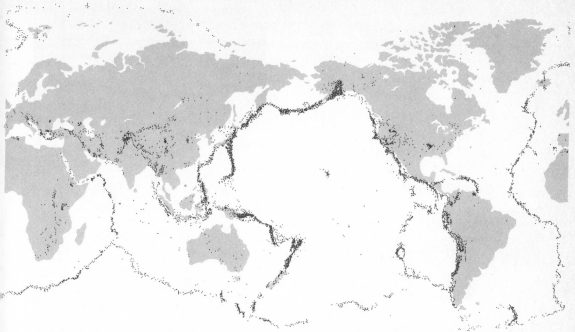

FIGURE 6.3
Locations of epicenters for all earthquakes recorded from 1961 through 1969. (From National Earthquake Information Center, U.S. Department of Commerce.)

tories containing such instruments, and these are the principal source of information on earthquake wave transmission.

Earthquake body waves, on which most of our knowledge of the earth's deep interior rests, can be subdivided into two types known

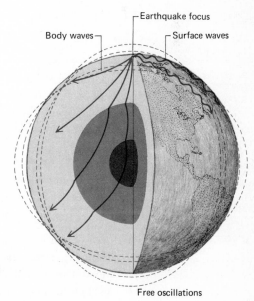

FIGURE 6.4
The three principal types of seismic waves: body waves, surface waves, and free oscillations.

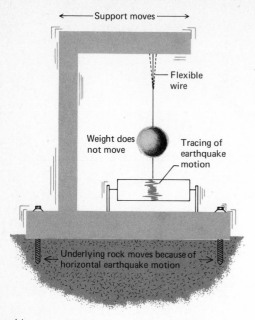

(a)

(b)

FIGURE 6.5
Seismographs. (a) Schematic diagram. Earthquake waves cause movements of the underlying rock and attached support; the weight tends to remain stationary while the pen records the motion. (b) A seismograph registering a large, distant earthquake. (United Kingdom Atomic Energy Authority.)

as **primary** (or merely *P*) and **secondary** (*S*) waves. In *P* waves the individual particles of the material through which the wave travels are elastically vibrated *back and forth* in the direction of wave movement; in *S* waves the particles are sheared *sideways*, at right angles to the direction of wave motion (Figure 6.6). Because of these different motions, the two waves move at different speeds: in the same material, *S* waves travel only about half as fast as *P* waves. The names reflect this fact, for "primary" *P* waves arrive at distant seismic observatories before "secondary" *S* waves from the same earthquake. In addition to these inherent velocity differences, the speed at which *both* waves move depends on the physical properties, particularly the density and elasticity, of the materials they are passing through. Both types move faster as the density decreases and the elasticity increases. Because of this influence on velocity, *P* and *S* waves are bent or *refracted* at the boundaries of layers with differing densities or elasticities, just as light waves are bent in passing from less dense air into more dense glass (Figure 6.7).

One easy way to visualize the bending of waves (seismic, light, water, etc.) is to employ a model proposed in the seventeenth century by the physicist Christian Huygens. Figure 6.7a depicts a seismic wave front striking a rock boundary on opposite sides of which the waves will move at different velocities. Refraction (bending) occurs because the end of the wave front that first reaches the boundary experiences

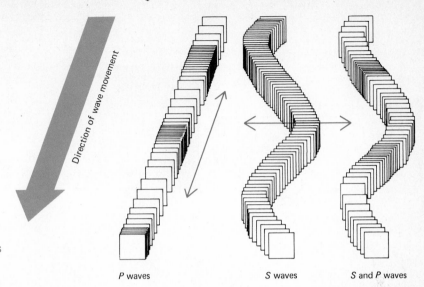

FIGURE 6.6
Schematic representations of the motions of P and S waves as they pass through the body of the earth.

Direction of wave movement

P waves   *S* waves   *S* and *P* waves

"drag," or slowing, while the other end is still moving at the original velocity. This is much like having your left front automobile tire catch in a deep puddle: it is dragged, and your car swerves to the left (Figure 6.7b. Like light waves, seismic waves may also be *reflected* from reflecting boundaries without penetration. Because of both reflections and refractions, the earthquake body waves reaching a distant seismic observatory consist not only of simple *P* and *S* waves that have traveled by the most direct route, but also of dozens of additional waves that have reached the station by more circuitous reflected and refracted paths.

How might we record and analyze these many seismic waves? If we want to make a permanent record of a visual experience, we use a special instrument—a camera—to do so. Once the photographic record has been made, it can be studied carefully to discover and interpret its contents. Similarly it is necessary to acquire recordings, called seismograms, of earthquakes in order to better "see" the earth's interior. Seismograms are also our principal means of locating the epicenter (the point on the earth's surface directly above the focus) and *focus depth* (Figure 6.2) for an individual earthquake. Figure 6.8a depicts a seismogram recording a distant earthquake. It can be used in the manner shown in Figure 6.8b to determine how far away from the seismographic station the earthquake occurred.

Once the distances from three or more widely spaced seismographic stations to an earthquake epicenter become known, a good approximation of the geographic location of that epicenter can be achieved in the manner shown in Figure 6.9.

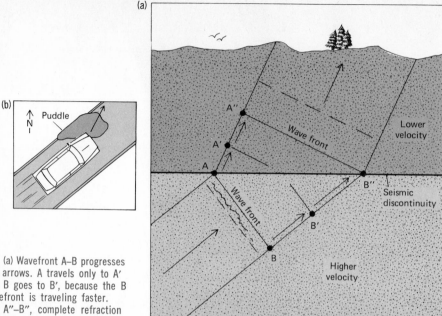

FIGURE 6.7
Wave refraction. (a) Wavefront A–B progresses
as indicated by arrows. A travels only to A′
during the time B goes to B′, because the B
end of the wavefront is traveling faster.
By the time of A″–B″, complete refraction
of this wavefront has occurred. (b) Sketch
showing a similar response as an automobile
strikes a mud puddle with one front tire
before the other.

6.2   Elastic Rebound Theory

We have described the properties of seismic waves and how their records have been analyzed to yield travel-time curves. In the following
section we shall show how such data can be interpreted to generate
a *model* for the earth's interior. Before doing that, however, one important question remains: What is the principal *cause* of earthquakes?
At the beginning of this chapter we suggested that "ruptures bounding large blocks of the solid earth" are the answer to that question. In
other words, the energy of most earthquakes is believed to come from
faulting. Early in this century, following a careful analysis of ground
survey data for the region around San Francisco from before and after
the great 1906 earthquake there, H. F. Reid proposed the **elastic rebound theory** as the mechanism for generating that earthquake. Since
then, most geologists have come to accept this model as essentially
valid for most earthquakes, although certain variations and exceptions
are still under study. Figure 6.10 shows a sequence of crustal configurations from an unstrained state, through the accumulation of elastic
strain, to the rupturing of the rock, allowing it to rebound to an unstrained state. Here, then, we have elastic rebound as the cause of
faulting, and the faulting as the cause of the earthquake. You can vis-

(a)

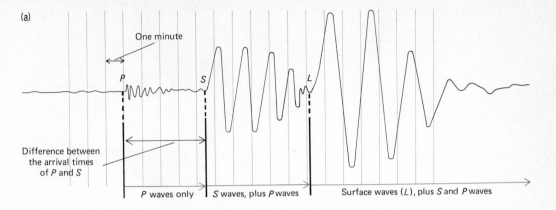

One minute

P

S

L

Difference between the arrival times of P and S

P waves only | S waves, plus P waves | Surface waves (L), plus S and P waves

FIGURE 6.8
(a) Seismogram showing the first arrivals of P, S, and L waves. (b) Travel time curves: these curves are compilations of many years experience relating the time taken by P, S, and L waves to travel known epicentral distances. It can be seen that the farther the seismic waves travel, the greater will be the time gap between the arrival of P and S waves. Thus, the epicentral distance is proportional to this time difference and can be determined from it.

(b)

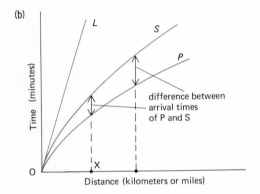

Time (minutes)

L

S

P

difference between arrival times of P and S

O

X

Distance (kilometers or miles)

ualize an analogy to this process by slowly stretching a rubber band until it breaks. The only rapid movement occurs when the two elastically strained pieces rebound back to equilibrium. You are made

FIGURE 6.9
Location of an epicenter. The epicentral distances from three seismographic stations (A, B, and C) are determined. Circles with radii of these distances are drawn around their respective stations. There is only one location that is correct for A, B, and C: the epicenter.

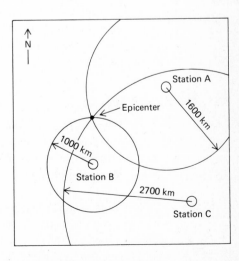

N

Station A

Epicenter

1600 km

1000 km

Station B

2700 km

Station C

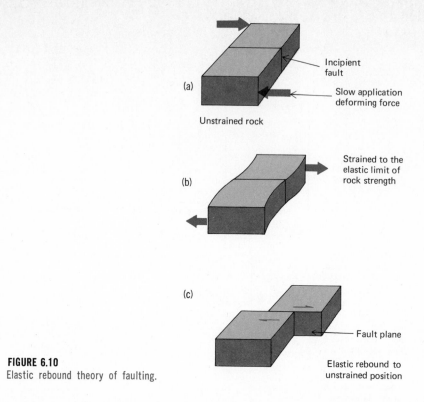

(a)   Incipient fault

Slow application deforming force

Unstrained rock

(b)   Strained to the elastic limit of rock strength

(c)

Fault plane

FIGURE 6.10
Elastic rebound theory of faulting.

Elastic rebound to unstrained position

aware that energy has been released by the accompanying sound and the sting of the pieces as they snap against your fingers.

We still, however, have not satisfactorily explained the cause of earthquakes in a fundamental sense. What accounts for the accumulation of strain in crustal rocks? Only in recent years has an answer to this question begun to emerge. It resides in the concepts of *plate tectonics*, our principal concern in Chapter 7. We will defer, therefore, our investigation into the fundamental cause of earthquakes until then.

6.3   The Earth's Interior

Observations on the arrival time of seismic waves coming from thousands of different earthquakes recorded at many observatories have also provided data for the calculation of **velocity-depth curves**, which show the change in speed of both P and S waves as they go deeper into the earth. These curves, shown in Figure 6.11, are the primary tools used for interpreting the earth's interior.

The velocity-depth curves of Figure 6.11 show two depths at which

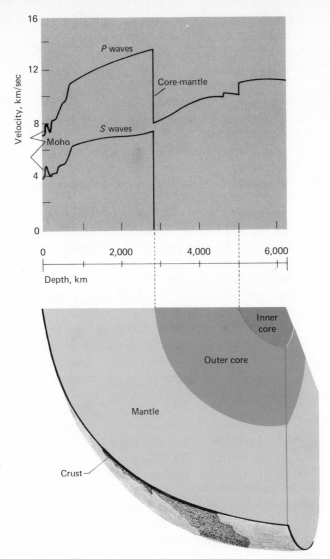

FIGURE 6.11
Velocity-depth curves for P and S waves (above) and the inferred internal structure of the earth (below). Major velocity changes occur at the Moho and core–mantle discontinuities. (Data from Anderson et al., "Science," vol. 171, 1971.)

earthquake wave velocities abruptly change: one at the relatively shallow depth of about 16 km (10 mi) (marked "Moho"); the other much deeper, at about 2,900 km (1,800 mi) (marked core-mantle), or about halfway to the earth's center. Such abrupt velocity changes are called seismic discontinuities: they clearly show that the materials of the earth's interior are not uniform in density and elasticity. Instead, the two major discontinuities define three concentric layers with differing properties: a thin outer **crust,** a central **core,** and a thick mantle lying between the crust and core. The mantle makes up about 80 percent of the earth's total volume, the core about 19 percent, and the crust only about 1 percent.

The outer discontinuity, that separating the crust and mantle, was discovered in 1909 by the Serbian geophysicist Andrija Mohorovičić. It has since come to be known as the Mohorovičić discontinuity, commonly abbreviated "Moho." The Moho seismic discontinuity provides the principal evidence that the rocks and minerals exposed on the earth's surface are but a thin veneer or "crust" overlying a much larger volume of materials with differing properties—the mantle and core.

The second major discontinuity, first recognized in 1924, is commonly referred to as the **core-mantle discontinuity.** Note in Figure 6.11 that not only does the velocity of *P* waves decrease dramatically at this discontinuity, but *S* waves *stop completely* at that depth—that is, they do not penetrate the core. This fact is of prime importance because *S* waves, you will recall, result from a *sidewise shearing* movement within the transmitting medium. Unlike the *pushing* motions of *P* waves, which can be transmitted through both solids and liquids, *S* waves can only be transmitted in solids, where the particles have enough adhesion to be pulled by one another. Liquids, of course, lack such adhesion and therefore do not transmit *S* waves. The lack of *S* wave transmission through the core provides strong evidence that the *core is liquid,* whereas the transmission of both *P* and *S* waves through the crust and mantle indicates that both the *crust and mantle are composed of solid materials.* To speak of "the solid earth" is therefore something of an oversimplification, since only the outer four-fifths of the earth's volume is made up of solid materials.

In addition to the two major seismic discontinuities defined by sharp breaks in the velocity-depth curves, minor discontinuities occur at several depths where the curves show less abrupt changes (Figure 6.11). These minor discontinuities provide additional clues to the structure of the core and mantle.

6.4 Internal Densities, Pressures, and Temperatures

We have noted that the velocity of *P* and *S* waves is dependent on both the *density* and the *elasticity* of the materials through which they pass. Neither of these properties can be directly measured for materials deep within the earth, and thus it is not completely certain whether the velocity changes seen in the velocity-depth curves result from changes in one or the other property, or both. Calculations based on the earth's gravity show, however, that the *average* density of the *entire* earth is about 5.5 grams per cubic centimeter (g/cm^3). This figure puts a restriction on the densities of the material in the mantle and core (because the crust makes up only about 1 percent of the earth's volume, it can be effectively ignored), for they must combine to have this average density. Because of this restriction and certain others caused by the manner in which the earth wobbles on its axis of rotation, it is possible to construct a reasonable model of the density distribution within the earth that is consistent with the velocity-depth curves. Such a model is shown in Figure 6.12a. Note that the densities of the mantle materials increase from the crustal value of 2.8 to about 5 at the base of the mantle. Then there is a sharp density increase to about 10 at the boundary of the core, and thereafter an increase from 10 to about 12 at the center of the earth. The average of these values, based on the different volumes of mantle and core material, is 5.5.

From the density curves, it is then possible to calculate the probable increase in pressure downward, which varies with both the amount and density of the overlying material. Such calculations show a steady pressure increase, from atmospheric pressure at the surface, to pressures about 1,000 times as great at the base of the crust and over *3 million* times as great at the earth's center (Figure 6.12b). These internal density and pressure curves are derived from the earthquake wave velocity-depth curves (Figure 6.11), but it should be stressed that these determinations also require educated guesses about the composition and elastic properties of the earth's interior materials and are therefore less certain than are the velocity-depth curves themselves, which are based on direct records of earthquake wave velocities.

Earthquake wave data provide little information about still another fundamental property of the earth's interior: its range of temperatures. The most obvious and important temperature inference that can be made from the velocity-depth curves is that the materials of the mantle, as solids, must be cooler than their melting point; whereas the materials of the core, as liquids, must be heated beyond their melting point.

Additional clues to the earth's internal temperatures are provided by still another physical property that can be measured at the surface: the manner in which heat is transmitted from the earth's warm interior to its cooler surface. Observations in mines and bore holes show that temperatures always increase rapidly downward through the outer few miles of the earth's crust. The rate of this increase varies from place to place, giving rise to differing *temperature gradients*; these range from about 9°C for each kilometer of depth (25°F/mi) to about 52°C/km (150°F/mi), with an average value of about 28°C/km (80°F/mi). Thus at the bottom of the deepest bore holes (about 8 km) temperatures are often far above the boiling point of water.

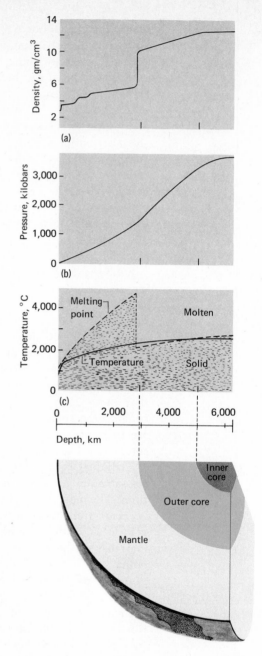

FIGURE 6.12
Inferred densities (a), pressures (b), and temperatures (c) of the earth's interior. (Data from Stacey, "Physics of the Earth," 1969; Verhoogen, "American Scientist," vol. 48, 1960.)

Below the very shallow zone of direct measurement in mines and bore holes, the earth's internal temperature gradient is poorly understood. The rapid rate of temperature increase seen in the outer crust probably does not persist farther downward, however, for if it did the materials making up much of the mantle would almost certainly be melted. (Figure 6.12c shows one estimated temperature-depth curve which fits this requirement).

THE CORE AND EARTH MAGNETISM

Studies of earthquake wave transmission show that the earth's interior has a basic threefold division into a thin solid crust, a massive solid mantle, and a central liquid core. As we have seen, earthquake wave data also permit the calculation of the probable densities and pressures within these fundamental units. There are, however, many other things that we would like to know about the earth's interior. For example, the composition and structure of the minerals that compose the mantle and core can only be inferred, for unlike the minerals of the outer crust these deeper materials cannot be directly sampled. Furthermore, the mantle and core are certainly not uniform and static masses, but dynamic units that show variations in their composition and structure. Although much remains to be learned about these variations, geophysicists have combined various indirect clues to arrive at reasonable hypotheses about some of them. For the remainder of the chapter we shall review these clues and hypotheses, beginning with the earth's central core and progressing outward to the mantle and, finally, to the deeper, hidden parts of the crust.

6.5   Composition of the Core

Seismic evidence indicates that the core is largely composed of a liquid whose probable density ranges from about $10g/cm^3$ at the outer core boundary to about 12 g/cm^3 at the earth's center. Most geophysicists now believe that this liquid is *molten iron*, with small amounts of other molten elements, particularly nickel and either sulfur or silicon, mixed in. Three principal lines of evidence suggest this.

Molten iron has a density of about 7.5 g/cm^3 at the earth's surface, but calculations show that at the extreme pressures encountered in the core, the iron atoms would be forced into a close-packed arrangement with a density slightly higher than that suggested for the core by seismic evidence. The slight discrepancy is believed to be due to small quantities of a lighter element, probably sulfur or silicon, in the liquid iron.

Additional evidence for an iron core is provided by the composition of meteorites, fragments of matter from interplanetary space that fall to the earth's surface. The materials in meteorites are believed to be similar to those that long ago consolidated to make the earth and other planets of our solar system (this will be discussed in more detail in Chapter 12). The composition of meteorites should, therefore, be similar to the bulk composition of the earth; that is, the most common elements in meteorites should also be the most common in the earth's core and mantle (recall that the highly variable rocks and minerals of the crust make up only 1 percent of the earth's volume and can be disregarded in considerations of the earth's bulk composition). Meteorites are of two principal kinds: iron meteorites, composed predominantly of iron, but containing between 5 and 20 percent nickel, and stony meteorites, composed of silicate minerals. The iron meteorites suggest that the earth's core may have a similar composition; in contrast, the stony meteorites may be similar in composition to the earth's mantle.

6.6   Earth Magnetism

The third reason for suggesting that the core is composed of molten iron is based on an important source of indirect evidence about the earth's interior that we have not yet introduced—that provided by surface measurements of the earth's magnetic field. The most familiar demonstration of the earth's magnetism is the common compass, the needle of which is merely a small bar magnet mounted on a swivel so that it may align itself with the earth's magnetic field. Compass observations have long shown that the earth behaves much as if it were a gigantic bar magnet with poles near the axis of rotation (Figure 6.13). Because the earth's magnetic poles are near the rotational poles, the compass needle always aligns itself in approximately a north-south direction.

The exact strength and direction of the earth's magnetic field at different points on the surface is measured by sensitive instruments called magnetometers, which can be used on land, at sea, or even in moving airplanes for rapid surveys of large areas (Figure 6.14). Such measurements show that the earth's overall magnetic field consists of two components which can be analyzed separately to yield different kinds of information. The largest component, the main magnetic field, originates deep within the earth and is a principal source of information about the interior. The second, much smaller component originates from the distribution of magnetic minerals, primarily magnetite (Fe_3O_4), in the earth's outer crust; it is of great value in understanding crustal composition and structure.

The origin of the main magnetic field is far more difficult to understand than is the simple magnetism caused by magnetic minerals in the outer crust. The main internal field cannot have such a source, for when magnetic minerals are heated they quickly reach a temperature, known as the Curie temperature, at which they lose their magnetism

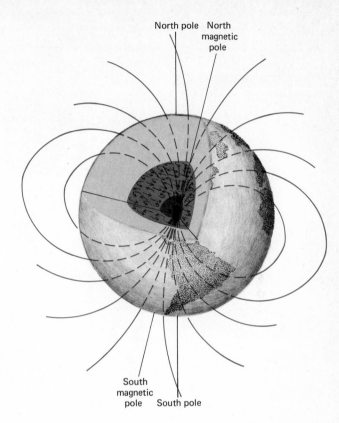

North pole   North
             magnetic
             pole

South
magnetic
pole   South pole

FIGURE 6.13
A generalized view of the earth's magnetic
field. The magnetism is believed to be
caused by motions within the liquid iron
core.

yet still remain solid. Curie temperatures are thus usually far lower
than the melting temperatures of minerals, and are also far lower than
the temperatures to be expected in the solid mantle. For these reasons,
any solid magnetic minerals present below the cool outer crust have
almost certainly lost their magnetism and cannot be a source of the
main magnetic field. There remains, however, one other potential
source of internal magnetism—that associated with electrical currents
(Figure 6.15).

   All electrical currents have associated magnetic fields, a fact first
established in 1819 by H. C. Oersted, who noticed that an electric current flowing through a wire will affect a compass needle placed near
the wire. Now if there were some means by which electrical currents
were generated deep within the earth, then the main magnetic field
could be understood as a direct natural consequence. The silicate minerals that make up the crust and, probably, most of the mantle are
poor electrical conductors and thus are unlikely sources of electric
currents and electromagnetism. Iron, on the other hand, is an excellent
conductor (in fact, the magnetic properties of iron can be traced to

Fixed eyepiece

Scale

Mirror

Balanced
magnet

−   ○   +

(a)

(b)

FIGURE 6.14
Magnetometers. (a)
Schematic diagram.
A balanced magnet
moves either up or
down with slight
changes in the
earth's magnetic
field; the position of
the magnet is read by
means of the eye-
piece, mirror, and
scale. More complex
electronic magneto-
meters are commonly
used for surveying
large areas. (b) A
simple magnetometer
in use. (Texas Instru-
ments.)

minute currents within its atoms), and thus motions in a core of fluid
iron (see Figure 6.13) might well create both electrical currents and
the main magnetic field. This possibility adds additional strength to
the suggestion that the core is, indeed, composed of molten iron.

FIGURE 6.15
The magnetic field produced by a loop
of wire carrying an electric current
is similar to that produced by a bar
magnet.

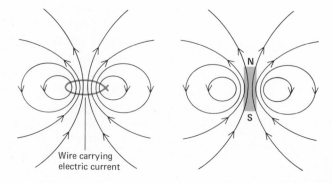

Wire carrying
electric current

6.7   Core Structure

Surface measurements of the main magnetic field show that most of its force could be produced by core-generated electrical currents that cause an **axial dipole magnetic field**—that is, a field with two opposing poles at opposite ends of an axis, much as in a simple bar magnet. After this large dipole field is subtracted from the main field there remains, however, a small **residual field** which, although originating in the core, is apparently not the result of the same current-producing core movements that cause the larger dipole field. Magnetic observations over the past hundred years show that the magnetic patterns of this residual field *move steadily westward* at a rate of about 19 km (12 mi) per year. The main dipole field, on the other hand, has remained rather constant in position. This westward drift of the residual field is of great importance, for it suggests that the core, at least in its outermost parts, is rotating less rapidly than the mantle and crust; only such a difference in speed of rotation would account for the steady change in position of the residual field with respect to the earth's surface where it is measured (Figure 6.16). This differential movement further suggests that the residual field may result from turbulent eddy currents caused by the drag of the solid mantle as it moves around the less rapidly rotating fluid core. Such eddy currents could produce small-scale electrical currents that might, in turn, give rise to the residual field.

Unfortunately, there is more difficulty in explaining the kinds of movements in the fluid core that might be responsible for the principal, dipole component of the main magnetic field. Not only are there theoretical problems in understanding how *any* motion in a fluid of uniform composition can produce magnetic fields, but there are also obstacles to explaining the underlying *causes* of the motions. Speculative suggestions for getting around these objections do exist, however, and most geophysicists now accept large-scale motions of the fluid core as the most probable source of the earth's main magnetic field. These internal motions are undoubtedly strongly influenced by the earth's overall rotational motion; this interaction probably causes the close coincidence of the magnetic and rotational poles.

One final bit of information about the structure of the core is provided not by magnetic data, but by earthquake wave transmission. The *P* wave velocity-depth curve of Figure 6.11 shows a relatively sharp velocity change within the core at a depth of about 5,000 km (3,200 mi). This discontinuity indicates a change in physical properties in the innermost part of the core; the small spherical inner core defined by this discontinuity is generally considered to be under such intense pressures that the iron atoms are closely compressed and respond as a solid, rather than as a liquid as does the bulk of the core.

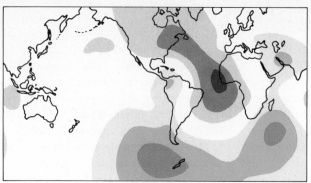

(a) 1922

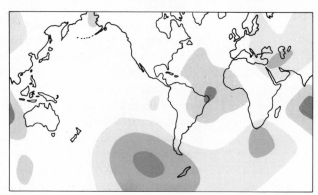

FIGURE 6.16
Westward drift of the residual field from
1922 to 1942. Strong positive values of
the residual field are shown in color;
strong negative values in gray. Note the
slight westward shift of the patterns during the 20-year interval. (From Vestine
et al., "Description of the Main Magnetic
Field," 1947.)

(b) 1942

Such a solid inner core would best account for the observed velocities
and patterns of wave transmission, but the evidence is still far from
conclusive.

THE MANTLE

The mantle makes up about 80 percent of the volume of the solid
earth, but because the average density of mantle material is about half
that of the underlying core, the mantle accounts for about 67 percent
of the earth's total mass. Besides occupying most of the earth's interior volume and contributing two-thirds of its total mass, the mantle
is of extraordinary importance because its properties have largely determined the nature of the crust which overlies it. Direct studies of
crustal rocks thus provide some clues to the hidden mantle but, as
with the core, most of our understanding of mantle composition and
structure is based on indirect evidence.

6.8 Mantle Composition

Unlike the liquid iron of the core, the solid mantle is believed to be composed predominantly of silicate minerals, just as are the more familiar rocks of the crust. This composition is indicated by three lines of evidence. The principal evidence of mantle composition is provided by meteorites which, we have seen, are of two principal types: iron meteorites, whose composition is probably similar to that of the core; and stony meteorites, composed of silicate minerals. Because there are compelling reasons for believing that meteorites are similar to the materials that originally consolidated to make the earth, stony meteorites may well reflect the general composition of the huge volume of material in the mantle. Stony meteorites do not in fact show the variety of compositions seen in crustal rocks; instead, they are composed largely of the silicate minerals *olivine* and *pyroxene* and are similar in composition to the relatively rare ultramafic rocks found in the crust. This suggests that the bulk of the earth's mantle may have an ultramafic composition.

The second kind of evidence of mantle composition is provided by laboratory experiments on the wave transmission characteristics of various silicate minerals at high temperatures and pressures. Such studies show that most crustal rocks transmit seismic waves too slowly to account for the velocities observed in the mantle (Figure 6.17a). There are, however, various mixtures of dense silicate minerals which have mantlelike wave transmission properties, and among these are mixtures dominated by olivine and pyroxene. The properties of one such mixture, called *pyrolite*, are shown in Figure 6.17b. This experimental work therefore supports the inference, based on examining the composition of meteorites, that olivine and pyroxene make up much of the mantle, at least in its outermost parts.

FIGURE 6.17

Mantle composition.(a) Seismic wave velocity (color) and inferred density (gray) of the mantle, at a depth of about 200 km, compared to those for the two principal crustal rocks—granite and basalt. (Data from Clark, Geological Society of America Memoir 97, 1966.) (b) Seismic wave velocity (color) and inferred density (gray) of the mantle, at a depth of about 200 km, are identical to those of laboratory mixtures of olivine, pyroxene, and garnet. Such mixtures, in the proportions shown, are called **pyrolite**. (Data from Anderson et al., "Science," vol. 171, 1971.)

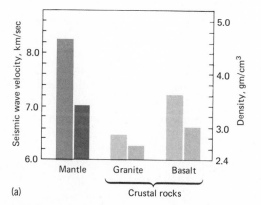

(a)

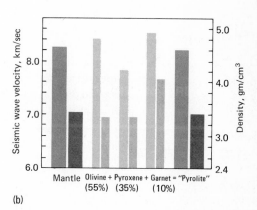

(b)

The final form of evidence of mantle composition comes from rare occurrences at the earth's surface of olivine- and pyroxene-rich ultramafic rocks which are believed to have crystallized in the upper mantle (Figure 6.18). These rocks contain small amounts of certain minerals (diamond, for example) which can form only under extremely high pressures—pressures greater than those normally encountered in the crust. Such minerals suggest that the ultramafic rocks originated in the upper mantle and were then somehow transported into the overlying crust, where they are exposed today on or near the earth's surface.

6.9 Mantle Structure

Recent seismic studies indicate that the mantle is not uniform throughout, but instead shows both horizontal and vertical changes in seismic properties. Figure 6.19 shows a modern survey of P wave velocities through the upper 800 km (500 mi) of the mantle, the region lying closest to the earth's surface which can be studied with high precision by seismic techniques. Note that in addition to the Moho marking the upper boundary of the mantle, there are three additional zones of very rapid increase in wave velocity bounded by regions of much less abrupt increase: these zones lie at depths of about 100, 400, and 650 km (approximately 60, 250, and 400 mi). There is also a single zone of sharp *decrease* in wave velocity lying not far below the crust, between depths of about 60 and 100 km (40 and 60 mi). These zones of rapid velocity change indicate levels at which the mantle materials undergo abrupt changes in their density, or elasticity, or both.

Some increase in density with depth is to be expected in the mantle because of the increasing pressures caused by the overlying materials, but calculations show this simple pressure increase to be both too small and too gradual to account for the sharp zones of velocity change. Instead, it is relatively certain that the lower two zones, at least, mark regions in which there is an *abrupt* change in the crystal structure of the silicate minerals that make up the mantle. At the temperatures and pressures occurring near the earth's surface, olivine and pyroxene are relatively stable minerals. Theoretical calculations and laboratory experiments show, however, that at higher temperatures and pressures olivine converts abruptly to a more dense and closely packed crystal structure (Figure 6.20). At still higher temperatures and pressures the silicon is forced from both olivine and pyroxene to yield a kind of quartz, while the remaining magnesium and iron convert to oxide minerals with very compact crystal structures and high densities (Figure 6.19). These transitions occur at about the

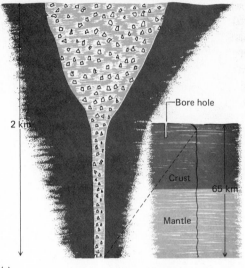

(a)

(b)

(c)

FIGURE 6.18

"Diamond pipes," cone-shaped masses of ultramafic rock that are believed to represent materials of the upper mantle forced upward into the crust and exposed today at the earth's surface. (a) Schematic cross section. (b) Ultramafic rock being mined near the surface in South Africa. (c) A deep diamond pipe, the abandoned Kimberly diamond mine in South Africa. The ultramafic rocks below the thick, overlying soil were dug out, crushed, and screened to remove the diamonds. (b: DeBeer Consolidated Mine; c: Bob Landry.)

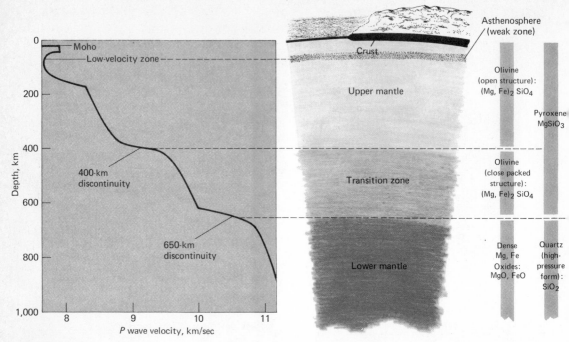

pressure–temperature conditions expected in the mantle at depths of 400 and 650 km, respectively, a finding that strongly suggests that the two lower zones of velocity increase are regions of abrupt structural changes in the minerals of the mantle. The region surrounding these changes is called the **transition zone** of the mantle; it separates the relatively thin and variable *upper mantle* from a much larger and probably more uniform mass of material composing the underlying *lower mantle* (Figure 6.19).

FIGURE 6.20
Diagrammatic structural change in olivine at high pressures. (a) Crystal structure of normal olivine as found in the crust and upper mantle. (b) More dense olivine with compacted crystal structure, inferred to be present in the transition zone of the mantle.

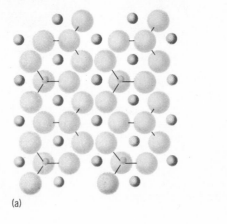

(a)

(b)

The velocity increase in the upper mantle at 100 km depth is more complex, because no transitions in the crystal structure of olivine and pyroxene should occur at such shallow depths. In addition, this zone of increase is closely associated with the overlying low-velocity zone, the single region of sharp velocity *decrease* in the mantle. Several lines of evidence suggest that the low-velocity zone is a region in which a *decrease in elasticity*, rather than a sharp increase in density, accounts for the low wave velocities; that is, it marks a zone where the minerals of the upper mantle are unusually warm, plastic, and mobile in contrast to the rocks above and below, which have the same composition but are cooler and more rigid (see Figure 6.19). Indeed, in some regions the low-velocity zone fails to transmit S waves, indicating that it is heated beyond the melting point to form local pockets of liquid magma. The velocity increase below the low-velocity zone apparently marks the zone in which the mantle materials become cooler and resume their normal rigidity. Because the warm low-velocity zone lies only slightly below the crust, it is probably the source for much of the molten magma that reaches the earth's surface to form volcanic rocks. In addition, the zone is intimately interrelated with movements of the crust.

The velocity-depth relations shown in Figure 6.19 are based on detailed seismic data for the upper mantle lying beneath western North America. Generally similar patterns have been found in other regions, but these differ enough in detail to suggest that the low-velocity zone, and the deeper zones of change in crystal structure, are not simple, earth-encircling spheres, but show lateral variations which, in part at least, reflect similar variations observed in the thin overlying crust.

THE DEEP CRUST AND EARTH GRAVITY

Only half of the crust's average 16-km (10-mi) thickness is accessible to direct sampling by even the deepest bore holes. The depths of the crust, like the mantle and core, are therefore known mostly from indirect physical measurements. As with the core and mantle, the principal evidence of deep crustal structure is provided by seismic wave velocities. Strong supporting evidence comes from an additional clue to the earth's interior that we have not yet introduced—that provided by gravity measurements.

6.10 Gravity

Gravity is a property of all bodies of matter by which they are attracted to other bodies by a force that increases as the mass (total amount of matter) of the bodies increases and as the distance between them grows smaller. This is the familiar force that causes unsupported objects to fall to the earth's surface; on a larger scale, the gravitational attraction of the sun, whose total mass is very great, holds the planets in their orbits.

If the earth were a perfectly spherical body with a symmetrical internal structure, then the gravitational force acting at all points on its surface would be exactly the same. Neither of these conditions is true: the earth's subsurface materials are not distributed with perfect symmetry, nor is the earth a perfect sphere, for its rotation causes it to bulge out slightly at the equator and become slightly flattened at the poles. Because of these variations in internal symmetry and external shape, measurements of the earth's gravitational force vary from place to place over its surface (Figure 6.21). These gravity measurements, made by sensitive instruments called **gravimeters** (Figure 6.22), can be considered to include three separate gravity components: a very large component resulting from the principal mass of the earth; and two smaller components, one resulting from local variations in the earth's shape and the other from local differences in the thicknesses and densities of the underlying rocks. It is the two small components that

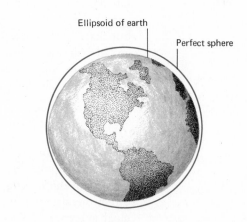

Ellipsoid of earth

Perfect sphere

FIGURE 6.21
Departures from uniform gravity caused by variation in the earth's external shape and internal symmetry. The earth is neither a perfect sphere nor are its surface and internal layers uniform in thickness and composition; these variations lead to differing values of the earth's gravitational force when measured at different points on the surface.

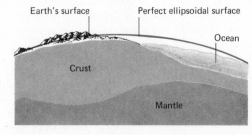

Earth's surface

Perfect ellipsoidal surface

Ocean

Crust

Mantle

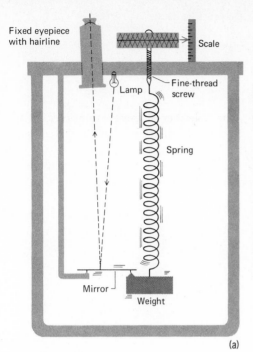

(a)   (b)

FIGURE 6.22
Gravimeters. (a)
Schematic diagram.
Gravity variations
cause the suspended
weight to move up or
down. The screw is
then adjusted to raise
or lower the weight
to a constant height
(mirror horizontal,
lamp centered in eye-
piece); the gravita-
tional force is read
from the height of
the screw. (b) A gravi-
meter in use. (Texas
Instruments.)

make gravity measurements useful for interpreting the earth's interior,
for they can be analyzed separately by first subtracting the large component contributed by the main mass of the earth and then making
certain simplifying assumptions about either the local rock distribution or the general shape of the earth, depending on which factor the
geophysicist wishes to measure. Tens of thousands of such gravity
measurements have been made on land and, more recently, from gravimeters on surface vessels at sea, and these have provided a wealth
of information about both the earth's shape and the varying densities
of rocks in the deep crust and upper mantle.

6.11   Isostasy

There is still another kind of information provided by gravity data.
If reasonable values are assigned to each of the three gravity components, then it should be possible to *predict* the force of gravity at any
point on the earth's surface. Such predictions have been made for
many years, and their careful correlation with actual gravity measurements shows that the two fit *only* over continental areas of low relief.

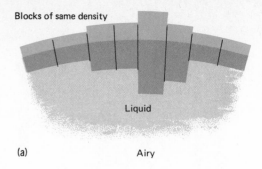

Blocks of same density

Liquid

(a)   Airy

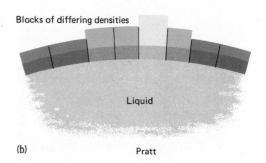

Blocks of differing densities

Liquid

(b)   Pratt

FIGURE 6.23
The Airy and Pratt isostatic hypotheses. In the Airy scheme (a) mountains and lowlands are composed of the same materials and mountains float higher because they have large roots; in the Pratt scheme (b) they float higher because they are composed of lighter materials.

In mountainous regions the force of gravity is normally *lower* than would be predicted, whereas over the deep ocean basins it is normally *higher*. These gravity anomalies, as they are called, are caused by variations in the rocks of the crust: apparently crustal rocks are relatively less massive under mountain ranges and more massive under ocean basins. This systematic variation of gravity with surface topography was discovered over a hundred years ago and has since been confirmed by thousands of gravity measurements.

The discovery that mountain ranges are underlain by relatively less massive rocks quickly led to the suggestion that mountains stand high because they are "floating" on the underlying rocks in much the same way that a block of copper or iron will float in heavier liquid mercury. The introduction of this concept of "floating," of the gravitational balance between light crustal rocks and denser mantle rocks—a concept termed isostasy (*iso* = equal, *stasy* = standing)—had far-reaching implications for the study of the deep structure of the crust. In particular, it was hypothesized that such floating could come about in two different ways. If all the rocks of the crust had approximately the same density, then mountains would require a large buried *root* of similar material to support them on the "fluid" subcrustal material. To provide buoyancy, this root would have to be much larger than the rocks exposed at the surface, just as the submerged part of a log or iceberg is larger than the part exposed above the surface (Figure 6.23a). But if there were density variations in the rocks of the crust, then mountains might "float" higher because they are composed of lighter materials than surrounding lowlands (Figure 6.23b). This second scheme was first proposed in 1854 by J. H. Pratt, an English geographer, and has come to be known as the *Pratt hypothesis*; the constant-density scheme was proposed the following year by another Englishman, the astronomer G. B. Airy, after whom it is called the *Airy hypothesis*. These isostatic hypotheses were the first clues to the deep structure of the earth's crust because, under either hypothesis, simple calculations showed that the rocks exposed in continental mountain ranges must extend far beneath the earth's surface and "float" in rocks of higher density.

6.12 Seismic Evidence of Deep Crustal Structure

It was not until 1909, when Mohorovičić discovered the seismic discontinuity at the base of the crust, that there was any additional evidence of deep crustal structure. Subsequent observations showed that the depth of the Moho discontinuity averages about 35 km (22 mi) under the continents but, in striking contrast, *is much shallower under the ocean floors*, where it has an average depth of only 8 km (5 mi). These observations were extremely important, for they showed that, on a continental scale, the base of the crust "mirrors" the topography of the earth's surface; that is, the crust is thickest under the high-standing continents and thinnest under the low-lying ocean basins (Figure 6.24)—as predicted by the Airy scheme of isostasy.

This is not the whole story, however, for seismic evidence *also* indicated that the continents are composed of less dense rocks than are those of the ocean floors, a finding that supported the Pratt iso-

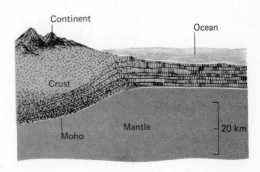

FIGURE 6.24
Moho depth and crustal thicknesses under continents and ocean basins.

static hypothesis. By making certain reasonable assumptions about the relationship between the density and elasticity of crustal rocks, it is possible to infer crustal densities directly from seismic velocities: faster velocities occur within denser rock. Early in the history of seismic studies it was noted that velocities in the continental crust are generally slower than in the oceanic crust, and these observations showed that continents are composed of lighter rocks than are the ocean floors. Seismic evidence thus indicates that *both* the Airy and Pratt schemes of isostasy play a role in the deep structure of the crust.

Early seismic studies also suggested that the crust itself is composed of "layers" of rock, each having a relatively constant wave velocity (and, by inference, a constant density), and each separated by a minor discontinuity in wave velocity. Continents were considered to have two principal layers: an upper, low-density layer and a lower layer of higher density. Oceans, on the other hand, showed evidence of only one principal layer whose density was about equal to the lower layer of the crust. These observations were interpreted to mean that the "oceanic crustal layer" is worldwide, extending under the lighter, upper continental layer as well as under the oceans.

Intensive modern seismic studies using precisely timed artificial explosions have shown that deep crustal structure, like that of the upper mantle, is more complex than this simple, layered model would suggest. Some continental regions show one or several small discontinuities within the crust, while in others the wave velocities increase uniformly with depth rather than changing abruptly (Figure 6.25). In all cases, however, there is a tendency for the velocities to increase downward, suggesting that although continental rocks may not necessarily be underlain by oceanic materials, they do become progressively denser with depth. In oceanic areas, the presence of a single principal layer with a relatively high and constant velocity has been confirmed. Velocities and inferred rock densities found throughout the oceanic layer are about as high as the maximum velocities and rock densities found deep under the continents, confirming once again the old suggestion that the varying heights of the crust's continents and ocean basins are related to large-scale differences in the deep crustal rocks which underlie them.

Finally, recent seismic studies also show that the Moho discontinuity, marking the base of the crust, can be clearly recognized in most regions, but that its depth varies in unexpected ways (see Figure 6.25). In some regions the crust thickens to a maximum of about 80 km (50 mi) under mountains, but other mountain ranges are underlain by crust of normal thickness. Conversely, abnormally thick crust sometimes occurs in regions of low surface relief. It should be stressed, however, that even though there is no simple, universal rela-

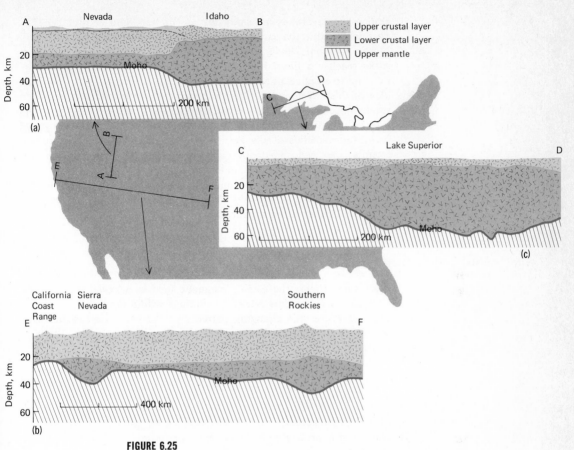

FIGURE 6.25
Variations in deep crustal structure and Moho depth, based on detailed seismic surveys of three U.S. regions. Note that the crust thickens under the Sierra Nevada and Rocky Mountains, but not under the California Coast Ranges (b). It is also unusually thick under the Lake Superior region (c), where there is little surface relief. Most regions show two crustal layers, but Nevada appears to have three (a) and much of California only one (b). (Modified from Bott, "The Interior of the Earth," 1971; after Hill and Pakiser.)

tion between crustal thickness and surface relief in local regions, on a much larger scale the difference between the thick continental crust and the much thinner oceanic crust has been abundantly confirmed.

SUMMARY OUTLINE

Earthquakes and the Earth's Interior

6.1   *Earthquake waves:* seismic (earthquake-shock) waves travel through the earth and provide the principal evidence of its internal structure;

among the several types, body waves (primary and secondary) that pass deep within the earth are the most instructive.

6.2   *Elastic rebound theory:* when the slow accumulation of strain in rocks reaches a limit, faulting is caused when the rocks elastically rebound to an unstrained condition.

6.3   *The earth's interior:* velocity-depth curves have provided a basis for subdividing the earth into a crust, mantle, and core.

6.4   *Internal densities, pressures and temperatures:* the probable pressures and densities of the earth's internal materials can be inferred from seismic velocity-depth curves; internal temperatures, on the other hand, are more poorly understood.

6.5   *Composition of the core:* seismic data indicate that the core is mostly liquid; this liquid is probably molten iron with small amounts of nickel and either sulfur or silicon.

The Core and Earth Magnetism

6.6   *Earth magnetism:* the earth's magnetic field is believed to originate from electrical currents caused by motions within the liquid iron core.

6.7   *Core structure:* the changing patterns of the earth's magnetic field, as measured at the surface, provide clues to the structure of the hidden core.

The Mantle

6.8   *Mantle composition:* the mantle occupies 80 percent of the earth's volume; it is probably composed of magnesium- and iron-rich silicate minerals, which are principally olivine and pyroxene in its outer regions.

6.9   *Mantle structure:* seismic discontinuities in the mantle indicate a warm, plastic zone just below the crust and two deeper, more rigid zones where olivine and pyroxene are believed to convert abruptly to denser mineral structures.

The Deep Crust and Earth Gravity

6.10   *Gravity:* the force of gravity varies at the earth's surface due to variations both in the earth's shape and in the characteristics of the underlying rock.

6.11   *Isostasy:* gravity measurements suggest that high-standing mountains and continents of the crust "float" on denser underlying rock.

6.12   *Seismic evidence of deep crustal structure:* the Moho seismic discontinuity, which separates the crust from the mantle, lies far deeper under continents than under ocean basins; the continental crust is therefore much thicker than the crust underlying the oceans.

ADDITIONAL READINGS

Bott, M. H. P.: *The Interior of the Earth*, St. Martin's Press, Inc., New York, 1971. An excellent intermediate text.

*Clark, S. P., Jr.: *Structure of the Earth*, Prentice-Hall, Inc., Englewood Cliffs, N.J., 1971. A brief, authoritative introduction, emphasizing the earth's interior.

Garland, G. D.: *Introduction to Geophysics*, W. B. Saunders Company, Philadelphia, 1971. An advanced text on the earth's interior.

Gaskell, T. F.: *Physics of the Earth*, Funk & Wagnalls, New York, 1970. A well-illustrated elementary introduction to the earth's interior.

*Hodgson, J. H., *Earthquakes and Earth Structure*, Prentice-Hall, Inc., Englewood Cliffs, N.J., 1964. A brief, popular introduction stressing the relation of earthquakes to man.

Phillips, O. M.: *The Heart of the Earth*, Freeman, Cooper and Co., San Francisco, 1968. A readable introduction to all aspects of the earth's interior.

Stacey, F. D.: *Physics of the Earth*, John Wiley & Sons, Inc., New York, 1969. An intermediate-level survey of the earth's interior.

*Available in paperback.

chapter 7

Mountains
and
Plate Tectonics

In Chapters 5 and 6 we described the geometry of the earth's rocks and raised unanswered questions about the possible causes of their structures. This chapter considers how localized regions of topographically high terrain, called mountains, have come into existence by deformational processes. Our discussion will lead to an introduction of plate tectonics, a powerful new model (paradigm) for the crustal movements that cause mountain building, earthquakes, and volcanism.

CONTINENTAL MOUNTAINS

7.1 Mountains and Geosynclines

On the largest scale, continental mountain ranges may be divided into two broad categories: those that originate largely from faulting, known as fault-block mountains, and those dominated by folding, which are called simply fold mountains* (Figure 7.1). Most of the large mountain ranges seen on earth today, such as the Himalayas, Alps, and large segments of the Rockies, are fold mountains. The Sierra Nevada, on the other hand (Figure 7.2), is a fine example of an exceptionally large fault-block range. Most mountains of this origin are somewhat less conspicious than fold mountains and are frequently associated with them in a secondary manner, having formed after a prior episode of strong folding. The Great Basin region of the western United States comprises many distinct basins and ranges, called grabens and horsts respectively, of classic fault-block origin (Figure 7.4).

Detailed studies of many folded mountain ranges have shown most of them to have similar patterns of rock deformation. Viewed in cross section, the outer flanks typically consist of thick sequences of sedimentary rocks that become progressively less folded and deformed away from the central axis of the range. Toward the axial regions, these deformed but otherwise unaltered sedimentary rocks give way first to metamorphosed sedimentary rocks and finally to large elongate masses of granite which form the central core of the range.

In many regions the typical fold mountain rock sequence, grading from relatively undeformed sediments near the flanks of the range to granite at the core, can be shown to have formed by the progressive alteration of an unusually thick accumulation of sediments deposited

---

*Still a third kind of mountain range originates not from deformation of preexisting solid rocks, but from volcanic lavas. Such *volcanic mountains* (see Figure 7.3) are common on the ocean floor and around the margin of the Pacific Ocean, but are less important within the continents than either fold or fault-block mountains. (See the discussion of volcanic landforms in Section 2.3.)

(a)

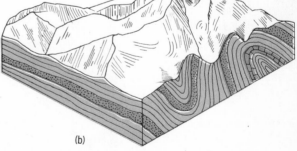

(b)

FIGURE 7.1
Fold mountains. (a) Southern Idaho. (b)
Schematic block diagram of fold moun-
tains. (a: John S. Shelton.)

near the margins of the continents. When the thin veneer of sedimentary rocks that covers much of the surface of the continents is traced
laterally into the flanks of the mountain range, the total thickness of
the original sediment can be seen to increase dramatically at just about
the point at which the mountain-building deformation begins. Furthermore, this rapid thickening continues toward the axis of the
mountain range, where the metamorphosed sediments give way to
granite. This general pattern was first described in 1859 by James

(a)

(b)

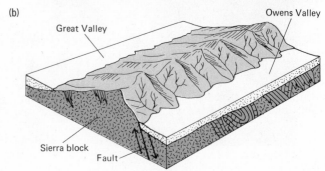

Great Valley

Owens Valley

Sierra block

Fault

FIGURE 7.2
Fault-block mountains. (a) Mt. Whitney region from over Owens Lake. (b) Block diagram of Sierran tilted fault-block. (a: U.S. Geological Survey.)

FIGURE 7.3
The Cascade Range, showing volcanic mountains. (U.S. Department of Agriculture.)

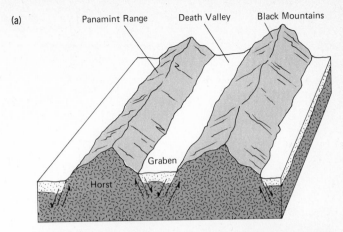

FIGURE 7.4
Fault-blocks forming basins and ranges. (a) Simplified diagram of the structure characterizing the Death Valley region, showing fault-block mountains and basins. (b) Photo of basins and ranges. (b: John S. Shelton.)

Hall, a geologist studying relationships between the Appalachians and the adjacent interior lowlands. Further investigations and theorizing gave rise to the concept of the geosyncline a few years later. Geosynclines were believed to have been formed where large volumes of relatively shallow-water sediments accumulated in subsiding elongate basins. According to the geosynclinal model, this thick accumulation of strata would eventually undergo intense compressional shortening, causing large-scale folding and reverse faulting, often combined with metamorphism and plutonism in the deep core of the newly formed mountain range due to the extreme pressures and temperatures that were reached (Figure 7.5). Although this model was generally accepted, it did not account for several important facets of the geosyncline's evolution: (1) What was the mechanism of basin subsidence? —It could be shown that sediment weight alone was not sufficient.

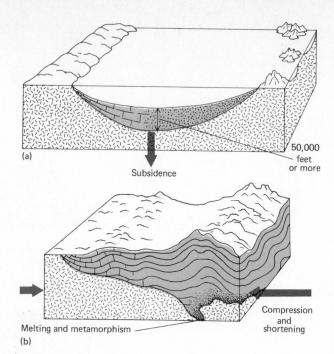

FIGURE 7.5
Model for the evolution of a geosyncline.
(a) Geosynclinal phase: subsidence and
sedimentation. (b) Mountain-building phase:
crustal shortening, folding and faulting,
and plutonism.

(2) Why does subsidence, which is a vertical movement, give way to horizontal compressional forces? (3) What forces within the earth are capable of causing tens, and even hundreds, of miles of crustal shortening? All of these questions were troublesome, but the evidence for geosynclines was so compelling that their existence in the past was not denied.

The conspicuous folded mountain ranges seen on the continents today have all formed relatively recently in earth history. Most older fold mountain ranges no longer stand high above the continental surface but have been worn down to zones of little or no relief through millions of years of erosion by water and wind. Such ancient mountain ranges are best exposed in regions known as **shield areas**, where they can still be recognized, even though they now lack mountainous relief, by their characteristic pattern of elongate granite masses flanked by progressively less deformed sedimentary rocks (Figures 7.6 and 7.7).

Over eons of geologic time the dimensions of the earth's continents are believed to have increased in the following manner: Sedimentary debris resulting from the erosion of old mountain ranges typically accumulated in geosynclines near the continental margins, where it ultimately became deformed to create new mountain ranges. The granitic cores of these newly formed mountains were welded to the silicic

FIGURE 7.6
The eroded roots of
an ancient mountain
range, exposed in
the central Canadian
shield area. (Canadian
Department of
Energy, Mines, and
Resources.)

nucleus of the "parent" continent. Through this process the rocks of
the continental crust were continually recycled, as granitic mountain
cores were eroded to form sediments which then melted deep within
the crust to form new granites which in turn became the cores of new
mountain ranges. In developing this model, the steps involving the
erosion of old mountains and the subsequent accumulation of sediments near the continental margins were more clearly understood
than were the forces that deform and uplift the geosynclinal sediments to produce new mountains. These complex forces are, however,
beginning to be recognized as but a side effect of still more fundamental motions of the earth's crust.

7.2 Continental Drift

One of the first suggestions that continental mountains might result
from larger-scale crustal motions was made in 1912 by Alfred Wegener, a German geophysicist. Wegener was impressed by the jigsaw
puzzle–like fit between the shapes of some of the continental margins. The most obvious are those between western Africa and eastern

FIGURE 7.7
The principal shield areas, which are regions where the igneous and metaphoric basement rocks of the continents are exposed at the earth's surface. Elsewhere on the continents the basement rocks are covered by a veneer of predominantly sedimentary rock. The arrow indicates the small area of basement exposure that is enlarged in Figure 5.2. (Modified from Kummel, "History of the Earth," 1970, after various sources.)

North and South America, but with suitable juggling rough fits can be made for other continental margins as well (Figure 7.8). A considerable amount of other evidence was advanced by Wegener supporting his view that our present continents have been horizontally displaced thousands of miles from the positions they occupied in the Mesozoic Era. Figure 7.9 summarizes only some of Wegener's evidence based on various fields of study: geologic similarities of fold belts, related fossil organisms that were incapable of swimming across deep ocean

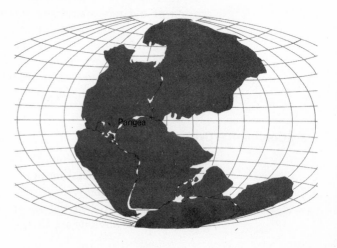

FIGURE 7.8
A reconstruction of the probable relationships of the continents before "continental drift." (After Dietz and Holden, "Scientific American," vol. 223, 1970.)

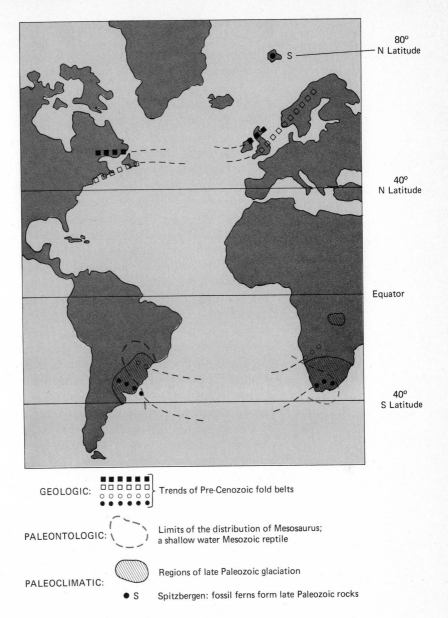

FIGURE 7.9
Evidence for continental drift as proposed by Alfred Wegener.

GEOLOGIC: ⬛⬛⬛⬛⬛ / ▫▫▫▫▫ / ○○○○○ / ●●●●● — Trends of Pre-Cenozoic fold belts

PALEONTOLOGIC: ⌒⌒ — Limits of the distribution of Mesosaurus; a shallow water Mesozoic reptile

PALEOCLIMATIC: ▨ — Regions of late Paleozoic glaciation

● S — Spitzbergen: fossil ferns form late Paleozoic rocks

basins, and paleoclimatic ("ancient climate") data that strongly suggest large-scale continental reorientations since late Paleozoic time.

All these associations led Wegener to suggest that the continents were originally joined in a single large landmass—which he called *Pangea* (*pan* = all, *gea* = earth)—that was subsequently broken apart, beginning in Mesozoic time, to create the separate continents which we see today. If entire continents were moving relative to one another, then mountain-producing deformations along their margins might be explained as some sort of relatively minor frictional or drag effect at the edges of the moving mass. Wegener's model explaining *how* the continents moved about and his unsuccessful efforts to provide a plausible force capable of causing these movements proved to be the undoing for continental drift, as his idea came to be called. Wegener envisioned masses of continental silicic rock "plowing" through oceanic mafic rock (Figure 7.10). This model was quickly rejected as unacceptable in light of the fact that mafic rock is *more* rigid than silicic rock: butter could hardly plow through steel! Geophysicists were among the most skeptical of continental drift because they (like Wegener) knew of no energy source capable of providing the immense mechanical work required to move continental masses. Primarily for these reasons, plus the natural conservatism of the scientific community, the notion of continental drift was largely dismissed.

In the Introduction (Section I.2) we discussed the concept of prematurity in science. Wegener's early work regarding continental drift seems clearly to be an example of a premature hypothesis. Even in the face of strong evidence that continents had once been connected and subsequently drifted apart, the idea was rejected for lack of a satisfactory mechanism; the science of Wegener's time provided no clues as to *how* it could happen—so it was concluded that it had not happened.

In the early 1950s new evidence derived from study of the earth's magnetic field began to suggest that large-scale continental motions had indeed taken place. When igneous rocks first crystallize, any magnetic, iron-bearing mineral that they contain becomes magnetized in the direction of the earth's magnetic field. Any subsequent change in the orientation of the mineral with respect to the earth's magnetic field also affects the mineral's magnetism; by using precise analytic techniques first developed in the 1950s, however, it is often possible to isolate the original magnetic component, which thus provides a rec-

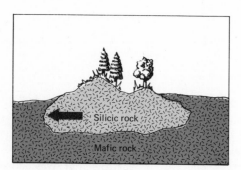

FIGURE 7.10
Wegener's continental drift: silicic rock "plowing" through mafic rock; quickly recognized as unacceptable, since mafic rock is more rigid than silicic rock.

ord of the direction of the earth's magnetic field at the time the rock crystallized. By combining data from many such "fossil magnets" taken from rocks of different ages found on each continent, it is possible to get strong clues about the changing positions of the continents relative to the magnetic pole (which is assumed to have always remained relatively near the earth's rotational pole, as it is today—see Figure 7.11). Such paleomagnetic studies of continental rocks provided additional evidence that the continents had moved relative to one another and led to a revival of interest in the idea of continental drift. General acceptance of the idea did not come, however, until the mid-1960s when it was discovered that still more fundamental crustal movements take place beneath the ocean floor.

THE OCEAN FLOOR

Among the most significant scientific advances of recent years was the discovery, first suggested as a hypothesis in the 1950s and then firmly supported in 1966, that the oceanic crust is in continuous movement like a giant conveyor belt. This discovery of what has come to be called *ocean-floor spreading* was a result of two principal kinds of research: one of these concerned the topography of the ocean floor, the other magnetism of the ocean floor. This section will first discuss these two research areas and their findings and will then move on to consider how ocean-floor spreading has provided a unifying principle that also helps to explain continental deformation.

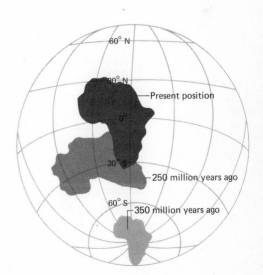

FIGURE 7.11
Paleomagnetic reconstruction of Africa's position, relative to the south pole, from 350 million years ago to the present. (After compilation by Clark, "Structure of the Earth," 1971, from various sources.)

7.3 Topography of the Ocean Floor

Because of the ease with which we can observe mountains, hills, and valleys on the continents, most of us are not accustomed to thinking of the extraordinary problems involved in learning whether similar topography exists on the ocean floor. Before 1920 the only method of determining ocean depths and, conversely, sea-floor topography was to laboriously lower a heavy weight, attached to thousands of feet of steel wire, over the side of a ship. Even though many such "soundings" were made, they were so scattered over the vast expanse of the oceans that they provided little evidence for ocean-bottom irregularities. The ocean floors were generally believed to be flat, featureless plains broken here and there by mountains that rose above the surface as islands.

In the 1920s the first *echo sounders* were developed for more rapid determination of ocean depths and bottom topography (Figure 7.12). These instruments measure depths by noting the time required for sound waves to reach the bottom and be reflected back to a shipboard recording device. The first systematic surveys of ocean-bottom topography using these instruments were undertaken in parts of the Pacific Ocean in the early 1940s. The principal result of these early surveys was the discovery of a great many rounded, flat-topped mountains, called guyots, on the floor of the northern Pacific (Figure 7.13). The tops of many guyots are covered by many thousands of feet of water, yet the only known mechanism for producing their broad, flat tops is the wearing action of waves at the surface of the sea. Because their flat tops could have been produced only at sea level, either sea level had risen many thousands of feet since they were formed or, more probably, the ocean floor on which they stand was somehow lowered several thousand feet. Guyots thus provided the first suggestion of large-scale movements of the ocean floor.

Another significant result of early ocean-bottom surveys was the location and mapping of several oceanic trenches—long, narrow zones where rocks of the ocean floor, investigators were puzzled to find, are depressed many thousands of meters *below* the surrounding terrain (Figure 7.14). Most of these trenches are parallel to continental margins and are usually bounded on their landward side by curving chains of offshore volcanic mountains that extend above the ocean surface to form island arcs (Figure 7.14). These unusually deep depressions immediately adjacent to high volcanic mountains also suggested that dynamic processes are taking place in the rocks of the ocean floor.

Since 1945 additional surveys of bottom topography have been undertaken in many areas: the Atlantic, Indian, Arctic, and eastern

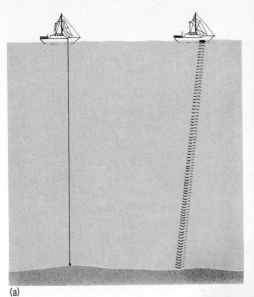

(a)

(b)

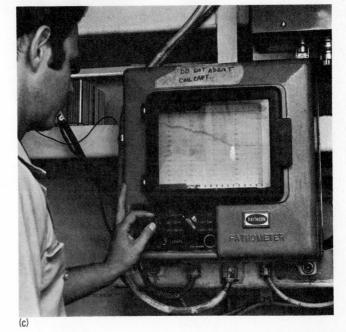

FIGURE 7.12
Methods of determining ocean-floor topography. (a) Before the 1920s the only method was the lowering of weighted lines (left); since then, increasingly sophisticated echo sounders (right) have been used to map the ocean floor. (b) A weighted line being used for sounding a shallow, uncharted area. (c) A recording echo-sounder showing a profile of the underlying topography. (b: U.S. Coast Guard; c: Ocean Science Laboratory.)

(c)

Pacific oceans have been intensely surveyed, and other oceans are known in less detail. These surveys have shown that the ocean floor, although relatively flat over large areas, also includes the longest,

FIGURE 7.13
Idealized drawing of a portion of the mid-Pacific ocean floor with water removed, showing two flat-topped guyots in the distance. The canyon in the foreground is cut into a guyot, most of which does not show. (Painting by Chesley Bonestell, from Hamilton, "Geological Society of America Memoir," 64, 1956.)

highest, and most continuous mountain ranges on earth. These ranges make up the **oceanic ridge-rise system,** a world-encircling belt of ridgelike mountains and broad rises that is of extraordinary importance for understanding movements of the ocean floor (Figure 7.15).

The most intensively studied part of the oceanic ridge-rise system is the *Midatlantic Ridge,* a chain of mountains lying almost exactly in the center of the Atlantic Ocean (Figure 7.16). The existence and general position of the ridge has been known for many years because it comes to the surface in several places in the form of oceanic islands such as Iceland and the Azores. Furthermore, it has long been known that the ridge overlies a zone of rather shallow earthquakes. In the 1950s detailed studies of the topography of the ridge and the distribution of earthquakes under it revealed two important relationships. First, although the total width of the ridge averages several hundred kilometers, the earthquakes are concentrated in a narrow band under

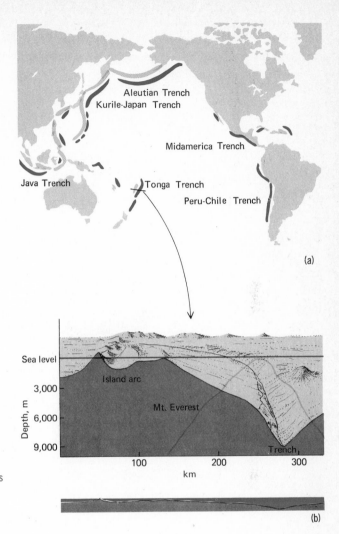

FIGURE 7.14
(a) The principal oceanic trenches (dark color) and island arcs (light color). (b) Cross section through the Tonga Trench, with Mt. Everest superimposed for scale. The vertical scale is exaggerated 10 times in the upper sketch; the true scale is shown below. (After Fisher and Revelle, "Scientific American," vol. 193, 1955.)

only the highest part or *crest* of the ridge. Second, the crest itself does not make up a single long ridge, but is instead composed of two parallel, steep-sided ridges separated by a deep valley. Similar but smaller and less continuous valleys, called **rifts**, are also seen on the surface of the continents, where they can be shown to have formed from normal faulting caused by a tension or *pulling apart* of the faulted rocks on each side of the valley (Figure 7.17). This immediately suggested that the Midatlantic Ridge might represent a similar but much larger zone of crustal tension in which the Atlantic Ocean floor on both sides

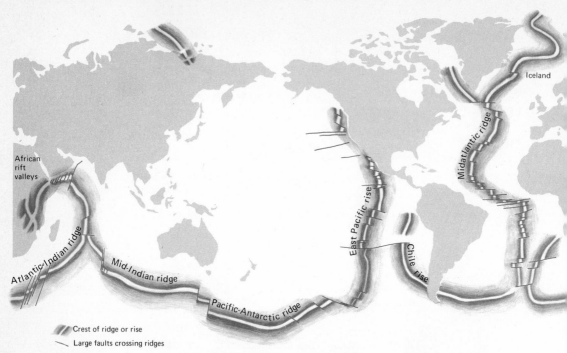

Crest of ridge or rise

Large faults crossing ridges

FIGURE 7.15
The oceanic ridge-rise system. (After Bullard, "Scientific American," vol. 221, 1969.)

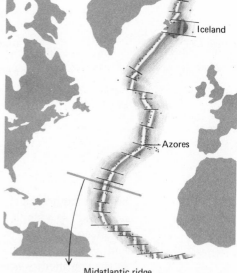

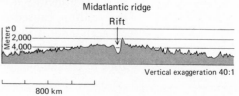

FIGURE 7.16
A portion of the Midatlantic Ridge showing the location of earthquake epicenters (black dots). The cross section below shows the rift valley at the crest of the ridge. (After Heezen, "Scientific American," vol. 203, 1960.)

of the ridge is moving away from the rifted ridge; the rift valley at the crest of the ridge and the earthquakes underlying it would mark the line of separation. Other lines of evidence also pointed to the same conclusion. Where the tops of the ridge are exposed as islands and the rocks of the ridge can thus be sampled, they were found everywhere to be basaltic or ultrabasic rocks, as would be expected if the ridge represented a fracture zone in which basaltic lavas had poured out from the underlying mantle to build up the ridge. Furthermore, the entire width of the rift valley itself is exposed above the sea in Iceland, where tension cracks and other evidence of movement can be directly observed (Figure 7.17a).

After the significance of the Midatlantic Ridge was discovered, interest quickly spread to similar features in other ocean basins. On the basis of topographic information combined with data on the distribution of oceanic earthquakes, it was proposed in 1956 that the Midatlantic Ridge is only one segment of a much longer world-encircling system of oceanic mountains which represents a fundamental zone of fracture and movement of the earth's crust. This is the oceanic ridge-rise system which extends, with several side branches, down the center of the Atlantic Ocean, through the Indian Ocean, south of Australia, and up the eastern Pacific Ocean. Although the system mostly occurs in ocean basins, it does extend into the continents in a few places such as the Red Sea, eastern Africa, and the Gulf of California.

7.4   Magnetism and Motions of the Ocean Floor

The discovery of the oceanic ridge-rise system, with its associated earthquakes and, in many areas, central rift valleys, coupled with the discovery of generally high heat flow to the earth's surface, strongly suggested that horizontal movements take place in the crustal rocks beneath the sea floor on a tremendous scale. The elevation of this hypothesis of ocean-floor spreading (as it came to be called) to a theory and then to scientific principle was accomplished by magnetic studies of the rocks of the ocean floor.

In Sections 6.6 and 6.7 we saw that measurements of the strength of the magnetic field at the earth's surface are affected by two components: one, the earth's main magnetic field, originates from motions in the liquid iron of the core; the other originates from the distribution of magnetic minerals in rocks of the crust. There are no magnetic effects from mantle rocks lying between the crust and the core because the minerals making up the mantle are heated beyond the Curie point (the temperature at which they lose their magnetism). By subtracting the large component contributed by the main magnetic field,

FIGURE 7.17
Rifts. (a) A part of the rift valley of central Iceland.
(b) A small rift in east Africa. (a: Sigurdur
Thorarinsson; b: Sidney P. Clark, Jr.)

it is possible to study differences in the magnetic minerals of crustal rocks by making magnetic surveys of large areas with shipboard or airborne magnetometers. Such surveys have been conducted over land areas for many years and have shown complex magnetic patterns that are caused primarily by differences in the amount of the mineral magnetite present in the underlying crustal rocks. The first systematic magnetic surveys of the ocean floor were undertaken off the coast of California in 1955, surprisingly, these showed patterns that are *far more regular than any from continental areas* (Figure 7.18). In particular, the oceanic magnetic patterns showed a series of extremely long, narrow bands running for hundreds of kilometers approximately parallel to the coastline. One additional feature of these banded magnetic patterns is of great significance: in certain areas the patterns are broken and offset for hundreds of kilometers along huge faults running at approximately right angles to the present coastline. Some of these faults were already known because they make large submarine cliffs that had been previously discovered by echo sounding. The fact that oceanic rocks had moved for hundreds of kilometers along these faults was first shown, however, by the offset patterns of magnetic measurements. Here again was strong evidence for large-scale movements of the ocean floor.

Once the linear oceanic magnetic pattern was discovered, the next question was: What caused it? In continental areas the much less reg-

FIGURE 7.18
Contrasting linear magnetic patterns of oceanic rocks and irregular patterns of continental rocks. (After Raff, "Journal of Geophysical Research," vol. 71, 1966; Sauck and Summer, "Aeromagnetic Map of Arizona," 1971.)

San Francisco

Los Angeles

0 100
km

ular magnetic pattern could be shown, by geologic mapping, to be caused by differing amounts of magnetic minerals in the underlying rocks. There was, however, another more likely cause for the regular oceanic patterns. This was related not to differing *amounts* of magnetic minerals but to differences in their *direction of magnetization*.

Rocks of the ocean floor are almost exclusively basalts containing a large and relatively constant amount of the mineral magnetite (Fe_3O_4), which makes them strongly magnetic. When lavas cool to form basalt, this magnetite crystallizes from the liquid lava and then eventually cools to the Curie temperature. As this temperature is reached the magnetic crystals become magnetized in the direction of the earth's magnetic field. Now suppose for a moment that the direction of the earth's magnetic field is not constant but shows regular changes. If this were the case, then lavas forming at different times would have different directions of magnetization, reflecting the earth's magnetic field at the time the lavas cooled. (The original patterns would dominate because magnetic minerals permanently retain much of their original magnetic orientation gained on passing the Curie point and do not become completely reoriented to later changes in the earth's field.) Most significantly, at about the same time that the puzzlingly regular magnetic patterns were discovered in the oceans, additional studies of vertical sequences of magnetic mineral orientation in lava flows on land and in samples of sediments from the sea floor (fine sedimentary grains of magnetic minerals also become oriented to the earth's magnetic field when they are deposited by water) showed that the earth's main magnetic field has undergone frequent and rapid reversals in polarity, the north magnetic pole becoming the south pole and vice versa (Figure 7.19). The cause of these sudden magnetic reversals is still unknown, but they must somehow be related to changes in flow patterns in the liquid iron of the earth's core, where the main magnetic field originates.

Soon after the regular oceanic magnetic patterns were discovered, it was suggested that they might reflect sudden reversals in the earth's magnetic field in the following way: if basaltic lavas poured onto the ocean floor from a long crack in the crust, then cooled to form basalt, and were then somehow separated down the middle and moved away in opposite directions from the crack before new lava was extruded, and futhermore, if a reversal in the earth's magnetic field took place between the lava outpourings, then the result would be *a series of parallel bands of basalt each having minerals with a different magnetic orientation* (Figure 7.20). These differences in magnetic direction might account for the regular, parallel patterns of ocean-floor magnetism, since the normally oriental basalt would *add* slightly to the

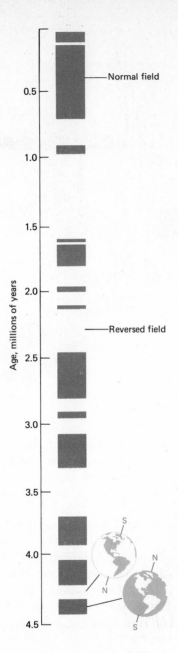

FIGURE 7.19
A time scale for magnetic reversals during the last 4.5 million years. Intervals of normal polarity are shown in color; reversed polarity in white. (Modified from Cox, "Science," vol. 168, 1969.)

present field of the earth, and the reversed basalt would *subtract*. Unfortunately, this idea still cannot be directly tested because it is impossible to determine the direction of magnetization without hav-

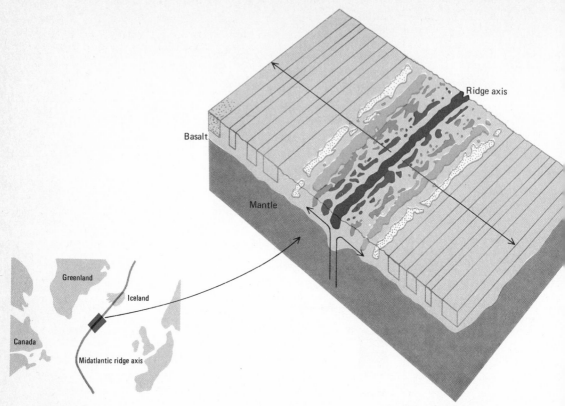

FIGURE 7.20
Interpretation of symmetrical magnetic patterns observed on a portion of the Midatlantic ridge south of Iceland (color). Lavas from the mantle pour out along the ridge axis and then move laterally outward on both sides of the ridge forming symmetrical bands that are progressively older (lighter color) away from the ridge. Bands of reversed polarity are shown in gray; normal polarity in color. (Magnetic pattern from Vine, in "The History of the Earth's Crust," 1968.)

ing carefully oriented samples of the magnetic rocks, and these are extraordinarily difficult to obtain for oceanic basalts covered by thousands of meters of water and hundreds of meters of sediments.

Strong evidence that spreading basaltic bands *are* the cause of the oceanic magnetic pattern was obtained, however, from another source in 1966, when the first detailed magnetic analysis was made of the Midatlantic Ridge. This study showed parallel magnetic patterns similar to those previously found off California, but this time the bands were found to be *identical on each side of the Midatlantic Ridge*, just as would be expected if lavas were pouring from the earthquake rift zone, solidifying into basalt, then being broken along the rift and moving away from it on both sides (Figure 7.20). Furthermore, the width of the parallel magnetic bands was found to closely match the relative times between reversals in the earth's magnetic field over the past several million years as established independently from magnetic studies of dated sedimentary rocks and lava flows. This discovery showed that the spreading basaltic rocks of the oceanic crust have behaved like a great moving magnetic recording tape which clearly reflects changes in the direction of the earth's magnetic field. Basaltic rocks of the oceanic crust must rise from the mantle along the oceanic ridge-rise system, solidify, and then be carried horizontally away from the ridges by powerful forces acting from below, within the hotter and less rigid rocks of the mantle. The oceanic crust thus becomes progressively older at farther distances from the ridge-rise system.

Following the dramatic indication of ocean-floor spreading from analyses of magnetic patterns, two other lines of evidence contributed further confirmation of the motions involved. One, a careful study of earthquake motions, has shown that large-scale crustal movements take place with precisely the patterns suggested by the magnetic evidence. The other, an important program of drilling from shipboard into the deep ocean floor, has revealed that the thin sediments lying upon the oceanic crust show a regular increase in age farther away from the ridge-rise system, as would be expected if the underlying basalts similarly varied in age (Figure 7.21).

PLATE TECTONICS

The confirmation of ocean-floor spreading quickly led to a reconsideration of *all* crustal motions, for the discovery of the movement of large blocks of the ocean floor indicated dynamic crustal processes on a scale that had never been obvious from studies of the more accessible continental crust. In particular, the position of earthquake belts and oceanic ridges now showed the crust to be divided into seven major moving **plates,** each relatively rigid within itself but bounded by zones of profound crustal movement (Figure 7.22). Differential motion of these huge plates is now believed to account for most of the dynamic features seen in the earth's crust, and their study has come to be called **plate tectonics** (*tectonics* means "land movement" and is applied to mountain-forming crustal deformations).

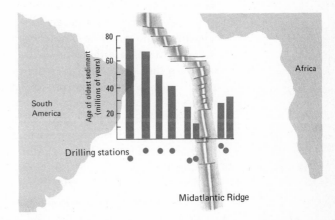

FIGURE 7.21
Confirmation of ocean-floor spreading from deep-sea drilling. Sediments lying just above the ocean-floor basalts become progressively older away from the axis of the Midatlantic Ridge. (After Maxwell et al., "Initial Reports of the Deep Sea Drilling Project," vol. 3, 1969.)

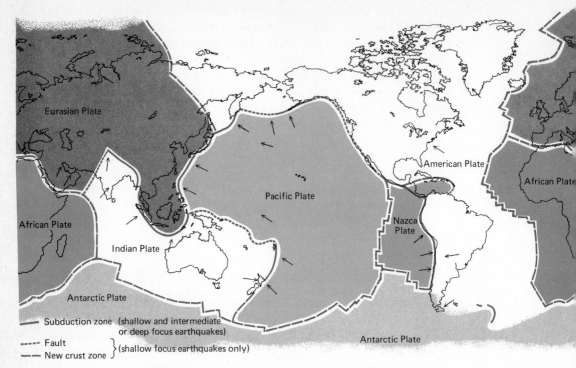

FIGURE 7.22
The major structural plates and their probable boundary types (the nature and position
of many of the boundaries are still uncertain). The arrows show the inferred relative
motions of the plates. Note that the plate boundaries generally follow volcanic and
earthquake zones as shown in Figures 2.18 and 6.3. Several small plates are omitted.
(After Morgan, "Journal of Geophysical Research," vol. 73, 1968.)

7.5   Plate Tectonics and Crustal Deformation

The boundary zones between crustal plates are usually marked by
earthquakes and volcanic activity, and are of three principal kinds.
The first kind of plate boundary occurs along the oceanic ridge-rise
system where basaltic lavas well up from within the mantle and pour
out onto the ocean floor to create new oceanic crust. This new oceanic
crust tends to move horizontally away from both sides of the ridge-
rise system and thus is added to separate moving plates on either side.
This plate boundary is a new crust zone (Figure 7.23). These zones,
commonly called **ridges** or **rises**, are characterized by shallow-focus
normal (tension) faults, as would be expected if crustal plates were
pulling apart.

The creation of new crust along plate boundaries must be balanced
by destruction of old crust at the other side of a moving plate unless

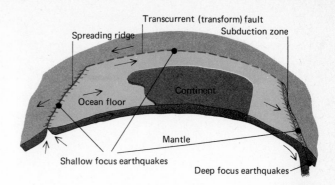

FIGURE 7.23
Idealized diagram of the three types of plate boundary: (1) new crust zones where volcanic crust is added to the plate along oceanic ridges; (2) subduction zones, usually marked by oceanic trenches and deep earthquakes, where the plate plunges deep into the mantle and is destroyed; and (3) transcurrent faults, where only horizontal motions between plates occur.

the earth is continuously expanding in diameter, a prospect with no convincing evidence to support it. Instead, it is clear that many plates *are* bounded by zones in which rigid crustal rocks abruptly descend deep into the mantle where they are ultimately destroyed by melting. Such regions, called subduction zones, are usually marked by deep oceanic trenches, volcanic island arcs, and deep-focus earthquakes caused by the motions of the descending plate margin (Figure 7.23).

If crust is created at one side of a moving plate and destroyed at the other, then there must be additional boundaries where compensating motions take place without either plate creation or destruction. This third type of plate boundary is marked by long transcurrent faults caused by horizontal motions between adjacent plates (Figure 7.23). These faults are now commonly called transform faults, a name suggested by J. Tuzo Wilson in 1965, since they occur where new crust zones (e.g., mid-ocean ridges) abruptly end, and so the fault is where one feature is "transformed" into another. At the time of Wilson's suggestion there was still a great deal of confusion surrounding these major ocean-floor faults. In Figure 7.24 a part of the Midatlantic Ridge is shown, with many offsets. Consider the fault that is labeled A–B: although there is a great deal of left–lateral transcurrent offset, the fault seems to terminate abruptly without continuing on to the South American or African continents. How can so much left–lateral offset disappear? Wilson noted that this problem would be solved if the ridges were centers of spreading that mark the *original fractures* between continents. Thus the left-lateral *offset* along A–B is *not* the result of left-lateral *slip* at all, but instead remains from very early in the separation of South America from Africa. Wilson predicted that careful seismic investigations of these faults would discover two surprising things: (1) faulting would be confined to the region *between spreading ridges only*, and (2) faulting would have lateral movement in directions *opposite* to the directions of ridge offset (Figure 7.24b). The almost immediate confirmation of these predictions did a great

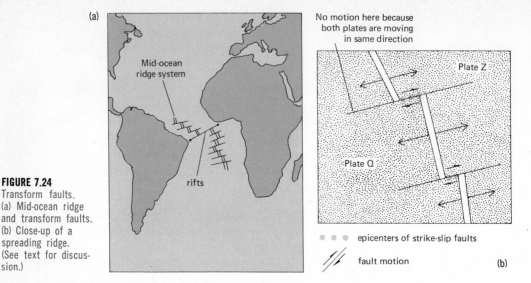

(a)

Mid-ocean ridge system

rifts

No motion here because both plates are moving in same direction

Plate Z

Plate Q

● ● ●   epicenters of strike-slip faults

⫽   fault motion

(b)

FIGURE 7.24
Transform faults.
(a) Mid-ocean ridge
and transform faults.
(b) Close-up of a
spreading ridge.
(See text for discus-
sion.)

deal to establish and develop the new plate tectonics model for crustal
movements.

A critical question involves the relation of continental masses to
movements of these crustal plates. Because continents cover only
about 30 percent of the earth's surface, much of the area of the seven
major plates is occupied by oceanic crust. Each plate except the Pacific
and Nazca also contains, however, large areas of continental crust
(Figures 7.22 and 7.23). It appears that the thick continental masses
are embedded in, and tend to move passively along with, the adjacent
oceanic rocks of each plate. (Note how this differs from Wegener's
idea of continental movement—see Figure 7.10.) Furthermore, continental blocks can apparently be split apart and moved away from one
another whenever they are located over a new crust zone. Such movement creates a new ocean basin with a mid-ocean ridge and causes
"continental drift," as had been previously postulated from other lines
of evidence (Figure 7.25a and b).

When plate motions bring a continent in contact with a subduction
zone, the results are more complex (Figure 7.25b and c). Because continental rocks are significantly lighter than both the moving oceanic
crust and the underlying mantle, continents are apparently seldom
plunged downward at subduction zones, but instead remain "floating"
at the earth's surface while the adjacent heavier oceanic crust descends
into the mantle. These movements of the adjacent oceanic crust may,
however, cause deformation of sediments accumulated along the continental margins and thus be responsible for the marginal fold mountain chains which are such a characteristic feature of the continents.

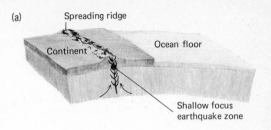

(a)

Spreading ridge

Continent

Ocean floor

Shallow focus
earthquake zone

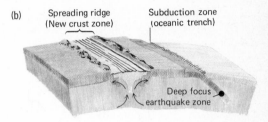

(b)

Spreading ridge
(New crust zone)

Subduction zone
(oceanic trench)

Deep focus
earthquake zone

FIGURE 7.25
Separation and deformation
of continental masses by plate
motions. Formations of a new
crust zone under a continent
(a), leads to rupture and
separation of the continent
(b). Upon reaching a subduction
zone the margins of the continent are deformed into
mountain ranges (c). (After
Dietz and Holden, "Scientific
American," vol. 223, 1970.)

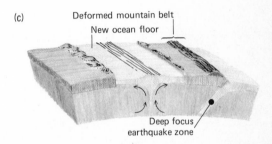

(c)

Deformed mountain belt
New ocean floor

Deep focus
earthquake zone

Furthermore, the descent of rigid oceanic crust and upper mantle into
a subduction zone accounts for the existence of the deep-focus earthquakes in these regions; prior to plate tectonics there was no satisfactory explanation for these deep (up to 700 km) seismic events.

In Section 7.1 we introduced the geosynclinal model for mountain
building and indicated that it suffers from an inability to account for
strong compressional forces and the scores of miles of crustal shortening that are known to have occurred. Using the plate tectonics model
these problems are resolved, since the dominant mechanism for deformation is the horizontal impinging of one plate against another,
thus deforming marginal sedimentary accumulations (previously called
geosynclines) into fold mountain ranges such as the Appalachians
(Figure 7.26). Fault-block mountains, such as those seen in the Great
Basin of the western United States, may also be explained as resulting
from large-scale tensional forces within a tectonic block that were
induced by transform faulting along one of its margins (Figure 7.27).
The structure of continental rocks is therefore far more complex than

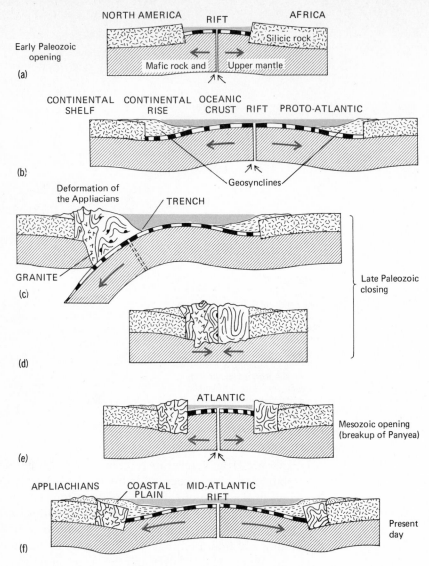

FIGURE 7.26
A model for the origin of geosynclines and fold mountains as a result of plate tectonics. (After Dietz, "Scientific American," vol. 226, 1972.)

that of the ocean floor, because the continents persist indefinitely and are repeatedly moved and deformed at the earth's surface. In contrast, the rocks of the ocean floor are created, moved, and destroyed relatively quickly and uniformly, and thus are both generally younger and less complex than continental rocks.

On a smaller scale, many of the complex structural features seen in continental deformation (discussed in Chapter 5) can be related to plate motions. Thus many *compressional* features such as folds, thrust

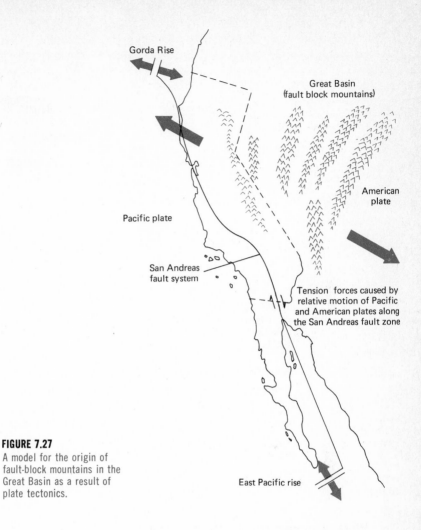

FIGURE 7.27
A model for the origin of
fault-block mountains in the
Great Basin as a result of
plate tectonics.

faults, and folded mountain ranges probably originate where a continent and an adjacent plate are pushed together at subduction zones
(Figure 7.28a). On the other hand, many *tensional* features, such as
large-scale normal faults and fault-block mountains, form when continents are stretched and broken over or near new crust zones or
transform faults (Figures 7.27 and 7.28b).

The discovery of plate motions clearly indicates that the basaltic
rocks of the oceanic crust originate from local melting of the upper
mantle, as described in Section 2.5, but still leaves unresolved the
question of the origin of the lighter, silicic rocks of the continents.
Because volcanic rocks associated with subduction zones along continental margins are usually silicic andesites rather than oceanic basalts,

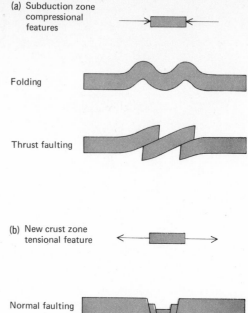

(a) Subduction zone
compressional
features

Folding

Thrust faulting

(b) New crust zone
tensional feature

FIGURE 7.28
The relation of the most common
structures of continental deformation to moving plate margins.

Normal faulting

some geologists have suggested that continents grow continuously along their margins by the addition of silicic materials that somehow separate from the mafic rocks of the descending plate or underlying mantle. An equally likely proposal suggests that the andesites result only from a deep melting of preexisting continental rocks, or of sediments derived from them, and thus shed no light on the *ultimate* source of the silicic continents. The question of continental origin thus remains a fundamental puzzle of the earth's crust.

7.6 Plate Tectonics and the Earth's Interior

Observation of crustal rocks exposed at the earth's surface, coupled with seismic data on the deep crust and upper mantle, have convincingly demonstrated that the earth's crust is formed and shaped by plate motions. A major uncertainty, however, concerns the underlying driving mechanism within the earth's interior which causes the creation, movement, and ultimate destruction of the plates.

Detailed studies of deep earthquakes caused by the descent of plates at subduction zones give an indication of the total thickness of the descending plates as well as of the overall depth to which they extend into the earth's interior. Such studies show that the plates include not only the 8-km (5-mi) thickness of the oceanic crust, but also a much greater thickness, up to about 80 km (50 mi), of the underlying mantle as well. This conclusion is confirmed by calculations of the strength of crustal rocks, which show that plate-sized units made up of only the thin crustal layer would tend to crush rather than to move as rigid bodies. Apparently, then, both the crust and the cool, uppermost mantle combine to form the earth's rigid outer skin. Lying beneath this rigid **lithosphere** (*litho* = rock, plus *sphere*), as it has come to be called, is the plastic, low-velocity mantle zone which is far warmer and less rigid than the overlying mantle and crust (see Section 6.9). This plastic mantle zone, sometimes called the **asthenosphere** (*astheno* = weak, plus *sphere*), evidently provides a lubricating layer over which the colder and more rigid lithosphere plates move (Figure 7.29).

Lithosphere plate motions must ultimately be driven by the heat energy concentrated in the underlying asthenosphere, but both the exact source of this heat energy and the way it is translated into horizontal plate movements remain uncertain. There are two likely sources for the earth's internal heat. One is *primordial heat* left over from the enormous frictional energy that is assumed to have been dissipated when the earth first consolidated from smaller masses of matter. Because of the insulating effect of the earth's great bulk, this initial heat would be lost very slowly and thus might still account for high internal temperatures even though the earth is very old. The second source, *radiogenic heat*, is provided continuously by the decay of radioactive nuclides within the silicate minerals of the crust and mantle. Although such nuclides make up only a tiny fraction of the earth's bulk, calculations show that their decay provides large quantities of heat to the earth's interior. Either primordial or radiogenic heat, or both, must in some fashion be concentrated in the warm asthenosphere, but there are many difficulties in understanding exactly how this takes place. Theoretical considerations indicate, for example, that a large fraction

FIGURE 7.29
Lithosphere and asthenosphere. Moving plates consist not only of the crust (black), but also of a much thicker layer of the underlying mantle (dark gray); together these units make up the rigid "lithosphere" which moves on the hotter, plastic "asthenosphere" or low-velocity layer of the mantle (color). The figure shows a descending plate at a subduction zone.

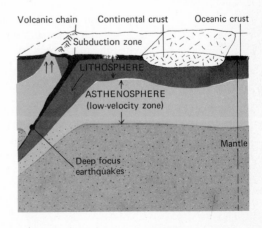

of the earth's heat-producing radioactive nuclides occur in the light silicic rocks of the continental crust rather than in the underlying mantle. Furthermore, there is no fully satisfactory explanation for the concentration of heat in a relatively narrow zone of the asthenosphere while the rocks both above and below remain cooler and more rigid.

Equally problematic is the question of exactly how the heat energy of the asthenosphere, whatever its origin, drives the regular horizontal plate motions of the overlying lithosphere. The most favored hypothesis relates plate motion to convection currents moving in either the underlying asthenosphere or, less probably, in a much greater thickness of the underlying mantle. Such convective movements, analogous to those seen in a pot of boiling water, occur whenever a fluid is heated from below in the earth's gravitational field. Convection currents always take the form of "cells" containing both rising and descending fluids; the rising areas might account for the upwelling of lavas along new crust zones, and the descending fluids might somehow pull down the oceanic crust at subduction zones. Although this hypothesis is perhaps the best yet advanced for the underlying cause of plate motion, convection currents are nevertheless difficult to explain in detail. How such regular fluid motions can account for the irregular outlines and changing configurations of the moving plates which shape the earth's crust remains a major challenge for this hypothesis.

SUMMARY OUTLINE

Continental Mountains

7.1 *Mountains and geosynclines:* most mountain ranges are made up largely of folded sedimentary rocks that originally accumulated along continental margins in thick sequences called geosynclines; near the axis of many mountain ranges the sediments have been melted to form elongate masses of granite.

7.2 *Continental drift:* the jigsaw puzzle–like fit of the continents and geologic, fossil animal, and paleoclimatic data combined with studies of "fossil rock" magnetism have suggested that the continents have moved or "drifted"; this phenomenon is now known to be related to still more fundamental movements involving the ocean floor.

The Ocean Floor

7.3 *Topography of the ocean floor:* the ocean floor is the site of the earth's longest and most continuous mountain chain, called the oceanic ridge-rise system.

7.4 *Magnetism and motions of the ocean floor:* the regular magnetic patterns of ocean-floor rocks show that basaltic lavas have been continuously poured out at the ridge-rise system and then moved laterally to create parallel bands of new crustal rocks; this process is called ocean-floor spreading.

Plate Tectonics

7.5 *Plate tectonics and crustal deformation:* as new basaltic crust is added at the margin of one of the large, rigid plates, it moves laterally, while the opposite plate margin may sink into the mantle and be destroyed; these plates include both continental and oceanic crust; mountain-building deformations of continental margins are caused by plate motions.

7.6 *Plate tectonics and the earth's interior:* moving lithosphere plates extend downward to include the rigid, uppermost mantle; they move upon the warmer and more fluid low-velocity mantle zone; the ultimate cause of plate motions is uncertain, but they may be related to thermal convection in the underlying mantle.

ADDITIONAL READINGS

*Clark, S. P., Jr.: *Structure of the Earth*, Prentice-Hall, Inc., Englewood Cliffs, N.J., 1971. Chapter 4 provides a good summary of plate tectonics.

*Cox, A. (ed.): *Plate Tectonics and Geomagnetic Reversals*, W. H. Freeman and Co., San Francisco, 1973. An anthology of research articles with excellent introductions by the editor.

Dickinson, W. R.: "Plate Tectonics in Geologic History," *Science*, vol. 174, pp. 107–13, 1971. A brief, authoritative review article.

*Keen, M. J.: *An Introduction to Marine Geology*, Pergamon Press, Oxford, 1968. An introductory survey with good discussions of ocean-floor topography, rocks and structures.

McKenzie, D. P.: "Plate Tectonics and Sea Floor Spreading," *American Scientist*, vol. 60, pp. 425–35, 1972. A popular review article.

*Takeuchi, H., S. Uyeda, and H. Kanamori: *Debate About the Earth*, Freeman, Cooper & Co., San Francisco, 1970. Excellent introduction to crustal magnetism, continental drift, and ocean-floor spreading.

*Wilson, J. T. (ed.): *Continents Adrift, Readings from Scientific American*, W. H. Freeman and Co., San Francisco, 1972. Popular review articles on earth structure, continental drift, and plate tectonics.

*Available in paperback.

chapter **8**

Stream Sculpture of the Land

In previous chapters we have considered the materials of the earth and the tectonic processes that cause them to move from their environments of origin into new positions, such as when a marine sandstone formation is folded into a mountainous region. We shall now discuss the destructional processes of weathering and erosion that operate on these rocks and minerals.

WEATHERING

8.1 Weathering Processes

Weathering concerns the static (in-place) alteration of rock material, an initial result of which is the production of soil. Erosion, on the other hand, involves the dynamic process of transporting rock debris (boulders, sand grains, soil, etc.) from one place, commonly called the *source terrane*, to another, the *environment of deposition* where sedimentation occurs. As a general rule, weathering precedes erosion.

There are two principal processes by which air and water act to change the structure and composition of rocks at the earth's surface. The first are mechanical weathering processes, the most important of which is known as frost wedging. Most rocks exposed near the surface have many small cracks and cavities filled with rainwater. Because water expands in volume as it freezes, it can exert tremendous pressures on the surrounding rock as temperatures fall below freezing. When freezing and thawing alternate, as often occurs as the land surface is warmed during the day and cooled at night in polar and cool-temperate regions, the repeated stresses of frost wedging are a significant cause of the breakup of massive rocks into smaller fragments (Figure 8.1).

Of still greater importance than mechanical weathering are the many processes, grouped together under the term chemical weathering, whereby both the structure *and* the chemical composition of rocks are changed. Most of these processes center on the chemical actions of the earth's surface waters. Absolutely pure water is chemically rather inactive, but natural surface waters are normally slightly acidic due to the presence of dissolved CO_2 from the atmosphere, some of which combines with hydrogen in the water to form a weak solution of carbonic acid. Dead and decaying plant and animal remains also contribute acids to surface waters. Like most weak acids, these surface waters are good solvents and slowly dissolve many minerals; they also enter directly into chemical reactions to form new minerals.

We saw in Chapter 2 that most of the earth's crust is composed of either silicon-rich granite or magnesium- and iron-rich basalt. The minerals of these two igneous rocks are therefore the primary mate-

FIGURE 8.1
The reduction of a massive limestone boulder to small fragments by mechanical frost wedging. (From Thornbury, "Principles of Geomorphology," 1969; photos by Alan Pratt.)

(a)

(b)

(c)

(d)

(e)

(f)

rials subjected to weathering as they are exposed to acid waters at the earth's surface. Such waters ultimately convert the feldspars of granite and basalt to the fine-grained sheet silicate *clay minerals*, which contain a relatively large amount of water in their crystal structures and are much more stable than are most igneous minerals under the temperatures and chemical environments of the earth's surface (Figure 8.2). Quartz, which is extremely resistant to chemical attack, is preserved without chemical alteration.

During alteration, much of the potassium, sodium, calcium, magnesium, and silicon of the original igneous minerals goes into solution and is removed from the place of origin by flowing water. Eventually, these elements are either deposited elsewhere in new sedimentary minerals or added by rivers to the oceans, which contain huge quantities of land-derived dissolved materials. Of the half-dozen or so common elements in igneous rocks, only iron and aluminum are relatively insoluble and tend to persist in solid form under the attack of chemical weathering. The iron is quickly converted to insoluble, rustlike oxide minerals, while the aluminum forms a principal constituent of clay

FIGURE 8.2
Schematic summary of the breakdown of granite and basalt by chemical weathering. Feldspar alters to clay minerals which under very intense weathering may, in turn, alter to aluminum oxides. Pyroxene and olivine alter to iron oxides. Quartz is preserved without chemical alteration.

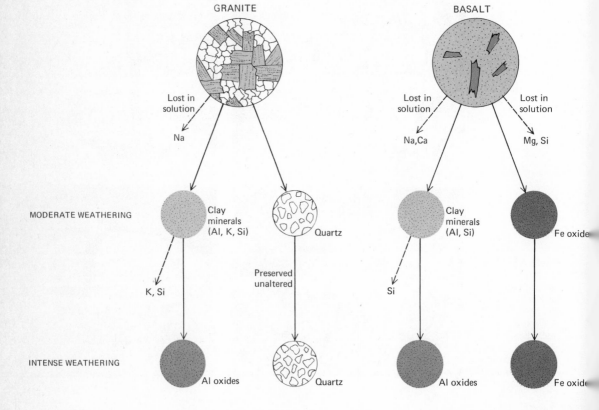

minerals. When chemical weathering is especially intense, even the silicon-oxygen tetrahedra of the clay minerals themselves may be removed by solution, leaving only a silicon-free residue of aluminum oxide (Figure 8.2).

The attack on fresh rock by weathering agents often creates rounded and smoothly curved surfaces, a process known as spheroidal weathering (Figure 8.3a). Chemical decomposition of rock-forming minerals proceeds from all joints and cracks in the rock where air and water are allowed to migrate. As can be seen in Figure 8.3b, this accentuates the depth and intensity of attack at any edge or corner, thus causing a spheroidal surface to evolve. On a much larger scale, the process of exfoliation yields similar-appearing curved surfaces of rock (Figure 8.3c). It is believed that the mechanism of exfoliation, however, is distinctly different from spheroidal weathering. Massive rocks, such as some sparsely jointed granites and sandstones, are believed to expand slightly as the pressure on them is reduced by erosional removal of overlying rocks, much in the manner a balloon expands as it rises into regions of less pressure. Expansion of the rock causes fractures to form that are approximately parallel to the surface of erosion, giving rise to the onionskinlike layered appearance known as exfoliation. Then weathering, both chemical and mechanical, begins to attack the rock along its newly formed fractures, causing slabs of rock to break away and yielding the layered look seen in Figure 8.3c.

8.2 Soil

The result of mechanical and chemical weathering is to cover much of the land surface with a thin veneer of altered rock debris known as soil. Soil, as we might predict from our discussion of the products of weathering, is made up primarily of tiny particles of various clay minerals mixed in varying proportions with larger particles of quartz. Because quartz is the only really stable igneous mineral under the action of chemical weathering, most of the gravel and sand-sized particles of the earth's surface are in large part composed of it.

In addition to the clay and quartz fragments produced by rock weathering, most soils also contain from 1 percent to as much as 50 percent organic carbon compounds derived from the decay of plants and animals supported by the soil. These organic constituents are extremely important components of soils, for they are closely interrelated with soil texture, structure, and composition. Plants, in particular, profoundly affect soils in many ways. By utilizing in their metabolic processes, dissolved elements left over from rock weathering—particularly potassium, sodium, and calcium—plants tend to retain these elements near their site of formation instead of allowing them to be

(a)

(c)

FIGURE 8.3

Spheroidal weathering and exfoliation. (a) Spheroidal weathering.
(b) Schematic diagram. (c) Exfoliating granite dome in Yosemite.
(a, c: E. A. Hay.)

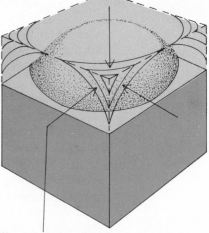

Chemical attack on all surfaces is deepest and most intense at a corner.

(b)

dispersed by moving water. These components are returned to the soil as the plants die and decay and are then utilized by new plants in an endless cycle. Only when this cycle is broken—for example, by harvesting the plants instead of allowing them to die and decay—are these nutrient elements permanently removed from the soil. It is for this reason, of course, that supplemental nutrients in the form of fertilizer are necessary for most cultivated soils.

Plant roots and burrowing animals also tend to facilitate weathering by making the soil porous so that surface water can penetrate to un-weathered, underlying rocks. In addition, the presence of organic acids from plant decay adds to the chemical activity of surface waters, as we have already noted. These processes tend, in general, to sharply increase the rate of chemical weathering on those portions of the land surface that are covered by plants. At the same time, plants facilitate the accumulation of weathered debris as soil by binding together the particles with their roots and by protecting the soil surface from the erosive effects of wind and water.

In those many parts of the land surface where plants are present and where the soil consequently contains significant organic matter, soils tend to have a definite cross-sectional structure known as a soil profile. Typically such profiles show three horizons of changing composition and structure with depth (Figure 8.4). The uppermost and thinnest horizon contains abundant decaying organic matter from the overlying plants and a minimum of inorganic constituents that are quickly removed by intense chemical weathering near the surface. Immediately below this is a thicker zone with less organic matter, where many of the weathered constituents, such as clay materials, quartz sand, and dissolved elements, are concentrated. Below this is a still thicker horizon of only partially weathered underlying rock that generally lacks organic matter but that is in the process of being converted to soil by slow chemical weathering. Superimposed on this basic threefold structure are a host of variations which permit soil scientists to define and classify seemingly endless minor soil types.

The kinds and intensity of weathering processes, and the soils that result from them, are closely controlled by climate. In the United States, for example, two major soil types are geographically distributed primarily as a function of average annual rainfall (Figure 8.5): pedalfers are relatively rich in aluminum and iron, whereas pedocals contain abundant calcium carbonate. The former is favored by a warm, humid climate, and the latter develops in regions of less rainfall. Yet a third major soil type develops in tropical climates where chemical weathering is increased by abundant surface water and high temperatures. Because of intense weathering, tropical regions commonly have silicon-poor soils composed mostly of iron and aluminum oxides (Figure 8.6). Where iron oxides predominate the soils are red and are known as laterite; where aluminum oxides dominate the soils are usually yellow or gray and are known as bauxite (Figure 8.7). The intense weathering and abundant vegetation of tropical regions cause

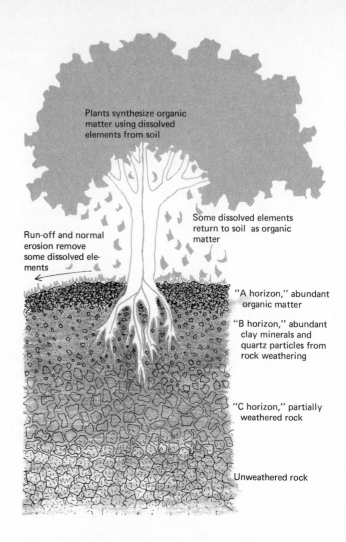

Plants synthesize organic matter using dissolved elements from soil

Some dissolved elements return to soil as organic matter

Run-off and normal erosion remove some dissolved elements

"A horizon," abundant organic matter

"B horizon," abundant clay minerals and quartz particles from rock weathering

"C horizon," partially weathered rock

Unweathered rock

FIGURE 8.4
A generalized soil profile.

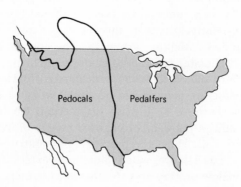

Pedocals   Pedalfers

FIGURE 8.5
Generalized distribution of major soil types in temperate North America. Pedalfers result from greater chemical weathering in regions of higher rainfall and temperature. (See text for discussion.)

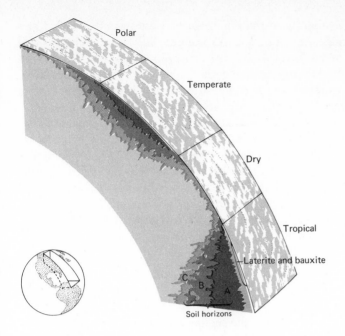

Polar

Temperate

Dry

Tropical

Laterite and bauxite

C
B
A

Soil horizons

FIGURE 8.6
Interrelations of soil and climate. Thick soils develop only under temperate and tropical climates where vegetation is abundant. The high temperature and rainfall of tropical regions lead to intense chemical weathering, deep soils, and residual iron and aluminum oxides (laterite and bauxite).

FIGURE 8.7
Bauxite: the result of intense chemical weathering. (Reynolds Aluminum Co.)

an extensive soil development: tropical soils commonly reach thicknesses exceeding 30 meters (90 feet) and have an average thickness of over 3 m (9 ft).

Chemical weathering is the dominant process in both tropical and temperate climates, but in the latter the generally lower rainfalls and temperatures make the process less intense. In temperate soils enough silicon normally remains to make clay minerals, rather than oxides, the dominant constituents. In contrast to tropical soils, temperate soils have an average thickness of only about 1 m (3 ft) (Figure 8.6).

In tropical and temperate regions the presence of a permanent protecting cover of vegetation permits a relatively thick soil layer to accumulate. In polar and dry climates, on the other hand, vegetation is sparse and the products of rock weathering are exposed to continuous removal by wind and water. Under such conditions only a very thin soil cover, with abundant exposure of bare rock surfaces, normally develops. In these regions mechanical frost wedging of the exposed rock supplements chemical breakdown as an important weathering process.

WATER ON THE LAND

Weathering is the first step in the sequence of events that shapes landscapes, for weathering reduces massive rock to small particles that can be removed from their place of origin by running water, wind, and ice. Because of weathering and removal of rock debris, the land surface is constantly being worn away. Indeed, were it not for the opposing internal forces of isostasy, mountain building, and volcanism, the land surface would have long ago been reduced to a flat, featureless plain by these processes of erosion.

Weathering, we have seen, is closely controlled by climate; the same is true of the other processes that shape the land surface. The most powerful agent of land sculpture is *moving water*, and under tropical and temperate climates it is the principal agent in shaping landscapes.

Each year, when averaged over the entire surface of the earth, a layer of water about 1 m (3.3 ft) deep is evaporated into the atmosphere, and each year the same meter of water is precipitated somewhere on the earth's surface as rain or snow. The amounts evaporated and precipitated differ, however, over the land and ocean. Over oceans, more water is evaporated than is precipitated; this excess falls on the land surface where, on the average, precipitation exceeds evaporation (Figure 8.8). In order to complete the cycle, this excess ocean-derived water dropped on land must ultimately return to the ocean; this occurs primarily as streams flow across the land and empty into the sea. The energy of this relatively small quantity of excess water, propelled by gravity as it moves over the land surface toward the sea, is the principal agent of land sculpture.

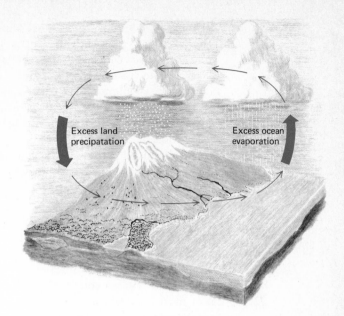

FIGURE 8.8
The hydrologic cycle. More water is evaporated than precipitated over the ocean; the excess is transported and precipitated over land and is ultimately returned to the ocean by streams and rivers. These relatively small quantities of water moving across the land are the principal agent of land sculpture.

Most of the water falling on land does not pass immediately into moving streams for the return trip to the ocean. Only about one-eighth reaches streams directly as runoff from the land surface; the remaining seven-eighths is absorbed into soil and rocks to become *groundwater*. Even this groundwater ultimately reaches streams and returns to the ocean, but its progress is far slower and less direct than that of surface runoff.

In addition to streams and groundwater, there is one additional form in which liquid water occurs on land—as lakes, relatively stationary concentrations of surface water. Neither lakes nor groundwater play a significant role in shaping the land surface when compared with the far more energetic action of moving streams. Consequently, we shall only briefly consider some of the principal characteristics of the former before looking more closely at streams.

8.3 Lakes

Viewed over millions of years of geologic history, lakes are extremely transient features of the earth's surface. They form when obstructions of rock or soil temporarily block the path of flowing water, and they disappear as the obstructions are worn away and as deposition of sediments fills the depression occupied by the lake. During their relatively short existence, lakes, particularly very large ones, manifest many of the same phenomena as the ocean—for example, surface waves, and temperatures that decrease downward due to surface warming and mixing. Most lakes differ profoundly from the ocean, however, in lacking a high concentration of dissolved materials.

Normal ocean water contains about 3.5 percent dissolved elements. A few "saline lakes" in dry regions where rains are infrequent and evaporation is intense—lakes such as the Dead and Caspian "seas" and the Great Salt Lake—also have high concentrations of dissolved materials, although not ordinarily in the same proportions as in the ocean. Most lakes, however, have only an insignificant fraction of 1 percent dissolved matter. Because of this negligible salinity, lakes differ from the ocean in several other important properties. In the ocean, differences in both temperature and salinity are important in causing density differences that lead to current movements. In lakes, only thermal differences occur, and density-induced currents are thus less pronounced than in the ocean. Another result of the lack of dissolved materials is that lake water freezes at 0°C (32°F) rather than having the indefinite freezing properties of seawater. As a consequence, lakes in very cold regions commonly freeze solid to depths of 100 m (330 ft) or more, whereas ice on the ocean (other than in land-derived icebergs) seldom exceeds 3 m (10 ft) in thickness.

8.4   Groundwater

Under the influence of gravity, most of the water falling on the land surface sinks into the underlying soil and rock to become ground-water. As a result, except for a relatively thin outer layer which dries between rainfalls, all cracks, voids, and pores in continental rocks that are located above the depth (commonly thousands of feet) where they are closed by pressure are permanently saturated with water. The upper surface of this zone of saturation, which separates it from the periodically dried layer above, is known as the water table. In any one region the depth of the water table below the surface varies with topography, which it follows by rising under hills and dropping along the sides of valleys (Figure 8.9a). When the water table intersects the land surface, *springs* result.

Groundwater is not usually stationary, but moves very slowly under the influence of gravity until it ultimately reaches the surface at some lower elevation. Permanently flowing streams of tropical and temperate climates normally move in valleys whose bottoms lie *below* the local water table (Figure 8.9a); much groundwater ultimately reaches the surface by slow seepage into these perennial streams. In dry climates, on the other hand, this process is reversed. There, temporary, ephemeral streams, formed after the infrequent rainfalls, lose water to the ground and ultimately disappear until the next rainfall.

Groundwater is a valuable natural resource in many parts of the

(a)

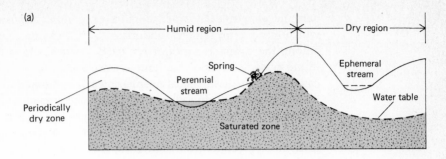

(b)

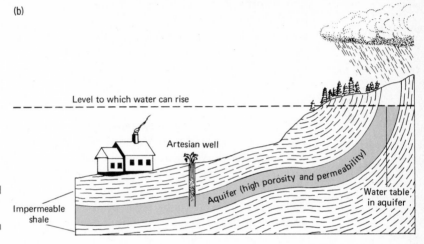

FIGURE 8.9
Ground water. (a) The relationship of ground water to climate and stream flow. (b) Conditions for an artesian well.

world. A body of rock that holds and can yield large volumes of water is known as an **aquifer** (Figure 8.9b). To serve as an aquifer a rock unit must be both porous and permeable—that is, it must have a significant volume of void space (porosity) and it must also have large enough openings between its grains to allow water to pass through (permeability). Sandstones whose voids between grains are not filled with cement or clay are very good aquifers, for example; shales, on the other hand, are poor sources of water even though they have high porosity (sometimes as much as 40 percent void), for in a shale the grains are so small that water molecules cannot easily pass between them.

In situations where a good aquifer is overlain by an impermeable rock unit, as shown in Figure 8.9b, free-flowing conditions may exist for wells that penetrate down to the aquifer: these are known as **artesian wells.**

Although groundwater has great importance to man, besides its crucial role in weathering and soil formation it plays a relatively minor part in shaping landscapes—with one very important exception, which occurs in regions underlain by limestone. Limestone alone among the common rocks of the earth's crust is readily dissolved by dilute acids. Waters of the land surface are usually slightly acidic because of both dissolved atmospheric carbon dioxide and the addition of acids from decaying plants. For this reason, groundwater commonly dissolves significant amounts of the underlying rock in regions where limestone is exposed at or near the earth's surface. This dissolution leads to large underground cavities, channels, and passageways which may open to the surface as *sinkholes*. Regions showing these distinctive features are said to exhibit Karst topography, named after the Karst region of Yugoslavia, where solution features are particularly well developed (Figure 8.10). Most of the earth's large dramatic underground caverns are developed in such regions.

8.5   Streams

The third, and geologically most important, form in which liquid water occurs on land is as streams, relatively long, narrow concentrations moving downward across the land surface under the influence of gravity. In everyday language we normally make a size distinction between smaller "streams" and larger "rivers," but geologists, recognizing that all gradations exist between the smallest flowing trickle and the largest rivers, generally refer to *all* flowing waters on the land surface as "streams."

Streams in tropical and temperate regions derive their water from two sources. Roughly half their flow comes directly from surface runoff of rain or of water from melting snow. The other half comes from slow seepage of groundwater into the stream channel, which normally lies below the local water table—at least in perennial, or permanently flowing, streams. It is this slow, continuous influx of groundwater along stream channels that maintains their flow between rainfalls. Conversely, stream flooding during periods of heavy rainfall is a result of the greatly increased surface runoff. In dry climates the water table is generally so deep that ephemeral streams receive none of their water from groundwater seepage. Instead, they are fed only from surface runoff after the infrequent rainfalls and are dry for much of the year (Figure 8.11).

Gravity is the principal driving force of stream (and groundwater) movements; it begins to act whenever solar energy moves water from the ocean into the atmosphere and deposits it above sea level on the land surface. The gravitational force then comes into play and moves the water downward, toward sea level, but is opposed by complex frictional forces as the water impinges on solid rock and sedimentary

(a)

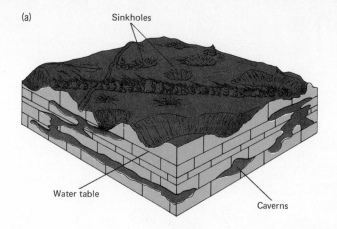

Sinkholes

Water table

Caverns

(b)

(c)

FIGURE 8.10
Karst topography. (a) Caverns, sinkholes,
and related features commonly form in
regions underlain by limestone, which is
readily dissolved by acidic ground water.
(b) Karst topography in central Kentucky.
(c) Close-up of a small, deep sinkhole in
northern England. (b: John S. Shelton;
c: Geological Survey of Great Britain.)

(a)

(b)

(c)

FIGURE 8.11
Varieties of streams.
(a) Perennial stream
in southern England.
(b) Ephemeral stream
in southern Arizona
in dry stage and
(c) in flood. (a: Geo-
logical Survey of
Great Britain; b, c:
Tad Nichols.)

particles. The net result of the impelling gravitational force and the opposing frictional forces is to impart a particular speed or *velocity* to the moving water.

In seeking to analyze the dynamics of stream flow, geologists in recent years have made intensive studies of many secondary factors that influence the two primary forces of gravity and friction in stream flow. The force of gravity is primarily a function of two such secondary factors: the volume of the flowing water, which is called the *stream discharge*; and the *slope* of the surface over which it moves. The opposing frictional forces, on the other hand, are primarily related to two other factors: the amount of sediment moved by the flowing water, known as the *sediment load*; and the *shape of the channel* through which the water flows. These four secondary factors—discharge, slope, sediment load, and channel shape—are closely and complexly interrelated with one another and with their ultimate result, the velocity of the moving water.

Analyses of records of velocity, discharge, slope, sediment load, and channel shape from hundreds of streams have shown that the interaction of these factors is not random, but varies in a systematic way; that is, the factors are so balanced that changes in one lead to surprisingly regular and predictable changes in the others (Figure 8.12). If, for example, the discharge at some point along a stream is doubled after a rainfall, then the stream velocity will increase by a factor of 1.3, the channel shape will change by a small but predictable amount as the water rises and the channel width and depth increase, and the amount of sediment carried by the stream will increase by a factor of 8. The mere fact that such increases take place is not surprising. Streams are easily seen to become wider and deeper when discharge increases during floods. Likewise, the fact that large quantities of sediment are moved during floods is clear because most streams flow on floodplains (Figure 8.13c and 8.14)—valleys partially filled with loose sand and clay that have been transported during periods of high discharge and dropped as the stream recedes into its normal channel. It is not that such changes in shape and sediment load take place that is

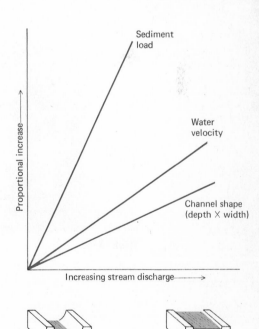

FIGURE 8.12
The dynamics of stream flow at a single point along a stream. Sediment load, water velocity, and channel shape all increase regularly, but at different rates, with increasing discharge. (Modified from Leopold et al., "Fluvial Processes in Geomorphology," 1964.)

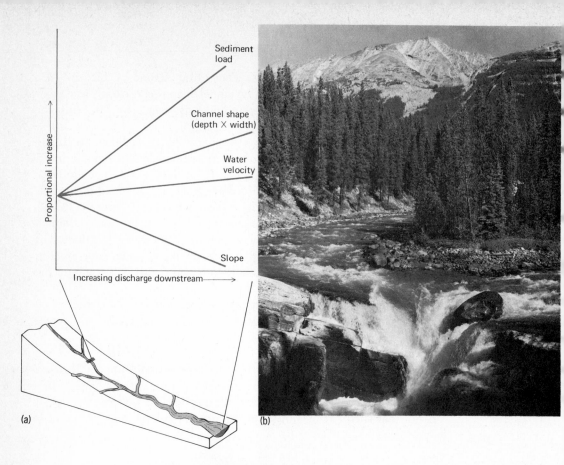

Sediment
load

Channel shape
(depth × width)

Water
velocity

Proportional increase →

Slope

Increasing discharge downstream →

(a)

(b)

(c)

FIGURE 8.13
Stream flow. (a) The dynamics of stream flow with increasing distance downstream.
Sediment load, channel shape, and water velocity increase, and slope decreases, with
increasing discharge. (b) Stream in western Alberta in its upper course, and (c) the lower
course of a stream in southern Wales. Although steep mountain streams appear to flow
faster than those in relatively flat alluvial valleys, careful measurements show that this
is not normally the case. (a: Modified from Leopold et al., "Fluvial Processes in Geomorphology," 1964; b: G. Hunter, "Information Canada Phototheque"; c: Geological Survey of
Great Britain.)

FIGURE 8.14
Typical meanders of a small stream across a floodplain in Colorado. (Tad Nichols.)

so astonishing, but that the *quantity* of such changes are regular and predictable, as established by careful observation and analysis.

So far we have considered only changes at a single point along the course of a stream. If, instead, we consider the interrelations of discharge, velocity, channel shape, and sediment load through the entire length of the stream, we find equally regular, but somewhat different, relationships (Figure 8.13a). In perennial streams, discharge normally increases downstream as water is added from runoff, groundwater seepage, and the inflow of tributary streams. Sediment load and channel width and depth increase regularly as discharge increases downstream. At the same time, slope—the one factor that (obviously) cannot be considered at a single point on the stream—decreases regularly downstream, so that in cross section the channel has a concave-up profile that flattens in the lower part of the stream course (Figure 8.13a). One of the most surprising results of modern stream analyses was the discovery that these factors interact to cause a small but regular velocity *increase* downstream. This conclusion had not been expected, since rushing mountain streams flowing on steep slopes *appear* to move faster than large rivers flowing on flat floodplains (Figure 8.13b and c). Measurements show, however, that this is not usually the case, for the downstream decrease in slope is more than compensated by the decreased friction of a larger channel.

Before moving on to consider how stream flow shapes the land, there is one rather puzzling characteristic of streams to be considered. When viewed from above, as from an airplane or on a map, large streams almost never flow in straight linear channels, but instead follow a regular zigzag series of loops known as *meanders* (Figure 8.14). These have the effect of increasing the channel length and thus adding to the distance through which frictional forces can act; they also decrease the channel slope, with the result that the gravitational impelling force is lessened. For these reasons meanders are believed to be somehow related to the dynamic balance of frictional and gravitational forces in stream flow, but as yet there is no satisfactory explanation of their exact cause.

LAND SCULPTURE BY LAND WATER

Shaping of the land surface takes place as rock debris produced by weathering is moved downward by gravity toward its ultimate resting place, the ocean. These movements of rock debris usually involve flowing water, but they commonly begin under the influence of gravity alone, as when soil slides down a steep hillside, or when a block of rock, loosened by weathering, falls from the face of a cliff. Such direct movements of weathered debris by gravity are called **mass wasting.** When mass wasting combines with weathering and stream transportation, the result is the formation of *valleys*, empty spaces formerly filled with rock, bounded by *hillslopes*, curved surfaces underlain by rock that is still undergoing weathering and erosion.

8.6   Mass Wasting

Soil and other rock debris formed by weathering move downward under the influence of gravity, just as water does, whenever the gravitational impelling force overcomes opposing frictional forces. On slopes steeper than about 40 degrees, gravity exceeds frictional resistance for *all* kinds of rock debris, and therefore only massive bedrock, exposed as cliffs, occurs on such steep slopes. Mass wasting also continuously takes place, however, on more gentle, soil-covered slopes. This mass wasting is of two sorts. Most commonly the frictional resistance is so great that downslope movements are very slow, from one to a few centimeters per year; such movements are known

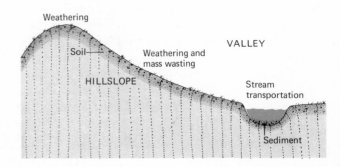

FIGURE 8.15
Erosion of the land surface. Weathering and mass wasting deliver rock debris to streams where it is transported as sediment.

FIGURE 8.16
Creep, slow downslope movements of soil and rock debris, and its effect on hillside objects.

as creep. All soil-covered slopes undergo creep, which is a principal cause of tilted fences, cracked sidewalks, bent tree trunks, and other common hillslope phenomena (Figures 8.16 and 8.17).

More rapid downslope movements by mass wasting are known collectively as landslides.* The speed at which rock debris moves in landslides varies from several meters per *day* to several meters per *second*, but is always much more rapid than the centimeters-per-year movement of creep. Rapid landslides are especially common on slopes where frictional forces are reduced by the lubricating effects of un-

*Chapter 11 contains a more complete discussion of landslides as they relate to man and his structures.

FIGURE 8.17
Soil creep. (William C. Bradley.)

usual amounts of groundwater. Such lubrication often occurs where soil is underlain by an impermeable layer of clay or, in polar climates, frozen groundwater, both of which limit and concentrate the downward penetration of surface water (Figure 8.18). Although landslides are often large and spectacular and can cause great destruction, they are less important in shaping landscapes than are the slower but more continuous movements of creep. Often, particularly in regions of strong mechanical weathering, rock fragments will loosen and fall a considerable distance under the influence of gravity alone. These fragments are called talus, and they sometimes form steep and jumbled talus slopes (Figure 8.18c).

8.7 Stream Erosion

Mass-wasting processes account for most of the *initial* movement of rock debris, but it is transportation of this debris by streams that normally carries it from its immediate area of origin and thus makes possible a continuous wearing away of the land surface. Solid debris particles are moved by flowing water in either of two ways: they may be completely surrounded and supported by the moving water as suspended load; or they may be rolled, slid, or bounced along the stream bottom as bed load (Figure 8.19). In general, larger and heavier particles move as bed load and smaller and lighter particles as suspended load. In addition to solid debris, streams also transport a dissolved load of ions weathered from rocks of the crust.

In the last section we noted that the total amount of sediment moved by a stream increases very rapidly as the discharge and velocity of the stream increase (Figures 8.12 and 8.13). For this reason, most sediment movement takes place at times of high water, when the transporting capacity of the stream is at a maximum. Between periods of flood, the sediment is temporarily deposited in the stream channel or adjacent floodplain. Most stream valleys are at least partially filled with such sediment, called alluvium, that is temporarily at rest on its downslope journey. The ultimate resting place of this alluvial debris is the ocean, for sea level marks the lowest point to which streams can ordinarily flow. On entering the ocean, the stream-transported sediments are moved, sorted, and reworked by waves and currents; ultimately they come to rest in beaches and other *depositional* landscape features.

The combined effect of mass wasting and stream transportation is to shape the land surface into valleys and hillslopes. Valleys result from the erosional action of the streams that flow through them. In general, two types of valley-stream relationships occur. In rugged mountainous regions where slopes are steep, the sedimentary debris

(a)

(b)

FIGURE 8.18
Examples of mass wasting. (a) Small landslides. (b) Large landslide. (c) Rockfall and talus. (a: Wisconsin Conservation Department; b: W. R. Cotton; c: Josef Muench.)

(c)

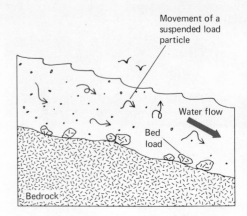

FIGURE 8.19
Transportation of debris by moving
water.

supplied by mass wasting is readily transported, and streams commonly flow directly on the underlying rock. When this occurs, chemical solution and mechanical abrasion cause the stream to cut actively into the rock to form steep-walled **V-shaped valleys** (Figures 8.20a and 8.21a).

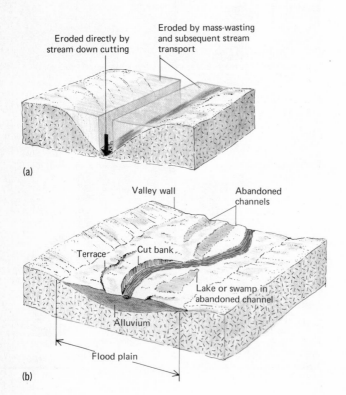

FIGURE 8.20
Types of valleys. (a) V-shaped and
(b) alluvial valleys. Alluvial valleys,
which characterize the lower reaches
of most streams, commonly show
abandoned channels, cut banks, terraces, and other features caused by
the shifting channel and moving
alluvium.

(a)

(b)

FIGURE 8.21
V-shaped and alluvial
valleys. (a) V-shaped
in southern British
Columbia. (b) Alluvial
valley in southern
England. (a: National
Film Board of Can-
ada; b: Geological
Survey of Great
Britain.)

In contrast, on the more gentle slopes that characterize the lower
reaches of most streams, valleys are floored by large quantities of
alluvial debris that make up the relatively flat floodplain over which
the stream flows. Even during flood, such streams generally erode
only these floodplain sediments rather than cutting downward into
the underlying rock. The resulting flat-floored **alluvial valleys** (Figures 8.20b and 8.21b) usually show characteristic patterns of temporary lakes, terraces, and other topographic features caused by the intermittent erosion and deposition of the underlying sediment. Some
of these features are summarized in Figure 8.20b.

V-shaped and alluvial valleys are easier to characterize than are the
hillslopes that bound them. When viewed in cross section, most hillslopes have a *profile* which is convex in the higher parts and concave
in the lower (Figure 8.22). The upper, convex slope is thought to be
shaped primarily by weathering and creep, whereas the lower, concave portion results from fluid transportation by concentrated surface
water runoff, leading to a more rapid removal of debris. Beyond this,
the exact shape and form of hillslopes is a result of a complex and
poorly understood interaction between the various processes of erosion, the structure and composition of the underlying bedrock, and
the length of time the rock has been exposed to erosion.

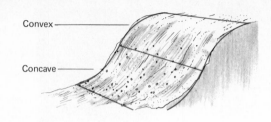

FIGURE 8.22
A typical hillslope profile.

When viewed from a larger perspective, streams develop drainage patterns that reflect regional structure and the relative erodability of the bedrock across which they flow. Contrast, for example, the three drainage patterns shown in Figure 8.23. What can each tell us about the underlying bedrock? The *trellis pattern* in Figure 8.23a suggests a strong *structural influence* on the stream's position. It has cut the

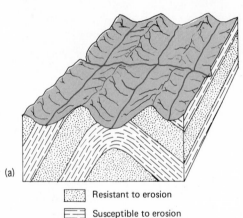

(a)

FIGURE 8.23
Stream drainage patterns. (a) Trellis pattern resulting from structural control. (b) Dendritic pattern, indicating a lack of structural control. (c) Radial pattern developed on a volcano.

⬚ Resistant to erosion

▤ Susceptible to erosion

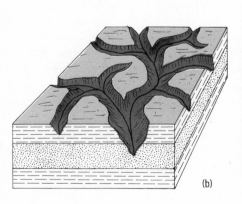

(b)

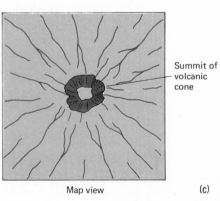

Map view   (c)

Summit of volcanic cone

deepest valleys along the trend (strike) of more easily eroded layered rocks; hence the trellislike pattern reflecting the regular geometry of the bedrock. The *dendritic pattern* in Figure 8.23b, in marked contrast, lacks geometric regularity altogether; it appears to be random, and therefore lacking in any structural influence on the part of the underlying rock. Flat-lying sedimentary or volcanic rocks, or nonlayered plutonic rocks, would all erode in this manner. Finally, a *radial pattern* (Figure 8.23c) is commonly developed on domal or cone-shaped structures. Large volcanoes provide our best example of radial drainage patterns.

8.8   The Graded Stream as an Equilibrium Concept

Let us now reconsider the factors controlling the effectiveness of stream erosion and analyze them in more detail in order to learn how changes in a stream's erosional activities can be predicted.

First, it can be noted that erosion, and hence transportation of debris, requires *work*; furthermore, work requires the expenditure of *energy*. There are two kinds of energy of primary concern to our understanding of stream erosion: potential energy and kinetic energy (potential energy is energy of position and kinetic energy is energy of motion). In a somewhat oversimplified manner, we may conceive of stream transportation as being a geologic process whereby the potential energy of water derived from rain and snow at high elevations is converted to a moving force (hence kinetic energy) capable of doing the work of rock transportation.

The amount of potential energy locked up in stream water is determined by two variables: mass and elevation. It is easy to see that the potential for water to do work will increase if (1) the amount, or mass, of water increases, and also if (2) a given amount is raised to a higher elevation so that it can fall farther. By way of analogy, consider the potential in an axe for doing the work of splitting logs: neglecting the force of the axe wielder, it is clear that the heavier-headed axe raised higher above the log can do the work most rapidly. This work on the log cannot actually be accomplished, however, until the force of gravity acts upon the axe, causing it to convert its energy of position into energy of motion. The work is achieved by the *moving* mass, not the *elevated* mass. By applying this example to stream water, we can see the importance of gravity acting on water masses at high elevations.

Imagine a mass of water perched at position A in Figure 8.24. By the time that water reaches sea level at position B, it will have expended all the potential energy it possessed by virtue of its position 600 m (2,000 ft) above sea level (at A). It is theoretically conceivable,

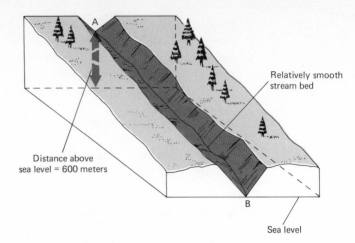

Relatively smooth
stream bed

Distance above
sea level = 600 meters

FIGURE 8.24
Block diagram of a stream valley.
(See text for discussion.)

B

Sea level

but not actually possible, that the water could convert all its potential
energy into downslope acceleration (velocity increase), in which case
the water emerging at B would be a smooth flow moving at an incred-
ible speed. Instead, what always happens in nature is that as the potential energy is converted into acceleration, the moving water soon
impacts the stream bed with enough force to dislodge fragments. This
hypothetical event brings about two highly significant changes in the
stream: first, the newly roughened surface causes turbulence; and
second, the fragments begin the process of being transported as part
of the *bed load* (Figure 8.19). The newly created turbulence, a condition where water is moving in random directions rather than smoothly
in the downslope direction, allows the stream now to transport small
particles in the *suspended load*. The overall effect of roughening the
channel, therefore, is to increase the work done by the stream at the
expense of decreasing the rate of downslope acceleration.

Let us now consider a stream that eventually achieves a condition
of equilibrium between its potential for erosion and its potential for
deposition, and what takes place when this equilibrium is disrupted.
Our model for such a stream in equilibrium is called a **graded stream**.
Even though it is unlikely that a perfectly graded stream can be found
in nature, the concept is nevertheless universally useful. Consider the
map in Figure 8.25, which shows a stream network of Lizard Creek
(A–B) and its tributaries. The long profile (side view of the straight-
ened-out stream bed) of Lizard Creek might have a generalized appearance as shown in Figure 8.26.

What does it mean to call Lizard Creek a graded stream? It means
that it is neither eroding nor depositing sediment—it is in *equilibrium*.
It has achieved a delicate adjustment of its channel shape, roughness,
and slope with the amount of water that flows through it so that all
the sediment that enters is carried through its entire length, with no

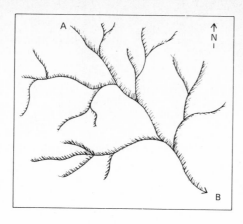

FIGURE 8.25
Map of Lizard Creek drainage system.
(See text.)

excess of energy available to erode additional bedrock. This concept of equilibrium carries with it the notion that any alteration of the equilibrium causes an immediate reaction in the stream in the direction of restoring equilibrium conditions. Referring back to Figures 8.25 and 8.26, let us consider how Lizard Creek would respond to local mountain-building forces tilting the earth's crust in a manner that steepened the long profile to the A'–B position shown in Figure 8.27.

Notice that A' is higher than A, and therefore water flowing in Lizard Creek now has more potential energy and hence a greater capacity to do the work of erosion. This can also be seen graphically by the steepening of A'–B with respect to the old profile A–B, causing the water to flow faster and cut down more at the stream bed. The effect of this increased energy will be to erode back toward A–B until the slope has been flattened enough to allow transportation of only the debris carried into A–B by its tributaries, thus restoring equilibrium. Although the stream is a nonthinking entity, it acts "as if" it said to itself, "Which must I do—erode or deposit—in order to achieve the ability to transport all the debris I receive, no more and no less? Well, I must flatten my slope, because I now have more velocity than is needed; and the way I can flatten my slope is to cut it down by erosion."

Thus any disruption of equilibrium for a stream segment will be met automatically with a response that works to reestablish the

FIGURE 8.26
Long profile of Lizard Creek. (See text.)

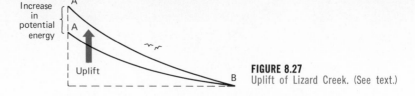

FIGURE 8.27
Uplift of Lizard Creek. (See text.)

graded condition: deposition will steepen the slope and erosion will flatten it.

The equilibrium of a stream segment may be altered by a variety of geologic processes. Climatic changes, for example, might cause downcutting or deposition in response to an increase or a decrease in rainfall, respectively. The degree of weathering and other elements of the bedrock's erodability must also be considered when analyzing a stream's equilibrium.

In truth, the delicate balance of stream erosion involves a complex interaction of all these variables and more. Throughout geologic time perhaps the most important one, though, has been that of upward crustal movement. Again and again we find evidence of stream **rejuvenations** (Figure 8.28), which appear to have resulted from raising of the land with respect to the sea, thereby endowing streams with increased potential energy, which they have expended in more vigorous downcutting.

8.9   Rates and Cycles of Water Sculpture

The landscapes of the earth's surface at any one time represent a balance between uplift of the land by the internal forces of crustal movement and destruction of the land by the external forces of erosion. Both of these processes act slowly—though at varying rates—over long periods of time. Therefore, in order to fully understand landscapes, we must know something of the historical sequence of uplift

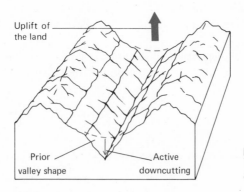

FIGURE 8.28
The process of rejuvenation as brought about by uplift of the land.

and erosion. Today, however, we can observe and study these processes at only a brief moment in the earth's long history, since interpretations of their interactions over millions of years involve many uncertainties.

The first and most basic question concerning the history of land sculpture is: How quickly does it take place? In other words, what is the rate at which the land surface is being lowered by erosion? Various schemes have been devised for estimating such rates, but none is fully satisfactory. Perhaps the most accurate is based on estimates of the amount of rock debris that is carried to the ocean each year by streams. Such estimates suggest that the United States, for example, is being lowered at an overall rate of about 8 cm (3 in.) per 1,000 years. Since the average height of the land surface on the entire earth is about 800 m (2,600 ft), such an erosion rate would reduce all the continents to sea level in about 10 million years. Geologically speaking, this is a very short time. There are, however, many difficulties in such estimates. One relates to the activities of man: the clearing of much of the land surface for agriculture during the last few hundred years has greatly increased rates of erosion—and by amounts that are difficult to estimate. In addition, the opposing internal forces that cause uplift of the land must be considered. Measurements of present-day rates of uplift suggest that the continents are now rising several times faster than they are being eroded. Moreover, the continents apparently stand higher today than they did through much of earth history; this suggests that rates of uplift and erosion may have been more nearly balanced in the past.

Perhaps the most serious problem in estimating erosion rates is caused by climatic change. Climate has a profound influence on erosion: under dry and polar climates, local erosion rates have been measured that are many times greater than the 8 cm/1,000 years average for the temperate United States. As yet, however, too little is known about overall erosion patterns under such conditions to permit close comparisons with rates derived from temperate regions.

Along with these difficulties in measuring merely the *rate* at which the land is being lowered, there are, as you might expect, even greater problems in reconstructing the exact *sequence* of landscape changes which result from the land being lowered by erosion and raised by internal forces over millions of years. Because no one has seen landscapes evolve over such a long period, all attempts at such reconstruction must be based on indirect evidence and intuition. One such intuitive scheme for reconstructing landscape history dominated the study of landscapes for many years. It was developed around the turn of the century by several American geologists, the most influential and articulate of whom was W. M. Davis of Harvard University. Davis visualized landscape development in temperate climates, where flowing water is the dominant erosional agent, as commencing on relatively flat surfaces, such as former sea floors, that were raised above sea level by tectonic forces (Figure 8.29). At first, he believed, streams would cut steep, V-shaped valleys without floodplains into the surface; later the valleys would widen, and broad floodplains separated by steep hillslopes would dominate; local relief would be at a maximum in this stage. Finally the hillslopes would be reduced by erosion to a relatively flat rolling surface which he called a **peneplane** (*pene* = almost, *plane* = flat). Landscapes showing each of these characteristics exist today in temperate zones, and Davis and his followers devoted much effort to classifying them according to their "stage of evolution" in his theoretical scheme of landscape development. This scheme, which involves such concepts as *youth*, *maturity*, and *old age*, described the degree to which the land had been destroyed by erosion. Most geologists now feel, however, that Davis's scheme was much oversimplified. Under certain conditions the sequence he envisioned might take place, but the complications of climatic change and the interaction of uplift and erosion probably lead to the formation of valleys and hillslopes by many complex sequences of events not included in Davis's appealingly simple analysis. There nevertheless remain general benefits to be gleaned from this simple "stage of landscape evolution" concept. High-standing crustal masses resulting from

FIGURE 8.29
The cycle of land sculpture as postulated by W. M. Davis.

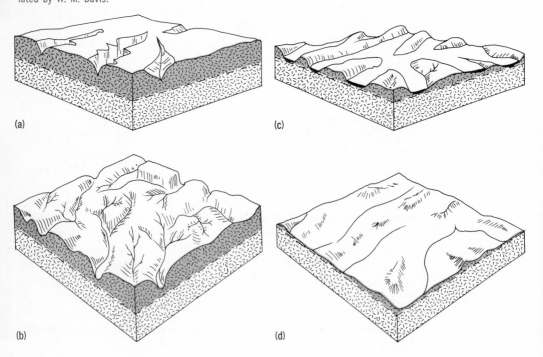

(a)

(b)

(c)

(d)

either deformation or volcanism and only slightly destroyed by stream erosion can correctly be recognized as "youthful"; others that display the effects of considerable denudation have greater maturity (Figure 8.30). It should be remembered, however, that such terms are relative and may be used only in a very broad sense.

(a)

FIGURE 8.30
Stages of landscape evolution. (a) Youthful (relatively little modification of terrain by stream erosion). (b) Early maturity (streams have fully established their drainage system; most of the original surface has been destroyed). (c) Late maturity (streams have widened valleys greatly by lateral erosion; topography is relatively subdued). (a: Tad Nichols; b: Santa Fe railway; c: Geological Survey of Great Britain.)

(b)

(c)

Weathering

8.1 *Weathering processes:* exposure to air and water at the earth's surface causes massive crustal rocks to break up into small particles and change composition by both mechanical and chemical processes.

8.2 *Soil:* weathering produces an accumulation of altered rock debris—mostly clay minerals, oxide minerals, and quartz fragments—that is thickest under tropical climates, thinnest under polar and dry climates, and of intermediate thickness under temperate climates.

Water on the Land

8.3 *Lakes:* lakes are temporary stationary accumulations of land water that show some properties of the ocean, but have relatively little influence on land sculpture.

8.4 *Groundwater:* much of the water on the land is in cracks, pores, and voids in soil and rocks; it directly influences landscapes only in limestone terrain

8.5 *Streams:* streams are the principal agent of land sculpture; the velocity and erosive effect of stream flow are controlled by a shifting balance among discharge, channel shape, channel slope, and sediment load.

Land Sculpture by Land Water

8.6 *Mass wasting:* soil and rock debris first move downslope under the direct influence of gravity—either slowly, by creep, or more rapidly, by various sorts of landslides.

8.7 *Stream erosion:* debris ultimately reaches flowing stream water that moves it from the area of origin; where slopes are steep, the debris is removed rapidly and the stream actively cuts a V-shaped valley; on more gentle slopes the debris accumulates to form an alluvial valley.

8.8 *The graded stream as an equilibrium concept:* the graded stream concept provides a basis for predicting changes in a stream's tendency to erode or deposit sediment in response to changes in its discharge or its channel shape, roughness, or slope.

8.9 *Rates and cycles of water sculpture:* stream erosion of the land surface is offset by mountain building and internal uplift; precise cycles of erosion and uplift are difficult to reconstruct.

ADDITIONAL READINGS

*Bloom, A. L.: *The Surface of the Earth*, Prentice-Hall, Inc., Englewood Cliffs, N.J., 1969. A brief introduction to landscapes.

*Gordon, R. B.: *Physics of the Earth*, Holt, Rinehart & Winston, Inc., New York, 1972. Chapter 5 provides a good introduction to land sculpture.

Leopold, L. B., M. G. Wolman, and J. P. Miller: *Fluvial Processes in Geomorphology*, W. H. Freeman and Co., San Francisco, 1964. An intermediate text on streams and their relation to landscapes.

*Morisawa, M.: *Streams: Their Dynamics and Morphology*, McGraw-Hill Book Company, New York, 1968. A brief introductory text.

Shelton, J. S.: *Geology Illustrated*, W. H. Freeman and Co., San Francisco, 1966. An excellent introductory text stressing landscapes; extraordinary photographs.

*Tuttle, S. D.: *Landforms and Landscapes*, William C. Brown Company, Dubuque, Iowa, 1970. A brief introductory text.

*Available in paperback.

chapter *9*

Oceans
and
Coastlines

As land-dwelling creatures, we tend to think of the dry surfaces of the continents as the feature most characteristic of the earth. An observer in space, however, is most impressed not with the earth's land areas, but with the enormous volume of ocean water covering more than seven-tenths of the earth's surface. Among the planets of the solar system, only the earth appears to have such huge quantities of liquid water. Water is present on earth not only in the ocean basins, but in smaller amounts on land, where it occurs in lakes, rivers, streams, and as groundwater saturating the uppermost levels of rock and soil. All of these waters make up a world-encircling sphere of liquid which is often referred to as the hydrosphere to distinguish it from the rocks of the underlying solid earth (Lithosphere) and the gases of the over-lying atmosphere. About 97 percent of the water in the hydrosphere occurs in the ocean, which, through evaporation and rainfall, is also the ultimate source of the 3 percent of water occurring on land.

OCEAN WATER

9.1 Chemical Composition and Physical Properties

In Chapter 1 we devoted many pages to describing the chemical substances that make up the solid earth, for even though only a few elements are common in minerals and rocks, these elements occur in many different proportions and combinations. In contrast, the elements making up ocean water do *not* occur in widely varying proportions and combinations, but are instead extremely limited in their variability. The most abundant elements of the ocean are shown in Figure 9.1. Hydrogen and oxygen, combined as water, are the most abundant, for the ocean generally consists of 96.5 percent pure water and only 3.5 percent of other elements, most of which are dissolved in the water to make a complex chemical solution. Although the 3.5 percent of dissolved material includes minute quantities of almost all the naturally occurring elements, chlorine and sodium predominate, accounting for 3 percent out of the total 3.5 percent.

The most important physical property of ocean water is its density (mass relative to volume). Density is extremely significant because the structure and movements of masses of ocean water are strongly influenced by slight density differences. The density of ocean water, in turn, is primarily dependent on three properties: salinity, temperature, and pressure.

Because most of the dissolved elements in ocean water are denser than pure water, the density of ocean water increases as the salinity increases. The effect of temperature on density is more complex. Most substances expand and become less dense when they are heated and, conversely, they contract and become denser when cooled. Pure water

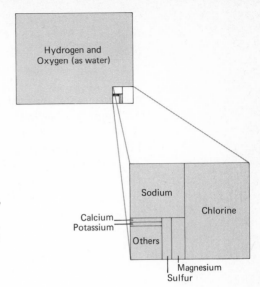

FIGURE 9.1
The most abundant elements in ocean water. Hydrogen and oxygen, combined as water, account for 96.5 percent of normal ocean water. The remaining 3.5 percent consists largely of elements dissolved in the water, particularly chlorine and sodium.

when cooled reacts in the normal way until it reaches a temperature of 4°C (39°F), just a few degrees above its freezing temperature. At that temperature, pure water reaches its maximum density; further cooling results in a *decrease* in density and an *expansion*, rather than contraction, of the water molecule. At the temperature of freezing, water (as ice) is about 10 percent less dense and occupies about 10 percent more space than at 4°C. This is why ice floats in water, rather than sinking, which is the fate of most substances that freeze from their own liquid.

Pressure, the final factor affecting the density of ocean water, is the least important of the three. Like all other substances, water increases in density as it is compressed into a smaller space by increasing pressure. Water, however, is a relatively incompressible substance; that is, increased pressure leads to very small decreases in volume and consequently to very small increases in density. The pressure in a vertical column of ocean water increases with depth because of the weight of the overlying water. Bottom waters in the deepest parts of the sea are under pressures 1,000 times as great as surface waters, yet water is so resistant to compression that the density is increased only a few percent. For this reason, pressure is far less significant in determining ocean water density than are temperature and salinity.

9.2 Ice in the Ocean

The formation of ice in seawater is an important structural property of the ocean. Ice crystals forming in the ocean are made up of almost pure water; their formation thus increases the salinity of the remaining water so that much lower temperatures are necessary for more freezing to take place. If this process continues, ice particles increase in abundance and freeze into a solid framework enclosing small cells of seawater which have too high a salinity to freeze at that temperature. The result is a frozen mixture consisting mostly of relatively pure ice crystals enclosing small quantities of salt and brine.

Complete freezing of ocean water in the manner just described sometimes takes place in shallow bays and lagoons in polar regions, but in areas of deeper water only a thin surface layer of sea ice, usually less than 3 m (10 ft) thick, forms in even the coldest polar seas (Figure 9.2). Such sea ice perpetually covers most of the Arctic Ocean and surrounds the ice-covered continent of Antarctica. It forms from the freezing of surface seawater and the accumulation of fresh water, as snow, on the surface. Sea ice seldom grows thicker than 3 m because as the waters beneath the ice are cooled, they increase in density, sink, and are replaced by warmer waters from below. Accumulation of snow and ice on the surface is also limited by the water temperature, for as snow accumulates on the surface, the ice below is depressed by the weight and an equivalent amount of ice is melted below by the warmer waters. Thus the thickness tends to remain constant even in areas of high snowfall.

In contrast to the thin sheets of sea ice that form on the oceans, ice on land, accumulating as snow on a solid surface with no water below to cause melting, can reach tremendous thicknesses. Most of Greenland and the continent of Antarctica are now covered by huge ice sheets that are more than a kilometer thick. In addition to these extensive ice sheets, most mountains of polar regions have valleys filled by smaller glaciers, thick rivers of ice that accumulate on land as snow. Where either ice sheets or glaciers touch the ocean, huge icebergs hundreds of meters thick may break away and float until they reach warmer water and melt. Icebergs (which along with sea ice form the two principal kinds of ice found in the ocean) therefore originate not from freezing of ocean water, but from the accumulation of fresh water, as snow, on the land surface. Most of the icebergs of the northern hemisphere occur in the North Atlantic and have their source in Greenland and arctic Canada. In the southern hemisphere, the huge Antarctic ice cap in some places flows beyond the boundaries of the continent as shelf ice floating on the surrounding seas. The largest icebergs of the ocean originate when huge masses of this ice break off and float northward (Figure 9.3). During the Antarctic winter, this thick, floating land ice is bounded on the seaward side by a much wider belt of thin sea ice similar to that which continually covers the Arctic Ocean.

(a)

FIGURE 9.2
Sea ice. (a) Off the coast of Antarctica;
note the ship breaking through thin,
transparent ice near the center of
the picture. (b) Close-up of Arctic sea
ice. (a: U.S. Navy; b: U.S. Coast Guard.)

(b)

(a)

(b)

FIGURE 9.3
Icebergs. (a) Surrounded by sea ice in the Arctic Ocean. (b) An enormous, tabular iceberg off the coast of Antarctica. (a: National Film Board of Canada; b: U.S. Navy.)

OCEAN MOVEMENTS

So far we have considered only the relatively stable, large-scale structure of the ocean, and have said little about more rapid and dynamic movements of ocean water. Such movements are of two sorts: **currents,** which are motions that transport water from one part of the ocean to another; and **waves,** moving disturbances traveling on or through the water, like wind ripples crossing a field of wheat, which do not directly transport water from one place to another. Currents will be discussed first; the following sections then treat ocean wave movements, tides, and the infrequent but devastating seismic sea waves that are sometimes (incorrectly) called "tidal waves."

9.3 Currents

Current movements are closely related to the ocean's layered structure. The most obvious take place in the surface layer, the upper hundred meters or so of water where solar energy and interaction with the atmosphere are concentrated. Most knowledge of **surface currents** —as horizontal water movements occurring in this surface layer are called—has come from records of their effect on the movement of ships. From thousands of such observations, navigators have compiled **current charts** showing the general patterns of surface currents around the world. The first comprehensive charts of this kind were made in the mid–nineteenth century as aids to navigation; a simplified modern version is shown in Figure 9.4.

An examination of the surface current patterns in Figure 9.4 shows certain similarities of the two largest oceans. The principal features are large circular patterns of flow called **gyres.** The Atlantic and Pacific oceans each have three such gyres: a small gyre flowing to the left (counterclockwise) in their northernmost extensions, and two much larger gyres on each side of the equator, the northern one flowing to the right (clockwise) and the southern one to the left (counter-

FIGURE 9.4
Principal surface currents of the ocean. Cold currents are shown in darker color, and areas of faster cold and warm currents by heavier shading.

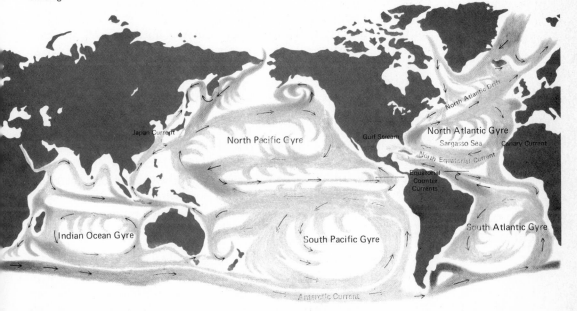

clockwise). Flowing along the equator between these large gyres in the opposite direction is a simple, one-directional current called the equatorial countercurrent. The Indian Ocean extends only a little bit north of the equator and has only the southernmost, left-flowing gyre well developed. The Antarctic Ocean has no gyres but merely a continuous eastward current, circling the entire earth around the continent of Antarctica. All of the major gyres persist in direction and undergo only relatively small-scale changes in position with the changing seasons; the smaller currents north of the equator in the Indian Ocean, however, completely reverse their directions in summer and winter.

The large gyres do not have formal names, but can be informally referred to as the "North Atlantic," "South Pacific," etc. gyres (Figure 9.4). Certain segments of each large gyre, however, have been given separate geographic names. For example, the western part of the large North Atlantic gyre is called the *Gulf Stream*, the northern part the *North Atlantic Drift*, the eastern part the *Canary Current*, and the southern part the *North Equatorial Current*. Usually only the parts with relatively rapid water movements are called *currents*; parts with slower movements are called drifts. Within each gyre is a large region of relatively stable and immobile surface water called an eddy. (The eddy within the North Atlantic gyre is called the *Sargasso Sea* because large quantities of the floating seaweed *sargassum* accumulate within its relatively quiet and stable waters; the eddies within the other major gyres do not have common names.) One other important fact concerning the ocean's surface currents is that all of the gyres tend to be asymmetrical to the west. Instead of having their centers in the middle of the ocean basins, the centers tend to be shifted to the left or westward in both the northern and southern hemisphere. As a result, the currents of each gyre tend to be narrowest and swiftest along their western sides. This is particularly evident in the Gulf Stream and the Japan Current—the *western boundary currents* (as they are called) of the North Atlantic and North Pacific gyres.

Of all the currents making up the gyres of the oceans, the Gulf Stream at the western edge of the North Atlantic has been by far the most extensively studied. Like other western boundary currents, it is made up of warm water flowing away from the equator at a relatively high speed, 6 km/hr (3.7 mi/hr); it somewhat resembles a "river" of warm water moving in a "channel" of cooler water. The "river" varies greatly in size but is typically very large, averaging about 80 km (50 mi) in width and several hundred meters in depth. Its boundary is often sharply marked by changes in water color or transparency. The Gulf Stream was long thought to be relatively constant in structure and position, but recent surveys have shown that, in fact, it is not a single mass of moving water, like a large river, but more like a series of small, rapidly flowing interconnected "streams" separated by stable water (Figure 9.5). The positions of the "streams" change rapidly, sometimes as much as 150 km in only a few days. It is the average flow of many such filamentlike streams over the wide region shown in Figure 9.4 that makes up the rapid northward movement of water in the Gulf Stream. Limited observations on the western boundary currents in other gyres suggest that they have structures similar to this.

9.4 Waves and Tides

Currents transporting water from one part of the ocean to another are not the only movements that take place in the ocean. Other movements occur which do *not* involve large-scale displacements of water, but merely the passage of moving patterns of disturbance, called **waves**, on or through the water. In watching surface waves on the sea, it at first appears that the water itself *does* move rapidly in the direction of wave travel. Closer observation of small objects floating in the water, such as a cork or a piece of seaweed, shows that it is merely the *form of the wave*, rather than the water itself, that progresses. As a wave passes, the cork and the individual water particles in which it floats move not forward, but in a vertically circular pattern that results in their ending up in approximately the same position as before the wave passed (Figure 9.6). Ocean wave movements therefore differ in a most fundamental way from the current movements we have already considered, for they do not involve large-scale transportation of water from one part of the ocean to another.

There are four principal ocean wave movements, each of which is caused by different dominant forces. The most familiar are *wind waves*, formed at the ocean surface by action of the wind, and *tides*, world-encircling waves caused by the gravitational attraction of the

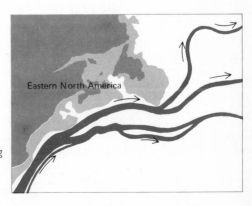

Eastern North America

FIGURE 9.5
Details of current flow in the northern Gulf Stream. The flow resembles a series of thin shifting "filaments" of north-flowing warm water. (After Munk, "Scientific American," vol. 193, 1955.)

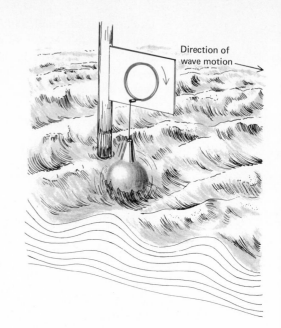

FIGURE 9.6
Water motion in wave passage.
An object floating in passing
waves does not progress with
the wave but inscribes an in-
place circle (color) with each
wave.

sun and moon. Somewhat less familiar are *seismic sea waves* (also
called *tsunamis* and, incorrectly, "tidal waves") caused by earthquakes, volcanic eruptions, or other disturbances of the ocean floor.
Internal waves, the final type, move beneath the surface of the ocean
and are the most difficult to observe and interpret: their exact cause
is still unknown.

Wind waves   The waves produced on the ocean's surface by the action of winds, called **wind waves**, are among the most fascinating and
easily observed phenomena of the ocean. Rather surprisingly, though,
they are still not fully understood. In the simplest case, any regular
sequence of moving waves can be described by three properties: the
length of the waves or *wavelength*; the *height* of the waves; and the
period of the waves, which is merely the length of time it takes one
complete wave to pass a fixed point (Figure 9.7a). From two of these
parameters, wavelength and period, the *velocity* of the waves can easily be computed. Likewise, the ratio of wavelength to wave height
determines the *steepness* of the wave (Figure 9.7b). Wind waves typically have wavelengths varying from less than 3 cm (1⅕ in.) to more
than 600 m (2,000 ft). The maximum steepness is reached when the
wavelength is about seven times as great as the wave height. At or
below this ratio the wave form becomes unstable and the water
plunges from the top to form a "whitecap" or "breaker."

Wavelength, height, and period are useful concepts for describing

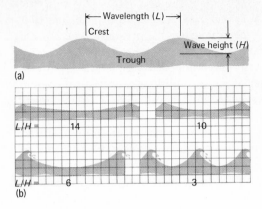

FIGURE 9.7
Wave properties. (a) Wavelength and wave height. (b) The ratio of wavelength to wave height determines the steepness of a wave. When the ratio is less than 7, the crest becomes unstable and plunges to form a "whitecap" or "breaker."

and discussing waves, but natural wind waves are usually far too complex and irregular to be understood by these properties alone. If you watch waves generated by a strong wind, you will see that it is impossible to follow single crests moving over long distances. Instead, the waves tend to grow and subside in complex patterns that defy simple recognition of wavelength and period. In considering such complex waves it is convenient to look at a completely smooth ocean surface over which a gentle wind has just begun to blow. The first effect is the production of very small surface irregularities less than 3 cm ($1\frac{1}{5}$ in.) across known as **capillary waves** (Figure 9.8). These capillary waves are so small that they quickly disappear if the wind subsides. If the wind speed increases and the winds become more turbulent and gusty, much larger and more irregular waves, ranging in length and height from several centimeters to a meter or more, begin to form (Figure 9.8). Unlike small capillary waves, these larger waves will continue to move under the influence of gravity even after the wind subsides; for this reason they are called **gravity waves.** With further increases of wind velocity the larger gravity waves, for reasons that are poorly understood, tend to gain energy at the expense of the smaller ones, with the result that the smaller waves are destroyed soon after they appear, whereas the larger waves continue to grow. The maximum height of these larger waves has been observed to depend on three characteristics of the generating wind: its *velocity;* its *duration,* or the length of time it blows; and its *fetch,* which is the distance over which it blows. The height increases as each parameter —wind velocity, duration, and fetch—increases. The maximum observed height of wind waves is rarely greater than 6 m (20 ft), although exceptional storm waves as high as 36 m (120 ft) have been reported (Figure 9.9). Even 6-m (20-ft) waves are unusual, for they require winds blowing at velocities greater than 80 km/hr (50 mi/hr), lasting for several days, and blowing over hundreds of kilometers of open water. It has been estimated that 45 percent of ocean waves are under 1 m (3.3 ft), 35 percent from 2 to 6 m (6.5 to 20 ft) high, and only 10 percent have heights greater than 6 m (20 ft).

Complex natural waves are considered to result from the interaction of many superimposed simple wave patterns (Figure 9.10). For a given set of wind conditions, a statistical combination of many such simple waves permits a calculation of the resulting distribution of energy in more complex waves of differing sizes. From this energy distribution, the general shape of the sea surface can then be estimated. These calculations agree rather closely with observations of natural waves and have formed the basis for an extremely useful scheme for predicting the maximum wave size to be expected under different conditions of wind velocity, duration, and fetch.

The complex wave patterns that we have been discussing occur in the **generating area**, the region where strong winds produce wave motions (Figure 9.11). Waves may, however, travel far beyond the generating area into calmer regions where large waves are no longer being produced by the wind. In this way, large storm waves often travel across hundreds or thousands of kilometers of open ocean with relatively little loss of energy. In the process of spreading from the generating area, the waves become more regular and are easier to describe by the simple properties of wavelength, height, and period. Such waves are called ocean **swell** to distinguish them from the more complex waves of the generating area, which are known as a **sea**.

Waves traveling across the open ocean as swell continue to move with little loss of energy until they reach shallow coastal waters, where their energy is dissipated by interactions with the ocean floor. There is a simple relationship for calculating the water depth at which surface waves begin to "feel bottom" and lose energy (Figure 9.12). As a wave passes a given point, the motion of water particles extends

(a)

(b)

(c)

FIGURE 9.8
Sequence of wind wave formation. (a) Small capillary waves; wind velocity 4 km/hr. (b-f) Gravity waves at increasing wind velocities: (b) 10 km/hr; (c) 30 km/hr; (d) 45 km/hr; (e) 80 km/hr; (f) 95 km/hr. (Photographs courtesy of Dr. F. Krugler, Hamburg.)

(d)

(e)

(f)

FIGURE 9.9
Unusually large waves in a storm off northern California. The ship, a Coast Guard Cutter on a rescue mission, is 100 m (327 ft) long. (U.S. Coast Guard.)

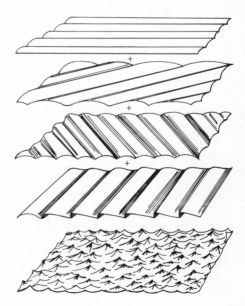

FIGURE 9.10
Wave averaging. A series of simple wave forms are statistically "added" to simulate complex natural waves (bottom).

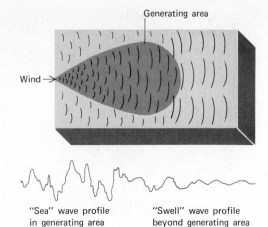

FIGURE 9.11
Development of waves in a "generating area" of strong winds. The waves may move far beyond the generating area; as they do they become more regular and are known as "swell."

"Sea" wave profile in generating area

"Swell" wave profile beyond generating area

downward, with decreasing energy, to **wave base**, a depth about half as great as the wavelength of the passing wave. Thus a wave 300 m (1,000 ft) long begins to "touch bottom" in water 150 m (500 ft) deep. When this happens, the form of the wave begins to change because the bottom restricts the movement of water particles in the wave: the velocity decreases, the wavelength becomes shorter, and the wave height increases. This steepening of the wave continues as the water becomes shallower until the wave reaches an unstable height and "breaks" to form surf, an action that dissipates most of its energy (Figure 9.13). Breaking normally occurs when the water depth is about $1\frac{1}{3}$ times the wave height.

Tides A second fundamental ocean wave motion is seen in the **tides**, world-encircling ocean movements caused by the gravitational attraction of the sun and moon. The rhythmic rise and fall of the tides seen along the coastlines of the earth are the most regular and, next to surface waves, the most easily observed of all ocean movements. Since at least the seventeenth century men have known that these move-

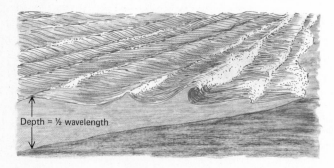

Depth = ½ wavelength

FIGURE 9.12
Wave motions in shallow water. At water depths less than ½ wavelength, the wave motion intersects the ocean floor causing the wave to move more slowly and, ultimately, to steepen and "break."

FIGURE 9.13
Waves steepening and breaking in shallow water along the central California coast. (Josef Muench.)

ments are related to astronomical forces, for they observed that regular tidal changes accompany the changing phases of the moon. Even though it has long been clear that tides are caused by the changing positions of the moon and sun, however the exact manner in which the gravitational forces produce varying tidal patterns is still not completely understood.

To visualize the origin of ocean tides it is helpful to think first of the effect of the moon alone on an idealized earth completely covered by oceans. On such an earth, ocean water should bulge slightly outward in the direction of the moon because of the moon's gravitational attraction. At first it would seem that only a single bulge would exist directly under the moon, and indeed this would be the case if gravity were the only force involved. The mutual gravitational attractions of the earth and moon, however, are balanced by *centrifugal* forces caused by slight motions of the earth around the mutual center of gravity of the earth-moon system. This centrifugal motion creates a force in a direction *opposite* to the moon and results in another tidal bulge on the side of the earth *farthest away from* the moon (Figure 9.14). The result of this balance of centrifugal and gravitational forces is to permanently deform the ocean surface into a slightly ovate (egg-like or football-like) shape, with the long axis extending in the direction of the moon. This deformation can be thought of as world-encircling waves with only two "crests," one at each tidal bulge, and two "troughs" in between. The forces involved in producing the tides are

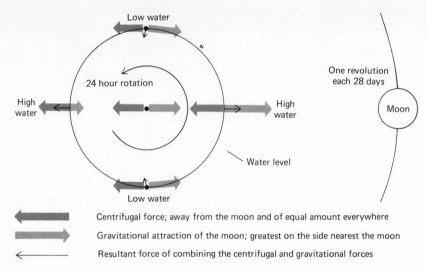

FIGURE 9.14
Origin of the tides. The gravitational interactions of the earth and moon produce a low, world-encircling wave with two crests (high water) and two troughs (low water). The earth rotates "under" these waves each 24 hours, producing two intervals of high water and two of low water at each point on the ocean surface.

so small that the height of these waves is usually less than 1 m (3.3 ft) in the open ocean, but their wavelength is extraordinarily long, for they extend halfway around the earth from one bulge to the other.

Daily tidal changes result from the fact that the tidal bulges remain in a relatively fixed position—that is, aligned with the moon—as the earth rotates on its axis every 24 hours. Thus, points on the earth's surface tend to rotate under both tidal bulges, and their intervening "troughs," every 24 hours. This is the reason that tidal records typically show two periods of high water and two intervening periods of low water each day. The tidal bulges themselves also move, however, as the moon rotates around the earth every 27 days. The effect of this movement is to make the tides, like the time of moonrise, come about 50 min later each day, thus completing one cycle and occurring at the same time once each 27 days (Figure 9.14).

This simple explanation of tides has neglected many complicating factors, the most important of which is the sun, whose gravitational attraction causes about *half* as much tidal deformation as does the moon. The principal effect of the sun is to cause tides to have their greatest range twice each month at **spring tides*** (near the times of full and new moon) when the sun and moon are aligned and their forces act together; conversely, tidal ranges are lowest at **neap tides** (near the first and third quarter of the moon) when the solar and lunar forces are in opposition and act at right angles to each other (Figure 9.15).

There are still other factors that complicate natural tidal patterns.

*Spring, in this usage, has nothing to do with the season of the year.

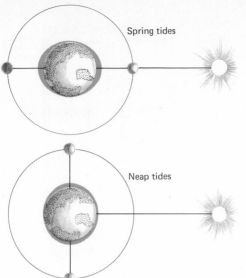

FIGURE 9.15
Spring and neap tides. Because of the gravitational attraction of the sun, tide ranges are greatest when the moon and sun are aligned (spring tides) and least when they are opposed (neap tides).

The earth is not, of course, completely covered by water, and thus the movement of the tidal "bulges" around the earth each day is strongly influenced by the shape and interconnections of the ocean basins. Furthermore, the extremely long wavelength of the tides causes them to "feel bottom" and behave as shallow water waves even in the deepest ocean basins. As a result, free movement of the tidal deformations are slowed down by frictional drag on the sea floor. Because of these and other complications, it has so far proved impossible to predict local tidal patterns from purely theoretical considerations. Instead, tidal predictions are based on long observation of local patterns, which tend to be repeated in regular ways along the coastlines of the earth.

Tidal observations taken over many years at thousands of stations around the world have shown that many regions, including most of the Atlantic Ocean, show a regular twice-daily tidal pattern with high and low tides of equal magnitude, just as predicted by our simple analysis. Such tides are called **semidiurnal** (Figure 9.16). In other areas, however, various complicating factors lead to two additional tidal patterns. Most common are *mixed tides*, having two highs and two lows of unequal magnitude each day. Such tides dominate most of the shores of the Pacific Ocean. In a few places, such as the Gulf of Mexico and Southeast Asia, there occur *diurnal tides*, with only one high and one low water per day.

In the open ocean, the rise and fall of water with the changing tides is difficult to measure, but tidal observations on isolated oceanic

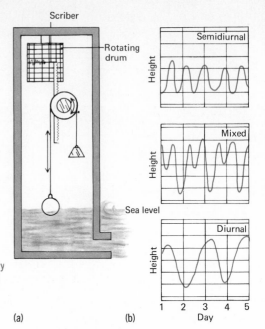

FIGURE 9.16
Tidal patterns. (a) Schematic diagram of a recording tide gauge.
(b) The three principal types of
tides, as recorded on a tide gauge.
Semidiurnal tides have two high
and two low waters of approximately
equal magnitude each day. In mixed
tides, some of the highs and lows
are unequal. Diurnal tides show
a single high and low each day.

islands suggest that the water seldom varies in height by more than
1 m (3.3 ft). In contrast, along the coasts of the continents much
greater tidal ranges may occur when water movements become concentrated and focused by the shape of the land and by interaction
with the shallow sea floor. The maximum tidal ranges so produced
are found on the shores of the North Atlantic, particularly in Nova
Scotia and around the English Channel, where the maximum range is
about 15 m (50 ft) (Figure 9.17). On most shores, however, the maximum tidal range is less than 3 m (10 ft).

9.5   Seismic Sea Waves

All of the ocean movements we have discussed so far have been
caused by either atmospheric forces (currents and wind waves) or external gravity forces (tides). Another ocean wave motion occurs that
derives energy directly from earthquake movements taking place in
the rocks of the solid earth. Such waves are commonly called seismic
sea waves, but are also often referred to by their Japanese name,
tsunamis, or by the misnomer "tidal waves." Seismic sea waves have
no relation to tide-producing forces; they do, however, resemble tides
in having very low heights and very long wavelengths. In the open
ocean they commonly have wavelengths of hundreds of *kilometers*

FIGURE 9.17
High and low tide in Nova Scotia, an area of extreme tide ranges. The wall in the right photo is about 8 m (24 ft) high. (G. Blouin, "Information Canada Photothèque.")

but heights of less than 3 m (10 ft), so low that the crests pass unnoticed by ships in mid-ocean. Unlike tides, however, the height of seismic sea waves may be greatly intensified as the waves move into shallow water. This topic will be developed more fully in our consideration of geologic hazards to man in Chapter 12.

COASTAL EROSION

While the properties of ocean water and of its waves, tides, and currents discussed so far in this chapter are important to an understanding of the earth, of primary concern to the geologist are the erosional landforms that develop along coastlines and the story they have to tell about crustal deformation and sea level changes. The principal agent in the production of coastal landforms is surf action.

9.6 Surf Action

As sea waves approach a shoreline where the water shallows, waves build up and eventually "break" to become a rush of water onto the land. Waves undergoing this process produce surf. *Surf action* is the dominant factor in coastal erosion. The principal mechanisms of surf erosion are the *abrasive action* of sand, silt, and pebbles against the coastal bedrock and the *hydraulic impact* of the water itself as it generates extreme pressures within the many small bedrock joints. Waves, to a lesser extent, contribute to the abrasional process by slowly grinding clastic grains back and forth in the zone where the wave base touches bottom.

This relentless attack upon the land by the sea causes any high-standing or emergent coastal terrain to undergo predictable erosional changes. First, a **marine terrace,** consisting of a surf-cut bench and a surf-cut cliff, forms rather rapidly (Figure 9.18a). Along coastlines

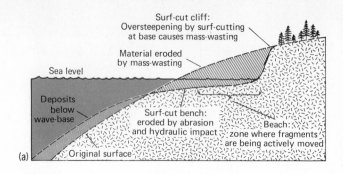

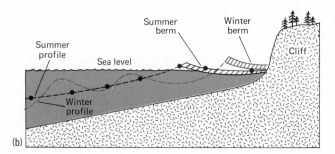

FIGURE 9.18
Shoreline profiles. (a) Marine terrace.
(b) Beach profile. Winter berm is narrower,
higher, and composed of larger fragments.
(After Bascom, 1964.)

where the land has recently undergone repeated upward movements
relative to the sea, several marine terraces can be seen extending hundreds of feet above the present surf zone (Figure 9.19). These relationships are common along the western coast of the United States.
How would it be possible to know if such elevated marine terraces

FIGURE 9.19
An elevated marine terrace (San Clemente
Island). (John S. Shelton.)

had resulted primarily from uplift (deformation) of the land rather than from sea level lowering? Although combinations of these two factors can give rise to complex relationships, there is one straightforward test that can be used to *exclude* a purely sea level–change mode of origin. When sea level lowers, it drops the same amount everywhere, so we would expect to find terraces formed by sea level lowering to be perfectly horizontal. Along the western coast of North America such horizontality is almost never observed. Instead the terraces are warped and in some cases severely deformed, thus proving the important role played by deformation in their origin (Figure 9.20).

9.7   The Beach and Longshore Transport

The beach is a coastal zone within which fragments are actively moved both onshore and offshore, as well as parallel to the shore. It extends from the base of the surf-cut cliff seaward to some variable position controlled by the wave base (Figure 9.18b). Because winter waves and surf are more energetic than summer waves, we should expect the beach's appearance to vary with the seasons. Indeed, not only is the summer berm (the layman's notion of "beach") wider and lower than in winter, as shown in Figure 9.18b, but also we frequently observe dramatic differences in fragment sizes exposed in the berm from season to season (Figure 9.21). High-energy winter surf not only moves more material offshore, it also selects all but the coarsest grain sizes for transport. During summer, the lower-energy waves and surf slowly push the fine-grained material back up toward and onto the shore.

FIGURE 9.20
A warped terrace, south of San Francisco. (N. T. Hall.)

(a)

(b)

FIGURE 9.21
La Jolla Beach. (a)
Summer. (b) Winter.
(Photos courtesy John
S. Shelton. From
"Geology Illustrated"
by John S. Shelton.
W. H. Freeman and
Company.Copyright ©
1966.)

In defining a beach, we suggested that its fragments move parallel
to the shoreline, as well as inward and outward from it. This **longshore transport,** as it is called, results from the fact that waves seldom
approach a coastline from straight out at sea. Wave fronts usually
approach the coast at an angle, and although they are typically refracted by being slowed in shallow water so as to become almost parallel, a small longshore-transport component persists (Figure 9.22).

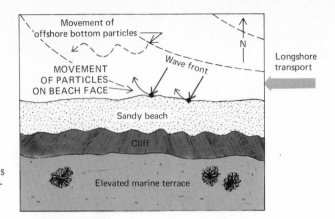

FIGURE 9.22
Map view showing longshore transport resulting from waves and surf with a westward component of movement.

Longshore transport often combines with local currents in shallow coastal waters to redistribute coastal sediments and to shape them into characteristic **barrier islands** (Figure 9.23).

The process of wave refraction contributes to another important aspect of a coastline's development. Suppose an irregular coastline

FIGURE 9.23
A barrier island along the Texas coast.
(John S. Shelton.)

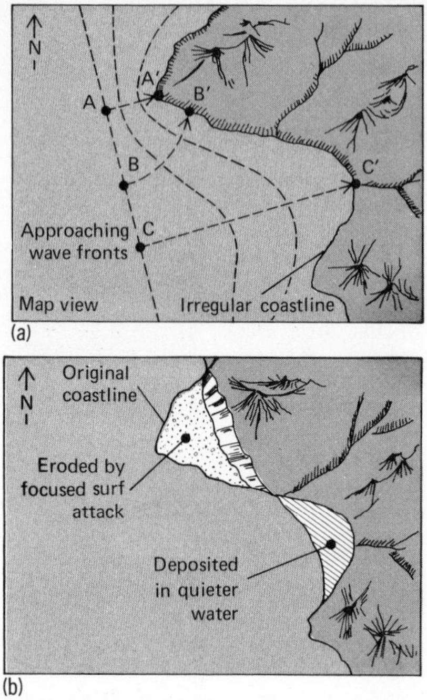

FIGURE 9.24

Effects of surf action. (a) Wave energy from AB is focused on the promontory at A'B', while BC energy is dispersed along B'C' segment. (b) The net effect of wave refraction plus time is to produce a straightened coastline.

comes into being, perhaps by deformation or sea level changes. How will such a coastline evolve? Will it become more or less irregular through surf attack? In view of the patterns of wave refraction shown in Figure 9.24a, it is apparent that the net effect will be to reduce the irregularities and straighten out the coastline by focusing energy at promontories and dispersing energy in embayments (Figure 9.24b). (Recall from Figure 6.7 that we can predict the direction of wave refraction in this manner: the wave will bend in the direction of whichever end of the wave front is slowed first.) Other aspects of longshore transport and surf attack that more directly affect man will be discussed in Chapter 12.

9.8 Continental Margins

Seaward from the narrow coastal zone, the continental margins slope gently downward and are covered by progressively deeper ocean water. At the edges of these submerged continental margins, which are called the continental shelves, there is usually a much steeper zone, called the continental slope, leading downward to the deep ocean basins (Figure 9.25). At their outermost edges, most stable continental

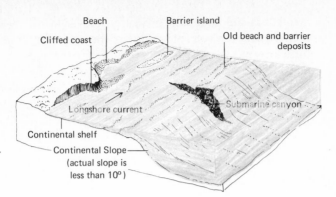

Beach   Barrier island

Cliffed coast   Old beach and barrier deposits

Longshore current   Submarine canyon

Continental shelf

Continental Slope
(actual slope is
less than 10°)

FIGURE 9.25
Characteristic landscape features of coasts and continental shelves. Vertical scale is greatly exaggerated.

shelves today are covered by about 200 m (660 ft) of ocean water; in tectonically active areas this depth is highly variable. Quite significantly, however, the size of the continental shelves has changed dramatically with the climatic fluctuations of Pleistocene time because the large continental ice sheets locked up enough water to lower worldwide sea levels by at least 100 m (330 ft) and, possibly, by as much as 200 m (660 ft). During intervals of maximum glaciation, much or all of the present continental shelves were therefore exposed above sea level, and the coastline lay offshore at points now deeply submerged. Conversely, it is estimated that if the present Greenland and Antarctic ice sheets were to melt, enough water would be added to the oceans to *raise* sea level about 50 m (165 ft), thus flooding much of the earth's present coastal regions. The ice sheet fluctuations of Pleistocene time have therefore led to numerous past migrations of the shoreline back and forth across the continental margins; the most important result has been to deposit beachlike sediments over much of the shelf region.

Although depositional processes have dominated the continental shelves, the outermost margins of many of them show dramatic erosional canyons that are deeper and steeper than any occurring above sea level (Figure 9.25). The exact origin of these submarine canyons is still uncertain; some geologists believe they were formed, in their uppermost parts at least, by streams flowing on dry land during intervals of maximum glaciation when sea level was much lower. Others think it more likely that they have been carved entirely by sediment-laden bottom currents of ocean water moving down the steep continental slope.

Beyond the steep margins of the continents, the deep ocean basins are regions in which the "land" surface is little affected by erosion. Instead, the construction of new crustal rocks by volcanic action and ocean-floor spreading, and the accumulation of land-derived sediments, are the dominant processes. Nevertheless, active erosion does occur whenever volcanic rocks of the ocean build above sea level to form oceanic islands. Over many years the tops of volcanic islands may be eroded and leveled by wave action. Such wave-eroded, flat-topped volcanic mountains, called *guyots* (see Section 7.3 and Figure 7.13), are a common feature of the ocean floor. The tops of the youngest guyots stand near present sea level, but older ones have been deeply submerged by the movement of ocean floor away from mid-ocean ridges.

SUMMARY OUTLINE

Ocean Water

9.1 *Chemical composition and physical properties:* ocean water is a uniform solution of many dissolved elements, of which sodium and chlorine are dominant; the concentration of dissolved elements defines the salinity of ocean water, which is normally $35^0/_{00}$ (3.5 percent); density, the most important physical property of seawater, increases with increasing salinity and pressure and with decreasing temperature.

9.2 *Ice in the ocean:* the dissolved matter in ocean water depresses the freezing point and prevents complete freezing; sea ice, formed from frozen seawater, is always relatively thin, even in polar regions.

Ocean Movements

9.3 *Currents:* currents are best known in the surface layer, where they move in large, circular gyres; the swiftest motions occur in the asymmetrical western portions of the gyres.

9.4 *Waves and tides:* wind waves are moving disturbances formed on the ocean surface by wind action; they increase in size as wind velocity, duration, and fetch increase and may move far beyond their generating areas to produce swell; tides are world-encircling waves caused by the gravitational attraction of moon and sun; points on the ocean surface normally have two high and two low tides daily, but this pattern is complicated by the shapes and limited interconnections of the ocean basins.

9.5 *Seismic sea waves:* these are large waves caused by undersea earthquakes.

Coastal Erosion

9.6 *Surf action:* surf action is the dominant factor in coastal erosion; marine terraces are formed by prolonged surf attack.

9.7 *The beach and longshore transport:* the beach is kept in constant motion by longshore transport.

9.8 *Continental margins:* continental shelves and slopes connect the beach with the deep ocean basin.

ADDITIONAL READINGS

*Bascom, W.: *Waves and Beaches*, Doubleday & Company, Inc., Garden City, N.Y., 1964. A popular and entertaining introduction.

Bird, E. C. F.: *Coasts*, The M.I.T. Press, Cambridge, Mass., 1969. A brief intermediate-level review of coastal land forms.

Ericson, D. B., and G. Wollin: *The Everchanging Sea*, Alfred A. Knopf, Inc., New York, 1967. A well-written popular survey of the study of the ocean.

*Gross, M. G.: *Oceanography*, Charles E. Merrill Publishing Co., Columbus, Ohio, 1971. A brief introductory survey.

Gross, M. G.: *Oceanography, A View of the Earth*, Prentice-Hall, Inc., Englewood Cliffs, N.J., 1972. A comprehensive and up-to-date intermediate text.

*Moore, J. R. (ed.): *Oceanography, Readings from Scientific American*, W. H. Freeman and Co., San Francisco, 1971. Many excellent popular articles.

*Turekian, K. K.: *Oceans*, Prentice-Hall, Inc., Englewood Cliffs, N.J., 1968. An authoritative and readable introduction, emphasizing ocean chemistry.

*Available in paperback.

chapter **10**

Landforms Resulting from Climatic Extremes

In Chapters 8 and 9 we discussed what might be called the "normal," or most common, modes of erosion and the landforms they create. Stream processes are indeed the principal agents of land sculpture, and surf and wave action take their toll wherever the land meets the sea. In this chapter we shall consider erosional processes and effects that are characteristic of two climatic extremes: deserts and glacial regions.

DESERTS

Worldwide belts of dry climate tend to develop in the zone of descending air that lies between the moist, rising air of the tropics and the turbulent mid-latitude belts of cyclones and anticyclones. These zones of dry climate are characterized by an annual amount of surface evaporation which equals or exceeds the annual amount of rainfall. Under these conditions, most of the scant precipitation returns rather quickly to the atmosphere, and there is no excess to form permanent streams or support permanent vegetation. The results are deserts, areas with little or no moisture or vegetation which today occupy about one-fourth of the land surface (Figure 10.1).

Although we normally think of deserts as being warm, high temperatures play a role in the origin of deserts only through increasing the annual amount of evaporation. For example, regions on the warm, equatorial side of the dry climate zone may receive 75 cm (30 in.) of rainfall a year, about as much as falls in Chicago or Seattle, and still remain deserts because of their high average temperatures and consequent high rates of evaporation. In contrast, on the cooler, poleward side of the dry zones, rainfall must normally be less than 35 cm (14 in.) per year to cause complete evaporation and desert conditions.

10.1 Desert Landscapes

Although the rain that falls in desert regions evaporates relatively quickly—usually in a few hours or days—it nevertheless plays a dominant role in shaping desert landscapes. Because there is little soil or vegetation to absorb the rainfall, it tends to remain on the surface where, particularly in desert mountains, it may be concentrated in normally dry channels to cause *flash floods* which subside in a few hours. Such floodwaters commonly have an enormous discharge and velocity, and thus transport large volumes of sedimentary debris to the base of desert mountains where it is deposited in fan-shaped wedges called alluvial fans.*

*This and other sedimentary deposits of desert origin were described in Section 3.6.

FIGURE 10.1
The principal present-day desert regions. ■ Deserts   ▨ Areas of sand accumulation

Figure 10.2 shows the relationship between an eroded fault scarp in Death Valley, California, and alluvial fans formed at its base. Several features typical of arid mountainous regions are apparent in this photograph: (1) little or no vegetation; (2) an absence of thick soils over the bedrock; (3) a sharp break in angle between the eroded bedrock and the surface of the overlapping fans, which are forming an alluvial apron, or bajada; and (4) white evaporite deposits in the valley bottom. Each of these features owes its existence, either directly or indirectly, to the arid condition of the terrain.

FIGURE 10.2
Alluvial fans and an eroded fault scarp. (John S. Shelton.)

The absence of moisture directly causes sparse vegetation and thinner soils (water is necessary to weather rock chemically). The reduced chemical breakdown of bedrock creates a situation where only relatively large mechanically weathered rock is exposed at the surface, which gives rise to the steep bedrock slopes. The evaporite deposits result from the absence of through-flowing streams which, in more humid regions, would carry dissolved ions to the oceans and allow them to concentrate there. In arid regions, to the contrary, intermittent **playa lakes** are formed by infrequent downpours and quickly dry up, leaving an accumulation of evaporites (Figure 10.3). It is noteworthy that such *interior drainage* is testimony to the fact that deformational forces have defeated erosional forces. It seems likely that rates of mountain building in more humid regions are not any slower than in deserts, but that more precipitation allows greater rates of erosion to carve through-going drainage systems.

Figure 10.4 shows what is believed to be the evolutionary sequence of landforms observed in the arid Great Basin of western United States. (These changes are reminiscent of the stages of landscape evolution by streams discussed in Section 8.9.) Figure 10.5 introduces terminology that is applied to many of the features typically seen in such regions. As is apparent in Figure 10.4 and 10.5, the **piedmont** (*pied* = foot, *mont* = mountain) slope is underlain by two distinctly different kinds of material, although their line of contact is frequently imperceptible: **pediment** is eroded bedrock, often covered by a very thin veneer of slope-washed alluvium; and the **bajada** is the joining together of many alluvial fans into a broad alluvial apron. The precise erosional mechanism responsible for cutting pediments and maintaining their abrupt angle with the mountain face is not well understood at this time.

FIGURE 10.3
Evaporite minerals on a dry playa lake bed. (J. Ramsey, Omikron.)

(a)

FIGURE 10.4
Arid cycle of erosion. (a) Youth (with
relatively fresh scarps and individual
fans). (b) Maturity (with mountain outliers
and well-developed bajada). (c) Old age
(with inselbergs). (a, b, c: John S. Shelton.)

(b)   (c)

10.2   The Role of Wind

Among the most distinctive characteristics of desert landscapes are
those caused not by running water, but by the wind. Winds in desert
areas are not stronger or more persistent than in most other regions;
but because of the lack of vegetation and water, exposed desert sand
and clay are more readily transported by wind action. Under such
circumstances, wind provides a more constant and widespread eroding
agent than does moving water. Just as in flowing water, sedimentary
particles may be carried by the wind as either suspended load or bed
load. Because air is a far less viscous fluid than water, however, only
very fine sand or clay particles can normally be carried in suspension.
Strong winds blowing across desert alluvium can carry fine sand and
clay several thousand meters into the air in *dust storms*, and later de-

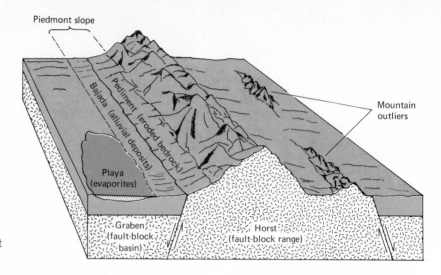

FIGURE 10.5
Desert landscape features in the Great Basin.

posit them hundreds of kilometers away as sheets of fine dust (Figure 10.6).

The removal of fine particles in this manner, called **deflation** (from Latin, meaning "to blow away"), leaves distinctive features behind.

FIGURE 10.6
Dust storm. (John S. Shelton.)

Figure 10.7 depicts the role played by deflation in the origin of **desert pavement** and the **ventifacts** (*venti* = wind, *fact* = face) that are commonly associated with it. **Sand dunes** are formed by the accumulation of clastic grains that are small enough to be moved, but too large to be blown away completely. Figure 10.8 displays some of the variety of dune forms and the conditions responsible for their origin. Although we normally think of sand dunes as particularly characteristic of deserts, they occupy a relatively small fraction of the earth's dry regions. Nevertheless, such sand concentrations can be impressive; the largest terrain of dunes, in the Sahara of North Africa, occupies an area larger than Texas.

(a)

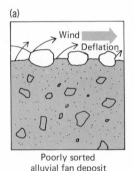

Poorly sorted
alluvial fan deposit

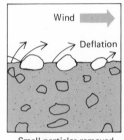

Small particles removed
and pebbles are "sand-
blasted" by wind blown
grains

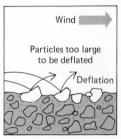

Surface is "armored":
desert pavement is formed.
Ventifacts are cut by
particle abrasion.

(b)

(c)

FIGURE 10.7
(a) Formation of desert pavement and ventifacts. (b) Desert pavement. (c) Ventifacts. (b: E. A. Hay; c: Sheldon Judson.)

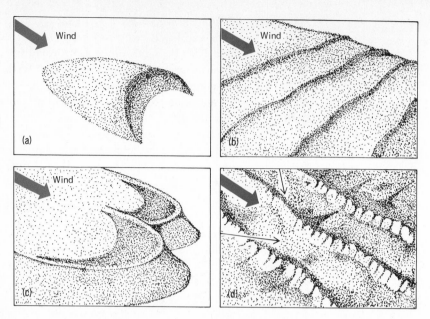

FIGURE 10.8
Sand dune forms.
(a) Barchans, which
occur on hard, flat
surfaces with limited
sand supply and a
steady wind direc-
tion. (b) Transverse
dunes, with abundant
sand supply and
constant wind direc-
tion. (c) Parabolic (U-
shaped) or "blowout"
dunes, which are
partially stabilized
by vegetation. (d)
Longitudinal or **seif**
dunes, which result
from variable wind
directions.

GLACIATION

Deserts form in regions where evaporation equals or exceeds precipitation. In tropical and temperate areas where evaporation does *not*
exceed precipitation, the yearly excess of water, being liquid, does not
accumulate, but moves downward to the ocean under the influence of
gravity. In polar regions, on the other hand, most of the precipitation
falls as solid water—snow; when the average yearly snowfall exceeds
the yearly loss from evaporation and summer melting, then the snow
continues to accumulate from year to year to form **glaciers** (Figure
10.9). Because of decreased temperature and increased precipitation,
glaciers can exist in high mountains even in the tropics. Mount Kili-
manjaro in East Africa rises to an elevation of 5,800 m (19,000 ft)
and has permanent glaciers even though located almost on the equator.

Although very high mountains everywhere and most mountains of
cool-temperate and polar regions today have extensive glaciers, it is
important to emphasize that much of the earth's polar regions is *not*
ice-covered. In many low-lying areas there is too little snowfall to accumulate even under extreme polar climates. For example, most of
cold, northern Alaska is ice-free, while glaciers are common in milder
southern Alaska, where there is more precipitation. In all, about 10
percent of the present land surface of the earth is covered by either
glaciers or ice sheets. During the maximum extent of Pleistocene ice,
about 30 percent of the land was so covered (Figure 10.10).

FIGURE 10.9
An ice sheet (background) on Baffin Island in the Canadian Arctic giving rise to two
converging valley glaciers (foreground). (Canadian Department of Energy, Mines, and
Resources.)

10.3   Glacial Movement

Glaciers are massive bodies of moving ice. As layers of snow accumulate to build glaciers, the structure and texture of the ice changes (Figure 10.11). Falling snow normally consists of delicate, lacy ice crystals; when these crystals first accumulate on the surface, they make up
a porous, powdery mass containing only about 20 percent ice and 80
percent air. Within a few days or weeks, the combined effects of evaporation from the surface of the snowflakes, daytime melting and night-
time refreezing, and compaction by the weight of overlying snow all
act to convert many powdery flakes into larger, roughly spherical ice
particles with the texture of coarse sand; these particles are known
as firn. At this stage the accumulated snow is concentrated to about
half ice and half air. As the firn is buried more deeply by subsequent

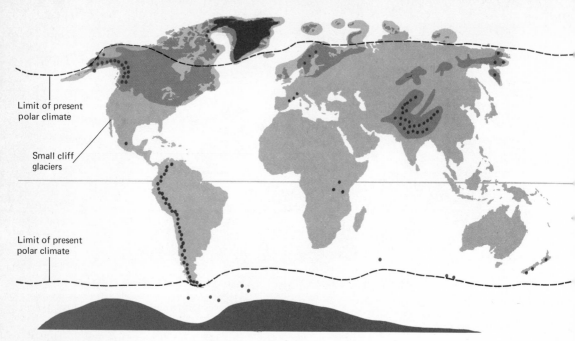

Limit of present
polar climate

Small cliff
glaciers

Limit of present
polar climate

FIGURE 10.10
Distribution of present-day valley glaciers (dots) and ice sheets (dark colored: Greenland and Antarctica). The maximum extent of Pleistocene glaciers is shown in lighter color.

snowfalls, the trapped air either escapes to the surface or is concentrated in large bubbles; at the same time the adjacent grains of firn tend to fuse. The final result is a solid mass of glacier ice containing only about 10 percent air. In addition to ice, most glaciers and ice sheets contain varying amounts of sedimentary particles; wind, mass wasting, and summer meltwater commonly deposit clay, sand, and gravel on the ice surface which then become buried by the next snowfall and incorporated, in layers, into the ice.

The mechanisms that permit solid glacier ice to flow like a very viscous fluid have been much investigated in recent years; these studies show that at least three separate mechanisms are important in glacier flow. Some movement is caused simply by *slipping* of the entire ice mass over the underlying bedrock. Because of the small but constant outflow of the earth's internal heat at the earth's surface, the basal portion of most glaciers is not solid ice, but a slightly melted mixture of ice and water. As a result of this basal melting, glaciers are seldom "frozen to" the underlying rock, but instead move over it on a lubricating film of water. A second factor in glacial movement is simple *fracturing* and *sliding* of large blocks of ice past one another. This fracturing takes place primarily near the surface, where the ice behaves as a brittle solid. Deep within the glacier the pressure of the overlying ice prevents brittle deformation; instead, the ice deforms

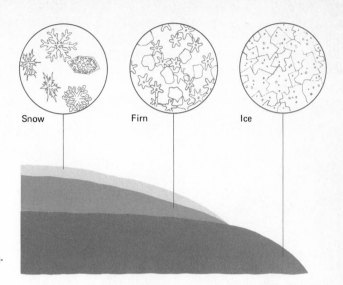

Snow   Firn   Ice

FIGURE 10.11
Structural changes during the transforma-
tion of snow to glacier ice.

by still a third process—*plastic flow*. We have seen that under high
pressures and over long periods of time the solid rocks of the earth's
crust slowly deform as if they were viscous liquids. Ice crystals under
pressure behave similarly, but have far less inherent strength than do
the silicate crystals of rocks; for this reason, ice flows even more
readily, and under far lower pressures, than does rock. As a rule, once
the ice attains a thickness of about 30 m (100 feet), flowage occurs.
This internal plastic flow is probably the most important factor in
glacial flow; it is known to involve several molecular mechanisms that
permit slow changes of shape in both single ice crystals and larger
ice crystal aggregates.

Careful measurements show that glaciers move at average rates of
a few centimeters to a few meters per day; the steeper the bedrock
gradient and the thicker the ice, the more rapid the flow. Under exceptional (and still poorly understood) circumstances the rate may be still
faster—the maximum observed movement is several hundred meters
per day. When glaciers occupy mountain valleys (Figure 10.12), rates
of movement are easily measured by observation of surveying mark-
ers placed on their surface; but huge **ice sheets,** such as those covering most of Greenland and Antarctica, are not confined by valleys,
and their flow is therefore not unidirectional but radially outward
from their high, dome-shaped centers. Because of their thickness (as
great as 4,000 or 13,000 feet) and size, flow measurements are far
more difficult to make on the Greenland and Antarctic ice sheets than
on valley glaciers, although it appears that the former flow much
more slowly.

FIGURE 10.12
Flow of glacier ice in mountains of northwestern Canada. Sedimentary debris on and within the moving ice clearly shows the patterns of glacial flow. (Bradford Washburn.)

FIGURE 10.13
Flow and dissipation of glacier ice. As the ice flows downward it eventually reaches lowlands where loss exceeds accumulation. Fresh snow and firn are absent and the older moving ice is dissipated by evaporation, melting, and runoff.

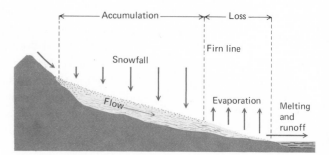

10.4 Valley Glaciers

Valley glaciers range widely in size from those fed by large ice sheets in arctic regions (Figure 10.9) to very small "glacierettes" or cliff glaciers found at high elevations in temperate zones (Figure 10.14).

All glacially modified valleys are U-shaped and headed by bowllike basins called cirques (Figure 10.15). These features owe their existence to the unique capabilities for erosion inherent in glaciers. Ice, being a solid, is able to *pluck* and *quarry* large rock fragments as it moves. Further, these fragments and other debris falling into the ice become powerful abrasives as a glacier scrapes them against the bedrock it is overriding, imparting a polished and grooved surface to it (Figure 10.16). Thus it is easy to visualize how a glacier is able to deepen and steepen the valley through which it moves, causing a typically V-shaped stream canyon to become U-shaped (Figure 10.17). But why do cirques have a steep headwall (upslope cliff)? Not much ice exists in that region to scour and cut such a precipice. The answer lies in the existence of the bergschrund, an air gap formed between the bedrock and the glacial ice due to the constant downslope movement of the ice (Figures 10.14 and 10.18). This is a zone of intense frost wedging, facilitated by the ready availability of water that seeps into rock fractures during the day and freezes at night. The headwall of the cirque is thus steepened by mass wasting into the bottom of the bergschrund, where the angular rock fragments are used to scoop out the depressions that, upon melting of the ice, are occupied by small lakes, called

FIGURE 10.14
Small glacier, showing bergschrund, in Sierra Nevada. (E. A. Hay.)

FIGURE 10.15
Cirques. The melting of valley glaciers exposes these bowllike
features of glacial sculpture. (John S. Shelton.)

FIGURE 10.16
Glacial polish and
grooves. (E. A. Hay.)

tarns, found in the bottoms of most cirques. Glacial **horns,** such as the famous Matterhorn in the Alps, are erosional remnants from cirque and U-shaped–valley development around the flanks of mountaintops (Figure 10.19).

We have seen that glacial modification of a stream valley imparts a U-shaped cross profile to it. A natural consequence of this valley wall steepening is the formation of **hanging valleys** and the impressive waterfalls that frequently pour over them (Figure 10.20). While the main valley glacier has the energy to effectively deepen, straighten, and steepen its valley, smaller tributaries are occupied by much thinner masses of ice, and are therefore unable to downcut at so great a rate. A comparison of the main glacier and its tributaries in Figure 10.12 makes this discrepancy apparent. When all the ice eventually melts away, the tributaries are left "hanging" above the major valley.

Waterfalls of another origin are also common in glaciated valleys. The waterfalls in Figure 10.21 occupy irregularities, or "steps," in the long profile of the glaciated valley, in contrast to hanging valleys which enter *into* such a valley. The origin of these irregularities resides in a glacier's unique ability to pluck and quarry large jointed fragments of bedrock in a manner beyond the capability of stream processes of erosion. Figure 10.22 depicts the manner in which jointing influences the glacier's excavation of a valley and the long profile steps that result. On a smaller scale, **roches moutonnées** (fleecy rocks) are asymmetric erosional remnants of similar origin that are frequently observed along the flanks and bottoms of glaciated valleys (Figure 10.23). The direction of glacial movement can be determined from their asymmetry because the upstream side is polished and the downstream side is plucked and steepened due to their different angles of exposure to the glacier's force.

10.5 Glacial Deposits

Material deposited directly from ice is called **till** (Figure 10.24a), and all glacially derived sediment—including stratified outwash from meltwater, lakes, and streams—is known as **drift,** a holdover term from early days when large glacial **erratics** (Figure 10.24b) were thought to have been floated (or drifted) over the central plains of Europe by icebergs and then deposited as the icebergs melted. An important and diagnostic characteristic of glacial till is its chaotic appearance. Glaciers possess so much energy that they transport every size of rock with equal ease, and when they melt all these fragments are immediately dumped. The result is, as can be seen in Figure 10.24a, a com-

(a)

(b)

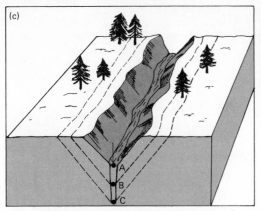

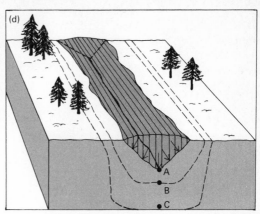

Stream erosion: down-cutting at arrow; mass wasting of sideslopes maintains Canyon shape as stream erosion deepens it from A to B to C.

Glacial erosion: all-around abrasion and plucking widens as well as deepens valley from A to B to C.

FIGURE 10.17
Conversion of U-shaped valleys by glaciation. (a) V-shaped valley. (b) U-shaped glaciated valley. Glaciers do not **initiate** valley cutting; they modify previously existing stream valleys. Note straightening of valley by truncating of ridge-ends. (c) Stream erosion. (d) Glacial erosion. (a: Geological Survey of Great Britain; b: Tad Nichols.)

pletely unsorted and unstratified pile of debris. When glacial deposits accumulate in various distinctive forms, the term **moraine**, used with an appropriate modifier, is applied: terminal moraines, lateral moraines, recessional moraines, and ground moraines are those most commonly seen (Figures 10.25 and 10.26). The ability of a glacier to be moving forward with a large sediment load and yet have its front margin (terminus) maintain a fixed position or even be retreating slowly (Figure 10.27) is what gives rise to the impressive ridgelike moraines shown in Figure 10.26.

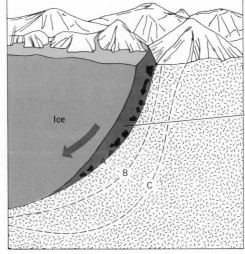

FIGURE 10.18
Bergschrund: the zone of intense frost-wedging activity at the head of a valley glacier. Valley head is modified from A to B to C.

FIGURE 10.19
Glaciated mountains with horns, cirques, tarns, U-shaped valleys. Wind River Range, Wyoming. (John S. Shelton.)

FIGURE 10.20
A hanging valley, Bridalveil falls in Yosemite. (E. A. Hay.)

FIGURE 10.21
Steps in the long pro-
file, Devil's Staircase
in Yosemite. (E. A.
Hay.)

Another group of stratified glacial deposits results from sediment transport and sedimentation by meltwater beneath a glacier and on its margins: eskers, kames, and outwash plains fall into this category (Figure 10.28). Two additional features of glaciated landscapes deserve mention: drumlins, which are mounds of morainal material

Smooth long profile resulting from stream erosion. Note variation in density of jointing in bedrock.

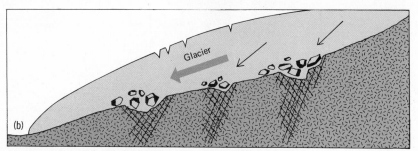

Where jointing is dense, blocks are small enough to be plucked out.

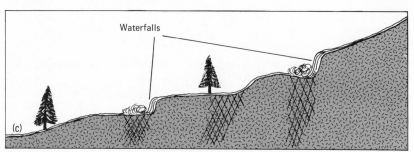

Melting of the glacier leaves stream valley with irregular long profile.

FIGURE 10.22
Development of steps in a long profile by glaciation.

that have been overridden and streamlined by ice; and **kettles,** formed by the melting of large blocks of ice that were incorporated with morainal debris (Figure 10.28).

THE GLACIAL EPOCH

Today glaciers are restricted to arctic regions and high elevations, but as recently as 20,000 years ago 30 percent of the earth's land surface was covered by huge ice sheets (Figure 10.10). The Pleistocene Epoch, which began about 2.5 million years ago, is also known as the glacial

(a)

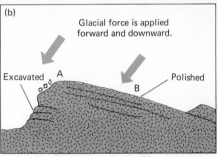

(b)

Glacial force is applied
forward and downward.

Excavated   A   Polished
             B

FIGURE 10.23
Roches moutonnées. (a) Lembert Dome in Yosemite. (b)
Diagrammatic sketch. At A, plucking and steepening occur
because the angle between the surface and the glacial
force is small and favors prying. At B, polishing occurs
because of the large angle between the surface and the
glacial force. (Photo by E. A. Hay.)

epoch, because during these past 2.5 million years the earth has experienced several extreme cycles of glaciation. Early study in Europe and
North America during the nineteenth and early twentieth centuries
suggested that there were four major advances and retreats of glacial
ice during the Pleistocene. These different glacial ages, as they were
called, were recognized primarily by analysis of different deposits of
glacial till that show greatly variable degrees of weathering, the most
highly weathered deposits obviously being older (Figure 10.29). More
recent investigations have used deep-sea cores that distinguish between cold-water and warm-water deposits, carbon-14 dating of
organic materials contained in various tills, and potassium-argon dating of volcanic rocks interlayered with morainal deposits in order to
establish a chronology of glacial and interglacial episodes. The results
have been startling: there have been many (perhaps eight or ten, or
more) major fluctuations of climate during the last two or three million years.

(a)   (b)

FIGURE 10.24
(a) Glacial till (near Lake Tahoe); note the poor stratification and sorting. (b) Glacial erratics (Yosemite): out-of-place boulders dropped by a glacier as it melted. (a: N. T. Hall; b: W. R. Cotton.)

It is only natural to wonder why these unusual conditions have occurred and what controls them. Before seeking a cause for the Pleistocene climates, however, let us inquire whether there were other such episodes in earth history? Have other major glaciations existed in the past? The answer is yes, but geologists are not sure exactly how many there were or precisely when they occurred. Recall from Section 7.2 and Figure 7.9 that there is excellent evidence for a period of late Paleozoic glaciation in the southern hemisphere. Somewhat sketchier indications of an Ordovician-Silurian glacial episode also exist in Africa. Many parts of the world have provided outstanding evidence for a late Precambrian glaciation, but without recourse to fossils geologists have found it difficult to determine if all these very old

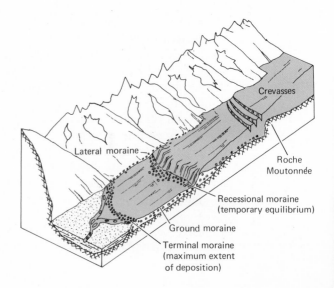

FIGURE 10.25
Origin of glacial moraines.

(a)

(b)

FIGURE 10.26
Glacial moraines. (a) Terminal moraine out of Bloody Canyon, Sierra Nevada (note its horseshoe-like shape). (b) Recessional moraine looking at Lyell Glacier, Sierra Nevada. (c) Dissipating margin of a glacier in the mountains of western Alberta. The ridge at left is a moraine; the sediments in the foreground form a small outwash plain. (a, b: E. A. Hay; c: National Film Board of Canada.)

(c)

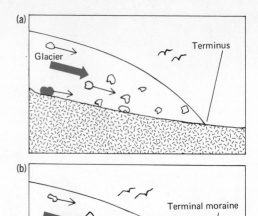

FIGURE 10.27
Development of a terminal or recessional moraine. (a) Equilibrium conditions: advance of ice equals rate of melting. (b) Note change in position of the dark boulder.

tillites (till that has been converted to solid rock) are of the same age: they could conceivably differ in age by 100 million years or more.

In any case, it is clear that glacial epochs have not been commonplace. They have occurred infrequently, without apparent periodicity, and were characterized by the affected areas undergoing several relatively abrupt temperature fluctuations between extremes varying from glacial conditions to environments perhaps somewhat warmer than those of today. What natural mechanism can account for these characteristics?

10.6 Hypotheses for Glacial Episodes

So many hypotheses have been proposed to explain glaciations that we shall not attempt to cite them all. The difficulty faced by all such efforts is that they deal with extremely long-term processes and with mechanisms that may or may not be operating today. In the Introduction we observed that geology is a historical science, and as such is confronted with problems that do not lend themselves to strategies of hypothesis testing by conducting controlled laboratory experiments. Our present problem is a classic example of this geologic condition. Since hypotheses concerning processes taking place over thousands or even millions of years cannot be easily tested, neither can they be easily rejected and dismissed. This is why so many of them "stay on the books," so to speak.

The most venerable of glaciation hypotheses are concerned with

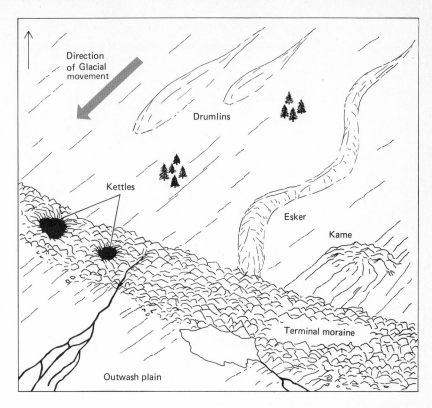

Direction
of Glacial
movement

Drumlins

Kettles

Esker

Kame

Terminal moraine

Outwash plain

FIGURE 10.28
Common features of
ice sheet glaciation.

past fluctuations of solar energy output and the possible effects of mountain building in redirecting patterns of atmospheric and oceanic circulations. The obvious difficulties with these views are that they are not subject to testing and do little to explain the many short-term variations that have been documented for the Pleistocene.

Other hypotheses, concerned with variations of atmospheric carbon dioxide, salinity changes in the upper oceans (thus changing the rate and amount of evaporation), and astronomical jugglings of earth motions and solar distances, are all stimulating, but untestable.

Even the new paradigm of plate tectonics has entered this arena. There are fairly good meteorological arguments suggesting that glacial advances are slightly favored if a geographic polar region is either wholly continental or wholly oceanic, a condition in which *both* polar regions found themselves throughout the Pleistocene. Might it be that the moving about of continents by plate tectonics positioned them in appropriate positions with respect to the poles during the late Precambrian and late Paleozoic glaciations? There are no reliable data available for a Precambrian reconstruction, but the late Paleozoic position of Pangea is compatible with this hypothesis (compare Figures 7.8 and 7.11). Even so, the only prudent statement we can make at

FIGURE 10.29
Sherwin Till, on the
east side of the
Sierra Nevada, with a
hammer imbedded in
a boulder, indicating
that the deposit is
relatively old.
(E. A. Hay.)

this time is that no *hypothesis* has thus far been tested sufficiently to be considered a satisfactory *theory* for glacial episodes.

10.7   Is the Pleistocene Over?

Let us close this chapter on climatic extremes by pointing out that we do not know whether our earth is about to embark soon on another glacial advance or retreat. Considerations of sea level change alone make this a very important question. There is good evidence that during the major advances of Pleistocene glaciers sea level was approximately 130 m (400 feet) below its present position. Crude estimates of the volume of water now locked up in the Greenland and Antarctica ice sheets indicate a likely rise of sea level of 40 to 50 m (130–165 feet) if this ice melts. Both of these potential changes, resulting either from glacial advances or retreats, present unsavory prospects to our coastal communities. The status quo is desirable, but any assurance that the status quo will persist, as well as the ability to influence the nature of any change in it, are beyond our present understanding.

SUMMARY OUTLINE

Deserts

10.1 *Desert landscapes:* periodic flash floods and the dominance of mechanical weathering give rise to distinctive desert landforms; alluvial
fans and playa lakes are common in the arid Great Basin of western
United States.

10.2 *The role of the wind:* constant wind erosion in the desert accumulates sand dunes and creates desert pavement by the process called
deflation.

Glaciation

10.3 *Glacial movement:* glaciers move by slipping over bedrock, by fracturing internally, and by plastic flow in response to high pressure
from overlying ice.

10.4 *Valley glaciers:* Many new landforms result when a stream valley
is glaciated: a U-shaped cross profile, hanging valleys, horns, cirques,
and irregular long-profiles are common.

10.5 *Glacial deposits:* deposits from a glacier's melting ice are typically
poorly sorted and poorly stratified; sometimes distinctive landforms
such as terminal and recessional moraines mark a position where the
rate of ice melting was the same as the glacier's rate of advance.

The Glacial Epoch

10.6 *Hypotheses for glacial episodes:* no wholly satisfactory hypothesis
has yet been proposed to account for the earth's history of glaciations.

10.7 *Is the Pleistocene over?:* worldwide sea level changes of plus or
minus 40 to 50 m could result from renewed melting or advancing
of present glaciers; either event would cause serious problems for
our coastal communities.

ADDITIONAL READINGS

*Bloom, A. L.: *The Surface of the Earth*, Prentice-Hall, Inc., Englewood
Cliffs, N.J., 1969. A brief introduction to landscapes.

Easterbrook, D. J.: *Principles of Geomorphology*, McGraw-Hill Book Company, New York, 1969.A standard intermediate text on landscapes.

Flint, R. F.: *Glacial and Quarternary Geology*, John Wiley and Sons, Inc.,
New York, 1971. A standard advanced text.

Shelton, J. S.: *Geology Illustrated*, W. H. Freeman and Co., San Francisco,
1966. An excellent introductory text stressing landscapes; extraordinary
photographs.

Thornbury, W. D.: *Principles of Geomorphology*, John Wiley & Sons, Inc.,
New York, 1969. A standard intermediate text on landscapes.

*Tuttle, S. D.: *Landforms and Landscapes*, William C. Brown Company,
Dubuque, Iowa, 1970. A brief introductory text.

*Available in paperback.

chapter **11**

Geology
and Man

As in all science, much of the geological knowledge summarized in this book was gathered to help quench man's intellectual curiosity, rather than to satisfy his physical needs. There are, nevertheless, two areas of vital human concern that draw heavily on geological information.

The first of these involves the discovery and utilization of mineral and water resources. Metals, fuels, fertilizers, industrial chemicals, and building materials are all found as local concentrations of specific minerals within the rocks of the earth's outermost crust. The search for and efficient exploitation of these resources requires sophisticated geological understanding and occupies the full attention of most of the world's trained geologists. Likewise, man's essential supplies of fresh water, whether occurring as streams, rivers, lakes, or buried groundwater, are a kind of special mineral resource whose proper utilization depends heavily on geological knowledge.

The second major area in which geology makes a direct contribution to man's well-being involves the recognition and prevention of potential hazards wherever the construction of houses and other buildings, roads, and other man-made structures takes place in regions subjected to earthquakes or unusually active erosional processes, such as landslides.

All these and related topics are the concern of *engineering geology,* an active and growing subdiscipline of the science. In this chapter we shall briefly review these two principal applied aspects of geology— the search for mineral resources and the prevention or reduction of geologic hazards.

MINERAL RESOURCES

Hundreds of different rocks and minerals are of direct value to man as mineral resources. These range from common sedimentary building materials, such as gravel and clay, which cost only a few dollars per ton, to rare igneous metals and gemstones, such as gold, platinum, diamonds, and emeralds, whose value may be hundreds or even thousands of dollars per *ounce.* Even though many such rocks and minerals are useful to man, only a relatively few are consumed in large quantities by industrialized societies. These, in turn, fall into three large categories: (1) *metals,* the most important of which are iron, manganese, aluminum, magnesium, copper, lead, and zinc; (2) *non-metallics,* principally fertilizers, industrial chemicals, and building materials; and (3) *fuels,* principally coal, petroleum, and uranium. The principal rocks and minerals from which these major resources are extracted are summarized in Figure 11.1.

Major Mineral Resources		Principal Rock or Mineral Sources
Metals	Copper	Rare igneous concentrations of native copper and copper-bearing sulfide minerals
	Lead	Rare igneous concentrations of lead-bearing sulfide minerals
	Zinc	Rare igneous concentrations of zinc-bearing sulfide minerals
	Iron	Sedimentary concentrations of hematite, magnetite, goethite; less common igneous or metamorphic concentrations of magnetite or hematite
	Manganese	Sedimentary concentrations of uncommon manganese oxide minerals
	Aluminum	Sedimentary aluminum oxide ("bauxite") deposits
	Magnesium	Extracted directly from sea water; also from sedimentary dolomites
Fuels	Coal	Sedimentary coal deposits interlayered with sandstones, shales, limestones
	Petroleum	Extracted from interstitial spaces in sedimentary sandstones, limestones
	Uranium	Rare sedimentary concentrations of uranium-bearing minerals
Building materials	Sand and gravel	Sedimentary deposits of quartz sand or pebbles
	Brick clay	Sedimentary deposits of mud, shale
	Limestone	Sedimentary limestones
	Gypsum	Sedimentary evaporite deposits
	Building stone	Sedimentary limestones, sandstones, igneous granites; metamorphic schists, marbles
Fertilizers	Potassium	Sedimentary evaporite deposits
	Phosphorus	Phosphorus-rich sedimentary rocks associated with limestones, dolomites
Industrial chemicals	Halite (common salt)	Sedimentary evaporite deposits
	Sulfur	Native sulfur formed by alteration of sulfate minerals in sedimentary evaporite deposits

FIGURE 11.1
Table of major mineral resources and their principal sources.

Continental rocks supply most of man's essential minerals. The only mineral removed in any quantity from the ocean floor is petroleum, and even that comes from wells drilled into rocks submerged under shallow waters along the continental margins rather than from the deep ocean basins. This is the case partially because of the inaccessibility of the ocean floor, which has precluded large-scale exploration for mineral resources there; but there are also good reasons for believing that the basaltic rocks of the ocean floor generally lack the large concentrations of easily extracted minerals that occur in the more complex silicic rocks of the continents. If so, then the continents will *always* supply the bulk of man's resources.

On the continents, the thin veneer of sedimentary rocks contributes a disproportionate share of minerals. Most of the building materials (other than granite used as building stone), the fuels, the fertilizers, and the minerals for industrial chemicals are extracted from sedimentary rocks, as are the principal supplies of iron, aluminum, and manganese. The much greater volume of continental igneous and metamorphic rock is of primary importance in the distribution of only three major metals, which in any event are scarce: copper, lead, and zinc.

11.1 Metals

Copper, Lead, and Zinc   These three important metals are closely related in their geologic occurrence, although they differ in their uses to man. (Copper is used primarily by the electrical industry and in the making of brass; the majority of lead is used for automobile storage batteries; zinc is used in various metallic mixtures and in anticorrosion treatment of iron and steel.) Each of the three metals occurs as a fractional trace element in many different minerals, but the quantities involved are normally too small for profitable extraction. Under exceptional circumstances, however, the three elements have become locally concentrated in rich deposits of otherwise rare sulfide minerals (Figure 11.2). These deposits are usually associated with present or former igneous activity and appear to form as the gases and liquids associated with molten magmas move through cracks in overlying solid rocks. There, the gases and liquids carrying volatile compounds of copper, lead, zinc, and sulfur cool and crystallize as solid sulfide minerals. Because both heat and water-rich gases and liquids appear to be involved in their origin, such concentrations are known as **hydrothermal ore deposits**.

Hydrothermal ore deposits do not occur at random on the continents, but tend to be concentrated in linear belts called **metallogenic provinces** (Figure 11.3). The exact cause of such concentration is uncertain, but the recent discovery of lithospheric plate motions suggests that metal-rich vapors should be most abundant at plate margins where volcanic magmas from the mantle pour out onto the earth's surface. When such a zone occurs under or immediately adjacent to continental rocks, then conditions should be ideal for the escape of gases and fluids into cool overlying rocks, where they might crystallize as hydrothermal ore deposits.

Iron   Iron has long been *the* essential resource of industrialized societies, where it is used principally to make steel for construction, trans-

FIGURE 11.2
Copper mining. (a) Drilling into copper ore in an underground mine, Montana. (b) An open pit copper mine in Arizona. (c) close-up view of an open pit copper mine. (a: The Anaconda Co.; b: Tad Nichols; c: Cities Service Co.)

(a)

(b)

(c)

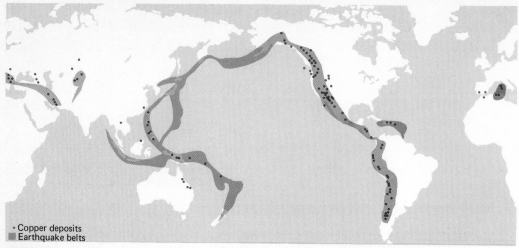

• Copper deposits
■ Earthquake belts

FIGURE 11.3
Distribution of principal world copper deposits (dots) and major earthquake zones (shaded). The close correspondence suggests that the copper originates in the mantle and is transported into the crust along plate margins. (Copper deposits from a compilation by Paul Eimon, 1970.)

portation equipment, and industrial machinery. Iron today accounts for about 95 percent of the total production of metals. Indeed, a significant fraction of the remaining 5 percent is composed of manganese and other metals produced principally to be added to iron in order to increase the strength and corrosion resistance of the resultant steel.

Iron is relatively easily extracted from any of its oxide minerals—such as magnetite, hematite, or goethite—by intense heating in the presence of carbon (charcoal or coke) (Figure 11.4). During this

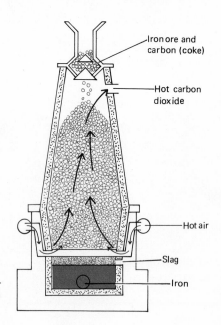

Iron ore and carbon (coke)

Hot carbon dioxide

Hot air

Slag

Iron

FIGURE 11.4
A modern blast furnace for the production of iron from iron oxide ore minerals.

"smelting" process, the carbon combines with the oxygen of the oxide mineral to form carbon dioxide gas, leaving behind an accumulation of relatively pure metallic iron. Man has produced iron by this process for at least 4,000 years, and modern steel mills differ from their earlier counterparts only in scale, efficiency, and refinements rather than in the fundamental process employed.

Because of the large volumes of both iron ore and coal (for coke) consumed in iron production, the traditional centers of iron and steel manufacture have been regions—such as the English Midlands, the Ruhr valley of Germany, or the American Great Lakes—where iron ore and coal are found in close proximity. In recent years, however, the development of cheap, large-scale ocean transport has upset this pattern by making it possible to ship the raw materials long distances to be processed. In this way such countries as Japan and the Netherlands have become important producers of iron and steel.

For many years iron was produced principally from ores that contain at least 50 percent iron—mostly relatively rare igneous concentrations of magnetite or goethite-rich soils. More recently, however, refinements in processing have made it profitable to mine hematite occurring in *banded iron formations*—thick sedimentary accumulations in which layers of iron-rich hematite alternate with layers of chert—most of which contain only 15 to 40 percent iron (Figure 11.5). This development has enormously increased man's reserves of relatively easily exploited iron, for great volumes of these enigmatic sediments, found only in ancient Precambrian rocks, are found on every continent (Figure 11.6). Because of these reserves, we are assured of an abundant supply of this essential metallic resource for many centuries to come.

Manganese, Aluminum, and Magnesium   Two of the three remaining primary metals have sedimentary sources, as does iron: manganese, which is mostly used in the production of steel, and aluminum, used principally as a structural metal and electrical conductor, are both mined principally from tropical soils from which intense chemical weathering has removed other rock constituents, leaving behind an insoluble concentrate of aluminum (or manganese) oxides (Figure 11.7). Such deposits are widespread in the tropics, including parts of Australia, the Caribbean area, northern South America, and central Africa. It has been estimated that such reserves will meet all demands for these important metals until well into the twenty-first century.

Magnesium, the lightest metal used industrially, is used principally in alloys to make light, corrosion-resistant structural materials. It has an entirely different principal source than all other metals, for it is the one major mineral resource that is extracted directly from ocean

(a)

(b)

FIGURE 11.5
Iron ores and iron mining. (a) Close-up
and (b) distant view of typical banded
iron formations, exposed in western
Australia. The ore mineral (hematite) is
concentrated in the darker layers. (c)
Mining iron ore in Minnesota. (a, b:
Brian J. Skinner; c: Standard Oil Co., N.J.)

(c)

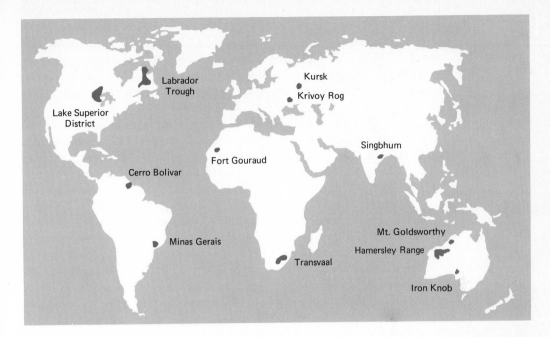

FIGURE 11.6
Location of the major Precambrian banded iron formations. (Modified from Skinner, "Earth Resources," 1969.)

water. Dissolved in the ocean are enormous quantities of many valuable elements, but the concentrations of most are so low that huge volumes of seawater must be processed to extract them, a procedure that has not yet proved generally profitable. Magnesium, the third most abundant dissolved element in ocean water (after sodium and chlorine), is an exception. It is extracted by adding to seawater dissolved calcium hydroxide, which causes the magnesium to precipitate as magnesium hydroxide (Figure 11.8). The oceans are so large that the potential supplies of magnesium and several other elements that might be extracted from them in the future are virtually inexhaustible.

FIGURE 11.7
Filling a pit from which aluminum oxide (bauxite) has been mined, Jamaica. (Reynolds Aluminum Co.)

11.2   Fuels

Mineral fuels are among the most crucial resources of all industrial-ized societies. Both coal and petroleum—the fuels that together supply the bulk of man's present energy needs—are the altered remains of ancient organisms that have accumulated in sedimentary rocks since the great expansion of life early in Cambrian time.

Coal deposits, which are the compacted and altered remains of an-cient land plants, were the first mineral fuels to be consumed on a large scale; they still provide more than half the world's energy (Fig-ure 11.9). Although enormous quantities have already been extracted, coal-bearing sediments are so extensive that remaining reserves could supply man's energy needs for at least another 250 years. Unfortu-nately, petroleum, which generally is a cleaner and more easily trans-ported fuel, is far less abundant.

(a)

Most petroleum is believed to have been derived from small marine animals and plants. When buried and compacted in sedimentary ac-cumulations, their organic remains become altered to less complex liquid or gaseous compounds, which may then be trapped in porous surrounding rocks—principally sandstones or limestones—to form petroleum reservoirs (Figure 11.10). Such reservoirs are most com-mon in Cenozoic rocks and decrease in abundance with increasing age.

(b)

Because petroleum accumulations usually are smaller, more deeply buried, and more erratically distributed than are coal deposits, they are much more difficult to find and evaluate. Careful estimates of both known and probable petroleum reserves, however, suggest that only a 50- to 100-year supply remains even if consumption were to be held to present-day levels from now on. For this reason, countries such as the United States that depend heavily on oil and gas as fuels are al-ready being forced to seek alternative energy sources (Figure 11.11). An increasing consumption of the more plentiful coal will undoubt-edly be one result of declining petroleum fuel production, but still other alternatives exist.

(c)

Among the more promising is "nuclear energy" derived from the heat produced by the fission of uranium and related elements. Because relatively small amounts of uranium fuel can be made to produce large amounts of energy, nuclear power plants are already replacing ones fueled by coal and petroleum and generating an increasing proportion of the world's electricity. The main difficulty with nuclear energy is

(d)

FIGURE 11.8
Magnesium extraction from sea water, Freeport, Texas. (a) Ocean water intake. (b) Large tanks in which calcium hydroxide is added to the ocean water to precipitate magnesium hydroxide. (c) Electrolytic cells which separate magnesium metal from other materials in solution. (d) Ingots of magnesium awaiting shipment. (Dow Chemical Co.)

FIGURE 11.9
A coal mine in Alaska. The coal beds are clearly visible as dark bands along the valley wall. (U.S. Bureau of Mines.)

disposing of nuclear plant waste products—particularly hot waters, which can endanger life in adjacent lakes and rivers, and radioactive wastes, which must usually be buried with the subsequent long-term risk of groundwater contamination.

Energy is also being obtained directly from the potential energy of waters on the earth's surface, both from man-made dams in streams and rivers as "hydroelectrical power" and along certain coasts where there are large tidal movements of ocean water. Furthermore, a few areas of hot springs are being directly tapped as "geothermal energy" for heating and for generating electricity, and much research is currently under way to harness more of the energy locked up in hot waters and rocks beneath the earth's surface. Careful projections show, however, that all of these potential sources combined are unlikely ever to account for more than a slight fraction of man's total energy needs.

There is, in addition, a far greater and as yet virtually untapped energy source: the direct radiation of the sun. Calculations show that the solar energy reaching a desert area only about one-tenth the size of Arizona could supply *all* the energy needs of the United States if it somehow could be trapped and stored. Many speculative schemes for such direct use of the sun's energy have been proposed, and some of these may become increasingly practical as petroleum reserves decline.

11.3   Building Materials, Fertilizers, and Industrial Chemicals

These remaining essential resources are mostly derived from sedimentary rocks, both detrital and chemical.

(a)

(b)

FIGURE 11.10
(a) Oil wells in central Kansas. (b) Liquid petroleum flowing from a "wellhead." (Standard Oil Co., N.J.)

Man's principal uses of *detrital* sediments are for building materials: sand and gravel are used for fill and for making concrete, and clay is used for making bricks (Figure 11.12). Such materials are consumed in enormous volumes, yet they are so universally abundant that there is no danger of their exhaustion. However, supplies conveniently close to the point of use are diminishing in many heavily populated areas, and so they must be brought from increasing distances at greater costs.

FIGURE 11.11
Past and projected energy sources in the United States. (Modified from Cook, "Scientific American," vol. 224, 1971.)

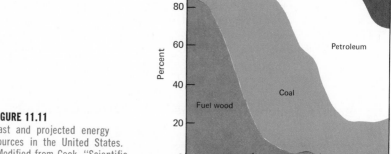

FIGURE 11.12
Sand and gravel being removed from under
water, Maryland. (U.S. Bureau of Mines.)

Of greater immediate concern than an ultimate shortage of these materials are the widespread, landscape-marring pits and quarries that result from their removal. With careful management such blight can be avoided and the areas reclaimed for constructive uses after the materials are removed. This reclamation work is often expensive, however, and relatively few cities or regions as yet require it.

In contrast to the comparatively restricted uses of detrital sediments, *chemical* sediments and the sedimentary rocks derived from them have many applications. Limestone, the most abundant chemical sedimentary rock, is used chiefly as the major ingredient of cement (Figure 11.13). By far the most versatile chemical sediments, however, are evaporite deposits which provide not only gypsum—the principal component of plaster and wallboard—but most mineral fertilizers and industrial chemicals as well.

As ocean or lake waters evaporate, the dissolved materials are precipitated principally as the minerals gypsum and halite. At several earlier intervals in earth history, large marine basins containing evaporite accumulations were far more extensive than they are today; as a result, the continents' veneer of sedimentary rocks contains abundant deposits of both gypsum, for building materials, and halite, which is used principally by the chemical industry (Figure 11.14).

Less common, but still adequate for any foreseeable demand, are potassium evaporite minerals and phosphorus deposits, both used mainly as fertilizers.

Still less common are deposits of native sulfur formed in only a few regions by the action of bacteria on the gypsum of certain evaporite deposits. Sulfur is a prime industrial chemical, used principally to make the sulfuric acid required for the processing of many other chemical products. For this reason, the relative scarcity of easily tapped native sulfur deposits has led to a search for alternatative sources. Fortunately, sulfur occurs in abundance in many other forms and is now being extracted in large quantities as a by-product of the processing of both natural gas and the sulfide ore minerals of copper, lead, and zinc.

11.4   Mineral Resources and the Future

Although much has been written about the need to conserve the earth's finite supply of mineral resources, many modern planners feel that concern for future shortages of most of these minerals has been premature. Such elements as iron, aluminum, magnesium, and potassium make up a large fraction of the earth's crust; with improved methods of extraction (and perhaps recycling), their sources will probably last indefinitely. There are also large potential sources of manganese, phosphorous, coal, building materials, halite, and sulfur.

(a)(b)

FIGURE 11.14
Evaporite minerals.
(a) Map showing the
areas known or
believed to be under-
lain by evaporite
deposits in the United
States and Canada.
(b) A salt mine in the
Dominican Republic.
(a: From Skinner,
"Earth Resources,"
1969, after U.S.
Geological Survey
Bulletin 1019-J and
A. D. Huffman; b:
United Press.)

Among the principal mineral resources, only copper, zinc, lead, and petroleum appear to be relatively limited. The history of technology suggests, however, that satisfactory alternative resources are usually developed as the supply of one resource drops, and thus even these seemingly essential materials may become obsolete in the future. Aluminum, for example, is already replacing copper in many applications. Similarly, coal and nuclear energy are beginning to replace petroleum as a principal source of electrical power.

Perhaps more serious than the ultimate shortage of essential mineral resources is the immediate problem of extracting and using them in ways that do not threaten other aspects of human existence. For example, unreclaimed quarries, strip mines, oil fields, and gravel pits blight large areas of the land surface, while even larger areas are scarred by the abandoned debris that result from processing and consuming mineral resources and that, unlike most animal and plant remains, do not quickly disappear through organic decay. Likewise, the potential hazard of long-term radioactive contamination will become increasingly serious as nuclear energy is increasingly exploited. In growing recognition of these problems, many conservationists now feel that the emphasis over the next decades should be shifted from merely searching for new mineral resources to perfecting techniques that minimize the damage of extracting and consuming them.

WATER RESOURCES

A second major area in which geology is of direct service to man concerns the discovery, use, and conservation of water resources. Most liquid water occurs on land as slowly moving groundwater: this source accounts for about 3,000 times the volume of water found in lakes and streams. As yet, however, less than one-third of the water used by man comes from wells tapping this large groundwater resource. The rest comes from the more easily exploited surface waters found in streams and in both natural and man-made lakes.

Most of the water used by man, whether for public water supply (about 10 percent of the total used), agriculture (about 45 percent), or industry (about 45 percent), is not actually *consumed* (Figure 11.15). Instead, it is returned with varying degrees of change to streams or the groundwater reservoir, where it may be reused again and again in a continuing cycle. For this reason, man's potential supply of fresh water is greater than any foreseeable need—with two important qualifications.

The first concerns the *location* of the water. Arid regions normally lack surface water in streams or lakes, and the deep-lying groundwater in such areas is often prohibitively expensive to exploit by wells (Figure 11.16). Thus neither agriculture nor industry with large water requirements is practical in most arid regions, and even municipal water supplies there may be limited and costly. In short, water resources are not evenly distributed on the land surface and must generally be exploited where they occur.

The second important reservation concerns not the location, but the

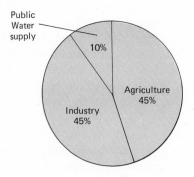

FIGURE 11.15
Principal uses of water in the United States.

FIGURE 11.16
Irrigation agriculture (young orange trees) based
on groundwater wells, southern Arizona. (U.S.
Department of Agriculture.)

quality of water resources. Most human uses require relatively pure,
clean waters; but wastes are frequently added that make these waters
impure. Traditionally, such "used" waters have been returned to
streams or the groundwater reservoir with little or no treatment, so
that contaminants were removed either slowly, by natural processes,
or more quickly by man himself before he used the waters again.
Along large rivers, for example, upstream cities and industries find
relatively pure water which they use and return to the river with im-
purities, which must then be removed by consumers farther downstream. The well-known results of this and similar practices are the
polluted lakes, rivers, and streams that now plague most industrialized
nations (Figure 11.17). Fortunately, there is increasing concern for
both the economic and aesthetic costs of such practices, and hopefully
the pattern will be reversed in the future by requirements that waters
be processed immediately *after* use so that they will be clean and pure
when returned to their natural sources.

Unlike waters of the land, ocean water is not a direct resource because large-scale removal of the ocean's dissolved salts has not yet
proved practical. Like lakes and streams, however, the ocean is used
by man as a receptacle for his waste products. Coastal cities normally
discharge sewage directly into the sea with little or no treatment; they
also dump enormous quanities of solid waste and debris onto the
ocean floor (Figure 11.18). Fortunately, the ocean is so large that
these processes have as yet had little apparent effect on the waters
and life of the open sea. In many partially enclosed near-shore waters,
however, such waste disposal has led to destruction of marine life,
including important fishery resources, as well as to pollution levels
that restrict recreational swimming and boating. Hopefully, increased

(a)   (b)

FIGURE 11.17
Water pollution. (a) Scene in a U.S. National Park. (b) Typical view of untreated wastewaters being returned to a stream. (United Press International.)

awareness of this condition may lead to corrective action before any large-scale damage is done to the overall balance of life in the ocean.

GEOLOGY AND ENGINEERING

Geology makes yet another important series of contributions to man through the field of engineering. Many natural hazards that may threaten homes, buildings, bridges, and other man-made structures can be recognized and anticipated through geologic study. Of major concern in such studies are problems involving earthquakes and slope failures or other erosional processes. This section considers these problems.

11.5   Seismic Hazards

In Chapter 6 we introduced the topic of earthquakes primarily in terms of their value in learning about the earth's interior. Of greater general concern to most humans are the seismic hazards that frequently accompany strong earthquakes. In an average year, ten earthquakes causing widespread death and devastation occur somewhere on the earth; about 100 others cause serious local destruction; 1,000

FIGURE 11.18
Solid wastes being hauled to ocean dumping grounds. (J. Paul Kirouac.)

do some damage; 100,000 are strong enough to be felt as tremors; and about 1 million can be detected by seismographs. Although earthquakes may occur anywhere on the earth's surface, they are primarily localized along plate boundaries, and the most frequent and destructive are concentrated in the seismically active zone around the margins of the Pacific Ocean (compare Figures 6.3 and 7.22).

The principal cause of natural earthquakes is believed to be faulting (see Section 6.2). There are four types of hazard associated with moderately strong to major seismic events produced by faulting (Figure 11.19):

1. Ground rupture If a structure is built across a plane of faulting that extends to the earth's surface, considerable damage can result.

2. Shaking This is the primary cause of earthquake damage. It may take several tens of seconds for the strong motions of P, S, and L (surface) waves to die out, and accelerations equal to the force of earth gravity have been measured near the epicenters of earthquakes.

3. Seismic sea waves (often called tsunamis or tidal waves, the latter being a misnomer) Vertical ocean-floor movements or large submarine landslides can generate large irregularities in water level that spread outward as waves. Upon reaching shallow coastal waters, the waves steepen and heighten until they break, much in the manner of normal surf—except that heights as great as 50 or 60 m (160–200 feet) have been reached. Although seismic sea waves travel at velocities up to 450 mi/hr across the Pacific Ocean, due to their low height and the great distance between successive wave crests while moving in the deep ocean environment they go unnoticed by ships at sea. They are dangerous, then, only when slowed and heightened in shallow water. One tsunami alone, for example, killed more than 27,000 Japanese in 1896.

(a)

(b)

(c)

FIGURE 11.19
Principal seismic hazards.
(a) Ground rupture.
(b-c) Before and after shaking.
(c) Seismic sea wave. (d) Slope
failure. (a: Karl V. Steinbrugge;
b: The American Iron and Steel
Institute; c: Mansell Collection;
d: J. R. Stacy, U.S. Geological
Survey.)

(d)

(e)

4. Ground failures Landslides, rock avalanches, severe settlement, and other forms of slope failure are frequently triggered by rapid ground accelerations and prolonged ground shaking.

Measuring Earthquake Strength   There are two approaches to measuring the strength of an earthquake. One involves instrumentally recorded seismograms, and the other depends on assessments of how much damage was done. Inasmuch as the amount of ground motion generated near the focus of an earthquake, and hence its potential for damage, dies out with distance traveled, two scales of measurement have been developed: one, called *magnitude*, describes maximum ground movement *at the focus*; the other, *intensity*, describes the amount of damage done at *any* geographical location. Magnitude is a measure of how much motion would be recorded by a "standard" seismograph at a fixed distance from an earthquake focus. Any seismogram can be interpreted to determine magnitude if the focal distance of the earthquake is known and if the recording instrument has been calibrated with respect to the standard instrument. Thus, although a seismograph in San Francisco would record less ground motion for an Alaskan earthquake than would appear on a seismogram recorded in Seattle, seismologists in both cities, using accepted procedure, would assign the same magnitude value to the earthquake. The relationship of less ground motion as a function of greater distance explains how seismographic stations all around the world can determine the same value for an earthquake's magnitude, even if the seismic vibrations are too faint to be felt by humans. Magnitude, then, is an absolute measure of the energy liberated by an earthquake, and its determination depends solely on the analysis of seismograms.

The most commonly used scale of magnitude was developed in 1935 by Charles Richter. The Richter magnitude scale is a range of numerical values used to describe the amount of ground motion (and energy) at the source of an earthquake. The great 1906 San Francisco earthquake, for example, was 8.3, while the Hebgen Lake temblor near Yellowstone Park in 1959 was 7.1 (Figure 11.19a and e). The largest earthquakes ever recorded had values of 8.6 on this scale, while the smallest that can be felt by humans are in the vicinity of 2.5.

The numerical values of the Richter scale are logarithms of the amount of ground motion. The significance of this is that the relationships between magnitude, ground motion, and energy generated at the focus are not simple geometrical proportions. Without pursuing the mathematics, we can summarize the relationships as follows:

Each Richter magnitude numeral represents 10 times the amount of ground motion and approximately 31.5 times the amount of energy as the next lower numeral. Thus an earthquake of magnitude 8 releases about 1,000 times as much energy (31.5 $\times$ 31.5) as a magnitude-6 event.

Prior to the development of an instrumental technique for assessing an earthquake's strength, the only widely used approach involved field evaluations of its effects: Was it felt by most people? Were they frightened? Did buildings collapse? Did water splash over river banks? These kinds of questions were asked in order to ascertain an earthquake's intensity. Today American geologists use the Modified Mercalli scale of intensity (Figure 11.20), adapted to North American conditions from a scale originally designed for use in Italy. Following an earthquake, investigators frequently compile data on its effects in order to prepare isoseismal maps (Figure 11.21), which are a graphic means of delineating regions of varying intensities. Under ordinary circumstances the epicenter of an earthquake is approximately coincident with the central region of highest intensity. Indeed, prior to instrumental techniques for locating epicenters (see Section 6.2 and Figure 6.9), using this rule of thumb was the only method for locating epicenters.

For most citizens the aspect of an earthquake that is of most concern is its intensity, especially as it pertains to their dwellings. What are the principal factors affecting how the building you live in will respond to an earthquake? In addition to the magnitude of an earthquake (over which you have no control) and the distance away from the epicenter (over which you have relatively little control) there are two other considerations of prime importance. One is that structures constructed on rigid bedrock suffer much less damage from shaking than do those built on poorly consolidated material. During the 1906 San Francisco earthquake, for example, buildings constructed on loose sandy soils and compacted but still watersoaked muds along the margins of San Francisco Bay were far more severely damaged than those resting on the bedrock of the city's higher hills. Another significant factor in a structure's ability to withstand strong shaking concerns its construction: design, materials, and quality of workmanship are all important. A fact that should be reassuring to most citizens is that a well-constructed one- or two-story frame house is one of the safest places to be during a strong earthquake, although special care must always be taken to avoid falling objects such as brick facings and chimneys.

Earthquake Prediction and Control   The United States government has recently launched extensive research efforts designed to increase our understanding of earthquakes. Though still highly tentative, the ultimate goal of these investigations is the prediction and possibly even the control of potentially devastating earthquakes. The only

The Modified Mercalli Scale

I Not felt except by a very few under especially favorable circumstances.

II Felt only by a few persons at rest, especially on upper floors of buildings. Delicately suspended objects may swing.

III Felt quite noticeably indoors, especially on upper floors of buildings, but many people do not recognize it as an earthquake. Standing motor cars may rock slightly. Vibration like passing truck. Duration estimated.

IV During the day felt indoors by many, outdoors by few. At night some awakened. Dishes, windows, doors disturbed; walls make creaking sound. Sensation like heavy truck striking building. Standing motor cars rocked noticeably.

V Felt by nearly everyone; many awakened. Some dishes, windows, etc., broken; a few instances of cracked plaster; unstable objects overturned. Disturbances of trees, poles, and other tall objects sometimes noticed. Pendulum clocks may stop.

VI Felt by all; many frightened and run outdoors. Some heavy furniture moved; a few instances of fallen plaster or damaged chimneys. Damage slight.

VII Everybody runs outdoors. Damage negligible in buildings of good design and construction; slight to moderate in well-built ordinary structures; considerable in poorly built or badly designed structures; some chimneys broken. Noticed by persons driving motor cars.

VIII Damage slight in specially designed structures; considerable in ordinary substantial buildings, with partial collapse; great in poorly built structures. Panel walls thrown out of frame structures. Fall of chimneys, factory stacks, columns, monuments, walls. Heavy furniture overturned. Sand and mud ejected in small amounts. Changes in well water. Disturbs persons driving motor cars.

IX Damage considerable in specially designed structures; well-designed frame structures thrown out of plumb; great in substantial buildings, with partial collapse. Buildings shifted off foundations. Ground cracked conspicuously. Underground pipes broken.

X Some well-built, wooden structures destroyed; most masonry and frame structures destroyed with foundations; ground badly cracked. Rails bent. Landslides considerable from river banks and steep slopes. Shifted sand and mud. Water splashed over banks.

XI Few, if any, (masonry) structures remain standing. Bridges destroyed. Broad fissures in ground. Underground pipelines completely out of service. Earth slumps and land slips in soft ground. Rails bent greatly.

XII Damage total. Waves seen on ground surfaces. Lines of sight and level distorted. Objects thrown upward into the air.

FIGURE 11.20
Modified Mercalli Scale of earthquake intensity. (After Hodgson, 1964.)

predictive capability that currently exists, however, is a long-established system for warning island and coastal communities around the Pacific of possible seismic sea waves. Because tsunamis travel 40 to

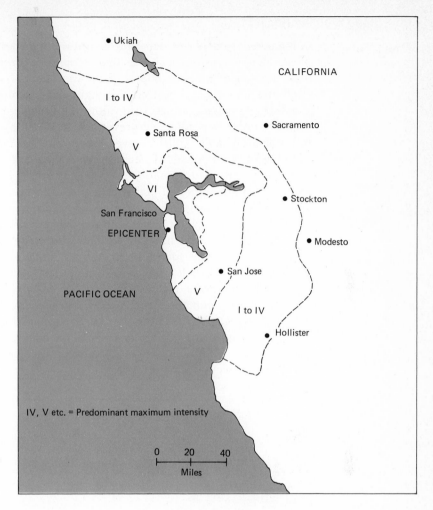

FIGURE 11.21
Isoseismal map of earthquake, San Francisco, 1957, Richter magnitude 5.3. (From California Division of Mines and Geology Special Report 57.)

50 times slower than earthquake waves, there is ample time to determine if an earthquake has occurred where a tsunami might be generated and, if it has, to accurately predict the time of its arrival at any geographic position.

More understanding of the underlying mechanism for earthquakes will be required, however, before predictions of major seismic events can be made. Perhaps we may never achieve this, but at present we are simply too ignorant to know. As for means of controlling an earthquake or reducing its strength, there are some encouraging models under development at present; but they, too, are in very preliminary form. The next ten years should see great progress in this area of geologic research.

11.6   Slope Stability Hazards

In the perspective of geologic time all natural hillsides are undergoing rapid change, but our concern in this section is with the much shorter time intervals of human life and of the useful existence of man-made structures. A principal concern of engineering geology is to understand the geologic factors controlling the development of natural slopes in order to predict and prevent those activities that cause them to become unstable. Hillsides everywhere are constantly undergoing adjustments to maintain a stable slope angle. Such **natural slope angles** are controlled primarily by three factors: (1) strength of rock and soil; (2) orientation of planar structures such as bedding, fracture, and foliation planes; and (3) climatic conditions, and especially the amount of water. Natural slope angles range from the vertical and even overhanging cliffs in regions such as Yosemite Valley, where the strong granite rock is sparsely jointed, to slopes of only a few degrees in weak shale deposits. Slopes with angles in excess of 40 degrees and still in equilibrium are rare.

Man's entry upon the scene is often attended by his residential or industrial development of the natural environment, which too often involves changes that artificially upset slope equilibrium and thereby trigger slope failure. Figure 11.22 graphically depicts examples of three such adverse changes: overloading, oversteepening by under-

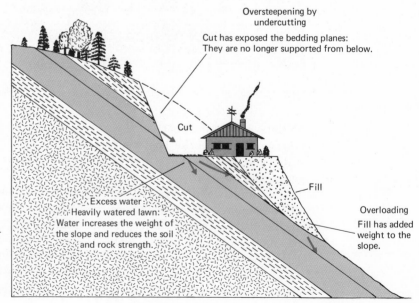

FIGURE 11.22
Composite diagram of factors contributing to slope failure. There are, of course, many other configurations whereby undercutting, overloading, and excess water contribute to slope failure.

Oversteepening by undercutting

Cut has exposed the bedding planes: They are no longer supported from below.

Cut

Excess water: Heavily watered lawn: Water increases the weight of the slope and reduces the soil and rock strength.

Fill

Overloading
Fill has added weight to the slope.

cutting, and introducing excess water. Seismic shaking is another important contributor to landslides. Figure 11.23 shows an unfortunate and dramatic result of failing to carefully evaluate a region's geologic setting before making cuts and fills for a subdivision.

11.7   Coastal Erosion

Many emergent coasts, such as those along the western coast of North America, display a variety of geologic hazards and inconveniences to man. Attack at the base of a sea cliff (Figure 9.18) during short episodes of high surf causes undercutting, thus oversteeping the cliff, which eventually causes landsliding and rock avalanching as a means of restoring the slope to equilibrium. Along the coast of central California, for example, rates of cliff retreat averaging 2 ft per year are not uncommon, and local occurrences of several times that rate are known. An unwary home builder, easily beguiled by the prospect of an attractive view from an ocean bluff, may fail to anticipate the precariousness of his setting in the not-too-distant future (Figure 11.24). In regions where the sea cliffs are composed of highly resistant rock, on the other hand, cliff retreat and landsliding may not pose any significant threat.

FIGURE 11.23
Homes destroyed by landsliding in Southern California. (W. R. Cotton.)

FIGURE 11.24
Home and cliff at Seal Cove, California. Home is already showing distress from slope creep; eventual destruction is inevitable unless home is removed. (W. R. Cotton.)

Along many coastlines man has interfered with the equilibrium of longshore beach transport (see Section 9.7) with disastrous results. In order to create a recreational facility such as a boat harbor or to enhance and protect a sandy beach, breakwaters and groins are often constructed to reduce wave and surf energy in the near-shore environment. More often than not, serious erosional activities are the undesirable side effects of such efforts. Consider the simple case shown in Figure 11.25: as the longshore transport moves sediment down the coast, an equilibrium is maintained as long as the backwash of the waves and surf is capable of carrying as many grains *out* as were brought *in* by the onwashing wave front. But what if the deposited grains are piled up in relatively quiet water created by a groin or breakwater? Not only is there insufficient energy to carry sand back out to deeper water, but down the beach from the groin the onrushing surf is without sediment to exchange as it strikes the beach. As a result, severe erosional changes are apt to occur.

11.8   The Need for Risk-Zoning

In addition to the more dramatic geologic hazards associated with earthquakes, slope failures, and coastal erosion, increased urbanization in many parts of our country has initiated or aggravated problems involved with flooding and soil erosion, expansive soils, adulteration of groundwater, and regional ground subsidence. All of these problems are avoidable if we apply current knowledge of engineering geology and soils engineering. Regrettably, however, a large communication gap exists between members of the geological and engi-

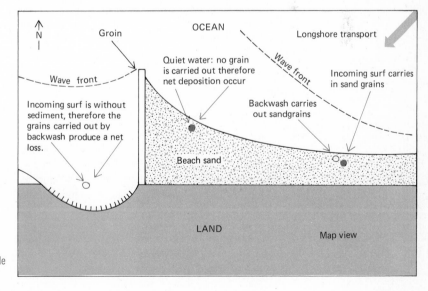

FIGURE 11.25
Disruption of longshore transport equilibrium. Net effect is deposition on the up-current side and erosion on the down-current side of an obstruction.

neering professions, on the one hand, and, on the other hand, most city and county officials who must provide for protection of the public. We are just now beginning to bridge this gap, and in the next few years most communities will acquire significant geologic inputs into their planning programs. Those cities and counties located in seismically active zones or containing hillslope areas subject to development will benefit measurably by developing thorough risk-zoning studies and appropriate codes and ordinances. Only in this way can the average citizen be protected against geologic hazards to the same degree that government protects him against inadequate electrical wiring and plumbing in the buildings he frequents and inhabits.

SUMMARY OUTLINE

Mineral Resources

11.1 *Metals:* these are principally copper, lead, and zinc extracted from minerals of igneous origin; iron, manganese, and aluminum extracted mainly from concentrations in sedimentary rocks and soils; and magnesium, extracted directly from ocean water.

11.2 *Fuels:* coal and petroleum are presently our principal energy resources, but replacements must be found for them since their deposits are limited and nonrenewable; nuclear and direct solar energy appear to be the only practical long-term alternatives.

11.3 *Building materials, fertilizers, and industrial chemicals:* these resources are mostly derived from sedimentary rocks, both chemical and detrital.

11.4 *Mineral resources and the future:* future concerns will be to provide substitutes for depleted resources and to minimize the detrimental environmental impact of extraction and utilization.

Water Resources

Fresh water is a valuable resource from both economic and aesthetic standpoints; energetic efforts to limit water pollution are in order.

Geology and Engineering

11.5 *Seismic hazards:* four types of hazards associated with earthquakes are ground rupture, shaking, seismic sea waves, and ground failure; the possibility of controlling major seismic events is still remote.

11.6 *Slope stability hazards:* hillsides generally stand at natural slope angles; their failure can be induced by overloading, oversteeping, or introducing excess water.

11.7 *Coastal erosion:* cliff retreat is an ongoing process that creates hazards in some areas; man's interference with the beach's longshore transport can also be disastrous.

11.8 *The need for risk-zoning:* risk-zoning studies and appropriate codes and ordinances are necessary steps toward protecting the public from potential geologic hazards.

ADDITIONAL READINGS

*Skinner, B. J.: *Earth Resources*, Prentice-Hall, Inc., Englewood Cliffs, N.J., 1969. An authoritative and readable introduction to mineral resources.

*Hodgson, J. H.: *Earthquakes and Earth Structure*, Prentice-Hall, Inc., Englewood Cliffs, N.J., 1964. A brief, popular introduction stressing the relation of earthquakes to man.

*Fagan, J. J.: *The Earth Environment*, Prentice-Hall, Inc., Englewood Cliffs, N.J., 1974. Chapters 2 and 4 provide good, brief reviews of water and energy resources.

*Available in paperback.

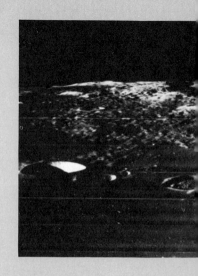

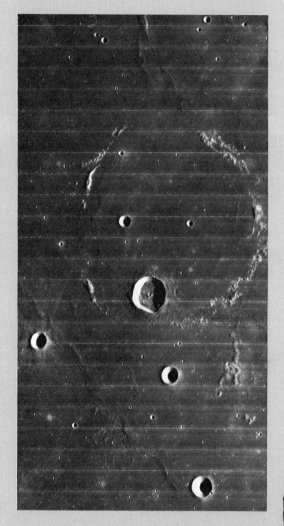

chapter **12**

Rocks Beyond the Earth

Although throughout this book we have investigated terrestrial rocks and minerals as the principal means for understanding the earth, rocks collected from beyond the earth also have an important contribution to make to our knowledge of how the earth was formed. In this chapter, then, we shall direct our attention to meteorites and to the moon—the only nonterrestrial elements of our solar system thus far sampled for careful study.

METEORITES: SAMPLES OF THE SOLAR SYSTEM

The "shooting stars" seen streaking across the sky on any dark night are caused by materials originally left in space during the formation of the solar system and now captured by the earth's gravity. The bright streaks are formed as the materials plunge into the dense atmosphere, where they are heated, and usually vaporized, by friction with gas particles at heights of 40 to 100 km (25 to 60 mi). Such objects are called meteors if they are completely vaporized and meteorites if a part of them survives the fall through the atmosphere and reaches the earth's surface.

12.1 Meteors and Meteorites

Telescopic analysis of the light produced by the fiery paths of meteors shows them to be of two sorts. Most are composed of low-density masses of very light material which leave behind relatively faint trails as they are vaporized in the atmosphere. This most common kind of meteor is believed to be composed of small frozen gas particles left behind by passing comets, a conclusion strengthened by the fact that such meteors are particularly abundant when the earth's orbit crosses that of a comet (Figure 12.1).

Meteors composed of frozen gases are readily vaporized in the atmosphere and are thus never preserved on the earth's surface as meteorites. The second kind of meteor appears to be caused by much more dense metallic or stony material entering the earth's atmosphere from space. This type of meteor leaves an intense "fireball" as it passes through the atmosphere. In contrast to the cometary meteors, a small number of these heavier meteors are not completely vaporized, but instead reach the earth's surface, at which point they are classified as meteorites. We have already made reference to meteorites because of their extraordinary significance in helping us understand many aspects of the earth: in Section 6.5 we saw that certain inferences about the nature of the earth's interior are based on the overall composition of meteorites, whereas in Section 4.7 we noted that radiometric dating of meteorite minerals gives important clues concerning

FIGURE 12.1

Meteors. (a) Meteors are most common when the earth's orbit crosses that of a comet. Most meteors are believed to be small, frozen masses of cometary debris that fall into, and are vaporized by, the upper atmosphere. (b) Photograph of a typical meteor. (b: Smithsonian Astrophysical Observatory.)

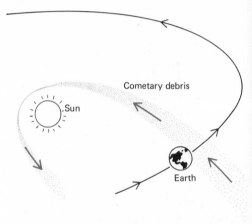

(a)

(b)

the age of the earth. In this section we shall review the characteristics of meteorites in more detail, emphasizing their importance as samples of the solar system beyond the earth.

About 500 meteorites the size of an orange or larger reach the earth's surface each year. Most fall in the ocean; only about 150 descend on land, and of these an average of only about 4 per year are recovered for study.

Although most meteorites are less than a meter in diameter, occasionally much larger masses, up to a kilometer or more in diameter, reach the earth's surface, where they make huge craterlike holes known as *astroblemes* (Figure 12.2). These are caused by the dissipation of their tremendous energy of motion as they collide with the

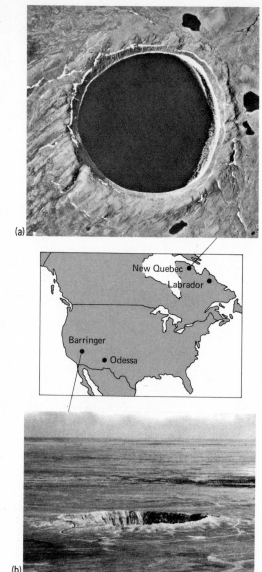

FIGURE 12.2
Astroblemes, the impact craters of large meteorites.
The map shows sites of four large North American
astroblemes that are believed to have been formed within
the past 2 million years. Many older craters, now much
modified by erosion, are also known. (a) Vertical aerial
view of the New Quebec crater, which is about 3 km
(1.8 mi) in diameter. (b) Oblique view of the Barringer
crater, which is about 1 km (0.6 mi) in diameter. (a:
Canadian Department of Energy, Mines, and Resources;
b: American Museum of Natural History.)

earth. It has been estimated that such large meteorites fall only about
once each 10,000 years.

At least 20 large astroblemes are known on the earth's surface today. Most of these probably formed in the last few million years of
earth history, whereas earlier ones most likely have been largely destroyed by erosion. In spite of the large masses of the meteorites that
form astroblemes, meteorite fragments are rare in and near the craters.
Apparently, such meteorites are largely vaporized by the enormous
heat created by their impact with the earth. It is therefore principally

the smaller meteorites which survive their fall to the earth's surface and become available for study.

12.2 Meteorite Composition and Origin

In overall composition, meteorites fall into three groups (Figure 12.3): some, called stony meteorites, are made up entirely of various silicate materials; others, called iron meteorites, are composed exclusively of metallic iron and nickel; still others, called stony-irons, contain mixtures of both metal and silicate. Most of the 1,600 or so meteorites recovered and preserved in the museums of the world are irons, which after they fall resist destruction by weathering far more readily than do the silicate minerals of stony meteorites. In addition, masses of

FIGURE 12.3

Meteorites: (a) Relative proportions of stony, stony-iron, and iron compositions among meteorites actually seen to fall. Most meteorites are stony "chondrites" containing small, glassy spheres of silicate minerals. (b-d) Cut and polished surfaces of typical meteorites. (b) Stony chondrite. The specimen, found in Mexico, is about 10 cm (4 in.) long. (c) Stony-iron. The dark masses are silicate minerals and the light matrix is iron. The specimen, found in Arizona, is about 5 cm (2 in.) long. (d) Iron. The specimen, found in Arizona, is about 25 cm (10) in. long. (b-d: Smithsonian Institution.)

metallic iron are far more likely to be recognized and preserved as exceptional materials than are stony meteorites, which are easily mistaken for ordinary rocks. For these reasons, it is impossible to estimate the relative proportion of meteorite types that reach the earth's surface by looking at *all* recovered meteorites. Instead, such estimates are based on studies of the much smaller number of meteorites, about 700 in all, that were *seen to fall* and then collected immediately. About 93 percent of such meteorite *falls*, as they are called, are of the stony type; irons make up most of the remainder, with stony-irons accounting for less than 2 percent (Figure 12.3). It thus appears that the material arriving at the earth's surface from elsewhere in the solar system is composed predominantly of silicate minerals, just as is the earth itself. The composition and structure of these meteoritic silicates differ, however, in many important respects from silicates found on earth.

Most stony meteorites are made up, at least in part, of small glassy spheres about 1 to 2 mm (millimeters) in diameter (Figure 12.4). These spheres, in turn, are composed of dark magnesium- and iron-rich silicate minerals, particularly olivine and pyroxene. These small silicate spheres are called *chondrules*, and the meteorites containing them are referred to as **chondrites.**

The origin of chondrules, which make up a large fraction of the material reaching the earth from space, is uncertain, because similar

FIGURE 12.4
Chondrites. (a) Microscopic view of a typical chondrite showing the characteristic rounded, glassy spheres ("chondrules") made up of silicate minerals. The spheres are 1 to 2 mm in diameter. (b) Mineral composition of a typical chondrite. Olivine and pyroxene predominate. (a: John A. Wood.)

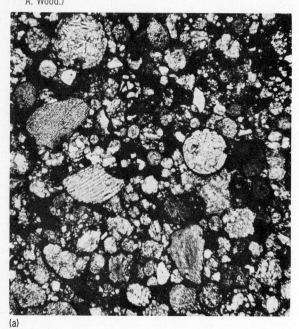

(a)

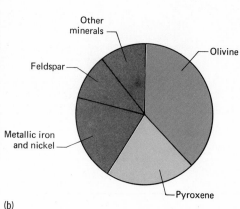

(b)

structures are not found among the silicate minerals of the earth. Chondrules look as if they formed from a heating-and-rapid-cooling process that first melted some preexisting material and then quickly "froze" small drops of the melt into spheres. Radiometric dating indicates that most of them solidified about 4.5 billion years ago, at the same time that the earth is believed to have consolidated. This observation suggests that they may be particles of matter that were melted by an initial solar heating and then "froze" as they moved away from the sun in the early history of the solar system. This suggestion is strengthened by the fact that their composition is remarkably uniform and shows about the same relative proportion of nonvolatile elements as does the sun (Figure 12.5). Indeed, one rare type of chondrule-bearing meteorite, called *carbonaceous chondrites*, even contains some of the more volatile solar elements, such as carbon, hydrogen, and oxygen, chemically combined into silicates and other solid minerals. Both because of their composition and their age, chondritic meteorites, particularly the carbonaceous chondrites, are believed to represent relatively unchanged samples of the primordial material that was initially dispersed from the sun to form the planets and other objects of the solar system.

Even if we knew that chondritic meteorites represented the primordial building materials of the solar system, we would still have to account for the small proportion of meteorites that do not contain chondrules (principally the irons and stony-irons, but also including a few non–chondrule-bearing stony meteorites called *achondrites*—see Figure 12.3). There are two general theories for the origin of these materials. Some workers believe that they condensed directly from solar gases, just as did the chondrites. Others feel that they have a secondary origin—from chondrites that at some time were swept to-

FIGURE 12.5
The relative compositions of chondritic meteorites and the sun. If the large volumes of solar hydrogen and helium are omitted, the remaining elements show almost identical abundance, a fact which supports the hypothesis that chondrites are fragments of solar matter. (From Glasstone, "Sourcebook on the Space Sciences," 1965, after Chapman and Larson.)

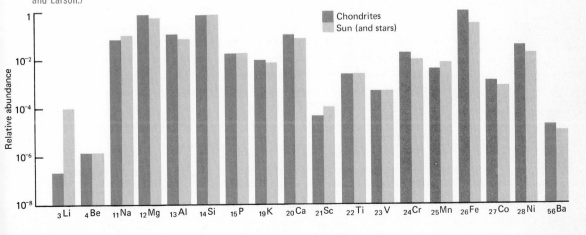

gether into large (several kilometers) bodies in which internal melting could take place. Under such circumstances, the heavier iron and nickel content of the silicates might have accumulated near the center of the bodies. If such differentiated bodies were later broken up—for example, by impact with each other—then the fragments might well resemble the small proportion of iron, stony-iron, and achondritic meteorites that reach the earth.

THE MOON

For many years meteorites provided the only materials from beyond the earth that were available for direct scientific study. Beginning in 1969, however, samples returned to the earth from the moon have added a new dimension to our understanding of the solar system. Before turning to the nature and implications of these dramatically important lunar materials, it will be useful to review some more general conclusions about the moon that have grown from many years of astronomical calculations and observations of its surface features.

Most of the nine planets have one or more satellites in orbit around them, yet the earth-moon system is unique because of the very large size of the satellite in comparison with its attracting body (Figure 12.6): The moon has a diameter about one-quarter as large as the earth's, whereas most planetary satellites have diameters less than one-twentieth the size of their respective planets. Indeed, the moon is half the size of Mars and two-thirds as large as Mercury; were it in orbit around the sun rather than the earth, it would qualify as a full-scale terrestrial planet.

By means of astronomical calculations the moon's diameter, total mass, and distance from the earth have long been established. The diameter and mass having been determined, the average density of lunar materials was easily computed to be 3.3 g/cm³, about the same as the materials of the earth's upper mantle but far lighter than the earth's *overall* density of 5.5 g/cm³. This was taken to suggest that the moon lacks the heavy iron core of the earth's interior, and is instead dominated by silicate minerals throughout. Because the moon is made up of lighter materials, its total *mass* is only about one-eightieth that of the earth, even though its diameter is one-fourth as great. The earth's far greater mass is the principal reason that a body as large as the moon can be held in orbit by the earth's gravitational attraction.

The moon completes one revolution around the earth every 27 days and, like the sun and planets, rotates about a central axis. Rather surprisingly, however, its period of rotation is the same as its period of revolution—about 27 days—so that the moon always shows the same

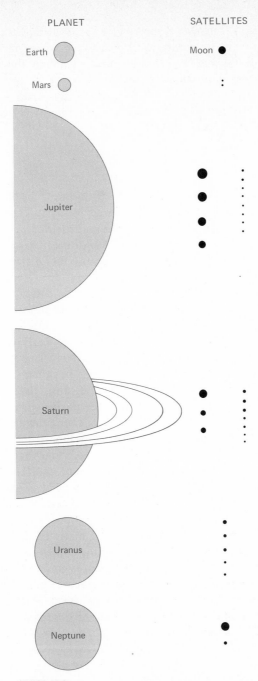

FIGURE 12.6
The relative sizes of the planets and their satellites. The moon is unique in having a diameter one-fourth as great as its associated planet; most satellites are proportionately far smaller than the planets they orbit. Mercury, Venus, and Pluto appear to lack satellites.

hemispherical face when viewed from the earth. Because of the tilt of the lunar axis of rotation, and because of minor wobbles in the moon's orbital motion, slightly more than half the lunar surface, or about 60 percent, can be observed at one time or another from the earth. The remaining 40 percent, called the farside—to distinguish it from the visible earthside—is forever hidden from earth-based observation.

12.3   The Lunar Surface

Before the recent flurry of space exploration, information about the lunar surface could only be gathered by earth-based telescopes. Today, however, artificial satellites have provided detailed photographs of the entire surface, including the long-hidden farside, and astronauts of the Apollo missions have sampled the moon's surface rocks.

Maria and Highlands   From afar, the most prominent features of the lunar earthside are large dark patches surrounded by lighter areas; both have been directly observed by everyone, for they are readily visible to the unaided eye (Figure 12.16). Early telescopic observations showed the dark areas to be relatively smooth and flat regions which contrast sharply with the more irregular and rugged terrain of the lighter regions around them. Because they bear a superficial resemblance to the earth's oceans, the dark, smooth areas were long ago named maria (Latin for "seas"; the singular is *mare*); the lighter regions of high relief are known as the lunar highlands. Maria cover about one-third of the moon's earthside hemisphere; one of the most surprising discoveries of recent lunar exploration was that maria are virtually absent from the long-hidden farside (Figure 12.7). For reasons that are still uncertain, the farside is covered almost entirely by rugged highlands.

On close inspection, the margins of many of the maria show traces of the rugged relief of the surrounding highlands protruding through the flat mare surface, giving the impression that the mare material rests on top of an older highland surface (Figure 12.8). For this reason, it has long been postulated that the dark maria are huge lava flows which spread to fill large rugged depressions on the highland surface. This idea has now been confirmed by samples from the mare surfaces, which show their dominant rocks to be dark, basaltic lavas. Except at the mare margins, the lavas completely cover the older highland surface which underlies them. Judging from the exposed highlands, this buried surface probably had mountains at least 6,000 m (20,000 ft) high. The complete burial of such rugged topography indicates that an enormous volume and thickness of lava was generated to form the maria at some stage in the moon's history.

(a)

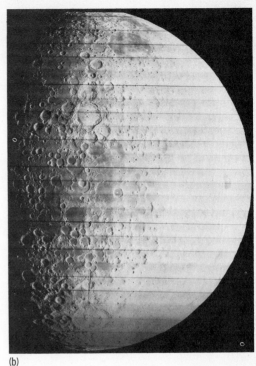

(b)

FIGURE 12.7

(a) Composite photograph of the lunar earthside, showing the distribution of maria (dark areas) and highlands (light areas). The dots indicate sites of samples returned to earth by the Apollo 11, 12, 14, 15, 16, 17 and Luna 16, 20 missions. (b) A portion of the lunar farside as photographed from an orbiting satellite. The farside is dominated by highlands and lacks the large maria of the nearside. (a: Lick Observatory; b: NASA.)

Craters The highland regions which cover about two-thirds of the lunar earthside and most of the farside are completely unlike the more familiar mountainous highlands of the earth. Instead of the narrow, linear belts of folded rock that make up the earth's mountainous regions, most lunar mountains appear to be the more-or-less-altered remains of circular depressions or craters which, next to the larger-scale maria and highlands themselves, are the most prominent features of the lunar surface (Figures 12.7 and 12.9).

The millions of craters visible in detailed photographs of the moon's surface come in all sizes and a variety of shapes. The smallest are only centimeters across, contributing to a highly pitted surface, and the largest have diameters of hundreds of kilometers. Many of the craters are fresh-looking and cup-shaped, whereas others are more dishlike, with flat bottoms and relatively low, irregular rims that give the appearance of having been modified by erosion. In general, craters with diameters of less than about 8 km (5 mi) tend to be cup-shaped and regular, while larger craters show a complete spectrum of shapes, from regular to strongly modified. The overall depths of the craters (distance between the crater floors and the tops of the surrounding rims) tend to vary with crater size; the largest craters, with diameters

(a)

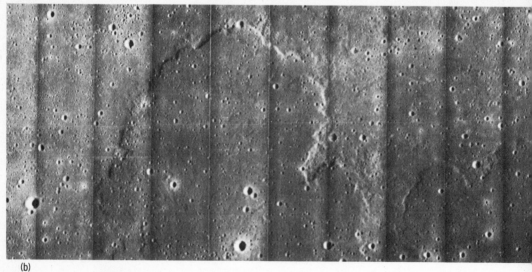

(b)

FIGURE 12.8
Mare lavas. (a) A mare margin, showing the partial filling and burial of highland
craters by flat, dark mare materials. Such evidence first suggested that the maria
might be dark lava flows that cover an older highland surface. (b) Lobe-shaped
pattern, suggesting lava flow, on a mare surface. (a, b: NASA.)

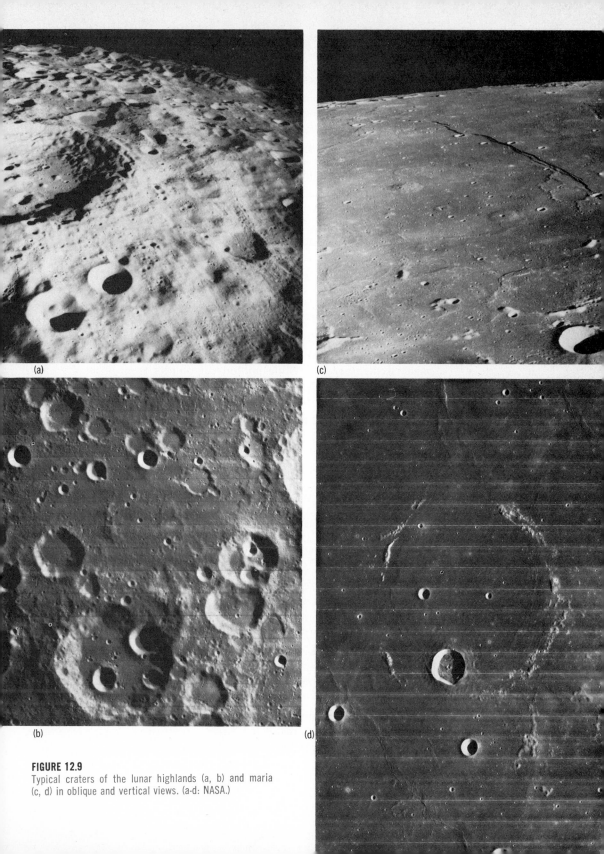

(a)

(c)

(b)

(d)

FIGURE 12.9
Typical craters of the lunar highlands (a, b) and maria
(c, d) in oblique and vertical views. (a-d: NASA.)

of tens of kilometers, have maximum depths of about 6,000 m (20,000 ft).

Craters occur on the surfaces of both the maria and highlands, but their size and distribution differ markedly in the two regions (Figure 12.9). Large craters, with diameters of several kilometers or more, are about 10 times more abundant in the highlands than on the maria. It is largely the modified rims of thousands of such large craters that give the highlands their rugged relief.

The origin of lunar craters has long been a subject of debate. On the earth, similar structures are known to form by either of two processes: on the one hand, circular, craterlike volcanoes result from surface outpourings of lava; on the other hand, the impact of large meteorites also leaves craters, called astroblemes, on the earth's surface (see Section 12.1). By analogy with the earth, both internal volcanism and external meteorite impacts have been cited to explain the craters of the moon.

Volcanism has undoubtedly produced some lunar craters, yet recent exploration has made it clear that most were formed by meteorite impacts. Remember that while millions of objects from space are captured each day by the earth's gravitational field and plunge toward its surface, most are vaporized as they enter the earth's dense lower atmosphere and only a few reach the surface. Although the moon, on the other hand, has far less mass and consequently exerts less gravitational attraction for such objects than does the earth, lunar gravity is still sufficiently strong to cause a continuous capture of materials. But more important, the moon lacks an atmosphere to cause melting and dissipation, and thus every captured object, both large and small, eventually plunges into its surface at high velocity. Countless millions of such impacts have probably given rise to most of the craters that cover the lunar surface.

Other Surface Features   In addition to maria, highlands, and craters, there are other, less common features observable on the lunar surface. Small linear faults can be seen to break the rocks in a few areas, yet the numerous large faults associated with the earth's moving crustal plates are absent. Likewise, the large-scale folded rocks that make up most of the earth's mountain ranges are not seen on the moon, although mare lavas sometimes show small-scale foldlike structures called wrinkle ridges (Figure 12.10). These are believed to have been caused by compression or collapse of the cooling lava.

Two additional features of the lunar surface are of interest because they have few analogues on the earth. The first are rays, thin coatings of light-colored dust that radiate for great distances from large, fresh

 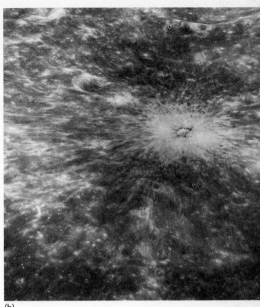

(a)   (b)

FIGURE 12.10
Lunar wrinkle ridges (a) and rays (b) (see text). (a, b: NASA.)

lunar craters (Figure 12.10). These rays, which can be seen from the earth only during a full moon, apparently form as lunar rocks are pulverized, and the resulting dust is scattered, by the force of large meteorite impacts. The second features are rilles, long, narrow, troughlike valleys that occur in and around many large craters on the surface of the maria (Figure 12.11). On the earth, somewhat similar valleys, but on a smaller scale, are caused by the shrinkage and collapse of tubular cavities in lava flows. Most lunar scientists believe that lunar rilles have a similar origin.

12.4 Lunar Materials

Telescopes and satellite cameras have provided a wealth of knowledge about features of the lunar surface, yet these instruments supply very little information about the *composition* of the materials that make up the maria, highlands, and other surface features. By employing an analogy with the earth's crust and by noting as well that the maria resemble huge lava flows, lunar scientists long assumed that the moon's surface is dominated by igneous rocks. This hypothesis has been abundantly confirmed, in the years since 1969, by several hundred kilograms of lunar materials that have been returned to earth and subjected to exhaustive analyses. Although generally similar to certain igneous rocks of the earth's crust, these lunar samples also have many distinctive characteristics.

FIGURE 12.11
Lunar rilles, troughlike valleys thought
to be due to the collapse of tubular
cavities in mare lava flows. Aerial view
of several rilles near a mare margin.
(NASA.)

Lunar Soil   Because the moon lacks water and an atmosphere to produce earth-style weathering, its surface was long believed to expose
much barren rock. One of the surprising initial findings of lunar exploration was that the moon's solid surface is almost everywhere covered by a deep layer of fine-grained "soil" particles, just as are most
regions of the earth (Figure 12.12). On the moon, however, this soil
debris originates from an entirely different kind of weathering than
do earth soils.

In the first place, it is clear that the moon has probably always
lacked the significant fluid cover which plays such a fundamental role
in weathering and shaping the earth's surface. Just as on the earth,
past volcanic outpourings from the interior of the moon undoubtedly
released large quantities of water vapor, carbon dioxide, and other
gaseous materials. Calculations show, however, that because of the
moon's relatively small mass, the normal molecular motions of even
the heavier gases would rather quickly cause them to escape its gravitational field and be lost to space. For this reason, it seems improbable
that the moon has *ever* had the extensive fluid cover required to cause
weathering of its surface rocks.

Instead of forming from weathering by water and atmosphere, the
debris that covers the lunar surface is the result of space weathering
—the continuous abrasion of the lunar surface by large and small
meteorite impacts. Acting in countless numbers over billions of years,
these impacts have pulverized the underlying bedrock and redistrib-

(a)

(b)

FIGURE 12.12
Lunar soil. (a) Typical view of the thick layer
of rock debris that covers the moon's surface.
A sample bag in the foreground and astronaut
in the background provide scale. (b) Close-up
view of a spacecraft imprint in fine lunar
soil material. (a, b: NASA.)

uted the resulting debris by "splashing" to produce the moon's soil cover.

The thickness of the lunar soil layer can be estimated from telescopic observations of lunar craters. Craters tend to show a somewhat different shape as they become deep enough to penetrate the underlying bedrock. Such observations show that the soils are thinner over the maria than over the highlands, which have presumably been subjected to a longer interval of space weathering. On the maria, the soil is usually between 2 and 10 m (33 ft) thick; in the highlands it may reach a maximum thickness of about 20 m (66 ft).

With one or two possible exceptions, all of the lunar materials so far returned to the earth have been pebble-, sand-, and dust-sized particles from the thick lunar soil, rather than direct samples of the underlying rock. All of this material shows evidence of igneous origin, yet much of it has distinctive secondary textures caused by space weathering within the soil layer (Figure 12.13).

When a meteorite strikes the lunar surface its impact energy melts some of the preexisting soil material at the point of impact and scatters this molten rock over a wide area, where it quickly cools to form small spheres of glass. These secondary spheres are a common constituent of lunar soils. Beyond the zone of fusion is a zone in which the impact energy does not melt the soil, but instead compresses and welds together small and large soil fragments to form a heterogeneous solid rock called **breccia.** Like the molten soil, breccia fragments are widely scattered by the energy of impact. Beyond the zone of breccia

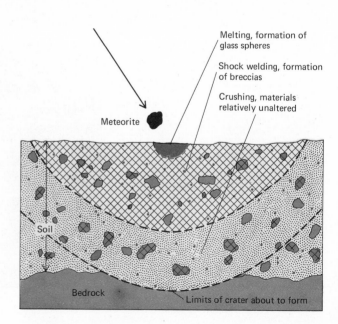

FIGURE 12.13
"Space weathering" and the origin of lunar soil materials. Continuous bombardment of the lunar surface by meteorites creates a soil layer composed of small glass spheres (dark color), composite breccia fragments (cross-hatched pattern), and fragments of relatively unaltered bedrock (gray). (After Wood, "Scientific American," vol. 223, 1970.)

Melting, formation of glass spheres

Shock welding, formation of breccias

Crushing, materials relatively unaltered

Meteorite

Soil

Bedrock

Limits of crater about to form

formation is a third zone in which soil materials—and, if the meteorite is a large one, the underlying bedrock—are crushed and scattered with no change in texture other than fragmentation into progressively smaller and smaller particles.

The meteorite impacts of space weathering thus produce three principal textures of materials in the lunar soil: small glass spheres, heterogeneous breccia fragments, and relatively unaltered rock fragments (Figure 12.14). The breccia and unaltered rock fragments may be of various sizes, ranging from dust and sand up to large boulders.

Lunar Rocks   The bulk of the larger rocks returned from the moon are heterogeneous breccias whose included particles show a complex history of fragmentation and welding by meteorite impacts. Analyses of these, supplemented by studies of the less common unaltered rock fragments and sand and dust particles, indicate that the lunar soil is derived from at least three fundamental kinds of underlying igneous rock: **mare basalts, premare basalts,** and **anorthosites** (Figure 12.15).

FIGURE 12.14
Lunar soil constituents. (a) Small glass spheres that average about 0.2 mm in diameter. (b) Breccia fragments of all sizes. Breccias are composed of heterogeneous rock particles welded together by meteorite impacts. The photograph shows a large breccia boulder, about 1 m (3.3 ft) across, made up of clearly visible light- and dark-colored rock components. (c) Relatively unaltered bedrock fragments of all sizes. Shown is a large boulder of mare basalt. (a: E. C. T. Chao; b, c: NASA.)

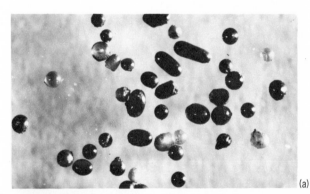

(a)

(c)

(b)

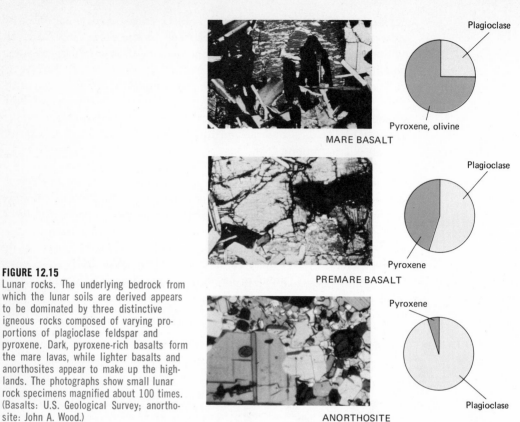

MARE BASALT

PREMARE BASALT

ANORTHOSITE

FIGURE 12.15
Lunar rocks. The underlying bedrock from which the lunar soils are derived appears to be dominated by three distinctive igneous rocks composed of varying proportions of plagioclase feldspar and pyroxene. Dark, pyroxene-rich basalts form the mare lavas, while lighter basalts and anorthosites appear to make up the highlands. The photographs show small lunar rock specimens magnified about 100 times. (Basalts: U.S. Geological Survey; anorthosite: John A. Wood.)

Mare Basalts. Most of the lunar rock samples, which were collected from the soils that cover the maria, are dominated by dark basalt fragments that apparently come from the underlying lavas. These are made up largely of plagioclase feldspar and pyroxene, as are earth basalts, but differ in containing more iron and far less sodium and certain other minor elements than do most basalts found on earth.

Premare Basalts. A second, quite different kind of basalt fragment is particularly abundant in soil debris believed to be derived from the highlands that bound and underlie the maria. Compared with the mare materials, these "premare" basalts have a higher proportion of plagioclase feldspar and less pyroxene and other dark minerals. As a result, they are relatively enriched in aluminum and silicon and depleted in iron and magnesium. They also contain much more radioactive potassium and uranium than do the mare basalts, but resemble them in having little sodium. Because of their association with the highlands, these basalts are assumed to antedate those of the maria.

Anorthosites. Unlike the basalts, the third principal moon rock is of a type that is rare on the earth. This rock, called *anorthosite*, is similar to gabbro, the coarse-grained version of basalt, except that it is light-colored and composed largely of plagioclase feldspar, with little additional pyroxene or other iron-bearing minerals. Because of the dominance of white feldspar crystals, anorthosites are very light in color and contrast sharply with the dark lunar basalts.

Like the premare basalts, lunar anorthosites appear to be derived from the highlands. Thus the three principal lunar rocks predominate in different regions. The well-sampled maria are composed of iron-rich basalts, while the less intensively sampled highlands apparently contain both aluminum-rich basalts and anorthosites. This general compositional difference, based on a relatively few sample locations, has been confirmed by sensitive instruments orbiting in lunar satellites, which show the highlands to be everywhere more aluminum-rich than the maria. Such instruments also suggest that the radioactive premare basalts may be confined to the northwest quarter of the earthside, with anorthosites dominating other highland regions.

The Moon's Interior   In addition to providing direct samples of the moon's surface rocks, the most recent lunar explorations began to investigate the materials of the moon's hidden interior. As on the earth, these investigations are based on surface measurements of the moon's internal wave propagation, gravity, and magnetism.

The most revealing glimpses of the lunar interior are provided by "moonquakes" recorded by sensitive seismographs placed on its surface (Figure 12.16). At least three widely spaced seismograph stations are required to accurately locate an earthquake or moonquake, and in 1971 the third such lunar station was established. The instruments have since shown that the moon is seismically very quiet compared with the earth. All recorded moonquakes are very small, and their total energy release is about a million times less than that of all earthquakes over a comparable period.

The most surprising result of lunar seismic measurements has been the discovery that most moonquakes are sharply concentrated in both time and space. They tend to occur twice each month, when the moon is farthest from and closest to the earth; thus they appear to be caused by tidal effects of the earth's gravity. In addition, about 80 percent of the moonquake energy appears to come from one very small region at a depth of about 800 km (Figure 12.16). The reason for this peculiar localization is unknown, but the mere presence of moonquakes at such a great depth—about halfway to the moon's center—indicates that the moon's deep interior is far more rigid than

(a)

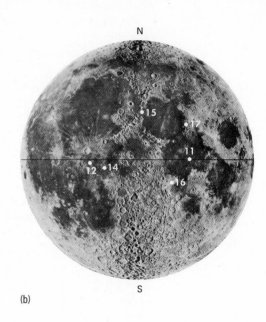

(b)

FIGURE 12.16
Moonquakes. (a) Astronaut setting up a seismograph at the Apollo 15 site. (b) Index map showing the location of lunar seismic stations established on the Apollo 12, 14, 15, 16, and 17 missions. Most moonquakes, which are far less energetic than earthquakes, appear to be concentrated at great depth beneath the small area shown by the color marker. (a: NASA.)

is that of the earth. Earthquakes are confined to the outermost one-tenth of the earth's radius.

Because of their small magnitude and their temporal and spatial concentration, moonquakes reveal relatively little about the moon's internal structure. Larger shock waves caused by the impact of abandoned spacecraft, however, have permitted the construction of crude seismic velocity-depth curves, which suggest that the lunar interior is layered, as is that of the earth: a rather sharp discontinuity at about 70 km (110 mi) appears to separate an overlying "crust" from an underlying "mantle." Wave velocities within the lunar crust are about what would be expected in basalts and anorthosites similar to those exposed on the lunar surface. Higher mantle velocities suggest denser underlying materials, perhaps similar to the ultrabasic materials of the earth's mantle. Little is known of the deeper interior because the relatively low energy of the spacecraft impacts, and the location of the seismic stations in relation to them, have so far made it difficult to measure deep wave velocities.

Satellite measurements of lunar gravity reveal only one unusual feature: anomalous concentrations of massive materials that floor several of the smaller maria. The exact nature of these surface concentrations of heavy materials is unknown, yet they confirm the earthquake evidence of a relatively strong and rigid lunar interior. On the more plastic earth, such masses would slowly sink into isostatic equilibrium rather than remain supported at the surface.

Unlike the earth, the moon today virtually lacks a magnetic field. This observation is in harmony with the moon's overall low density, which strongly suggests that the moon lacks the liquid iron core that causes the earth's magnetism. There is, however, one disquieting magnetic puzzle: all lunar rocks tend to show strong *remnant magnetism*, which indicates that a significant magnetic field was present when they formed. The nature and origin of this earlier magnetic field are but two of the many unsolved problems of the moon's history.

12.5   Lunar History

As with the earth, a principal reason for studying the present configuration of the moon is to understand its long and complex past. Likewise, our understanding of lunar history, like that of the earth, rests on two principal chronological techniques: *superposition*, or the relative vertical relations of its surface features; and *radiometric dating* of its minerals.

Surface Chronology   Long before the first lunar samples were returned to the earth in 1969, a large body of information about the moon's history had been compiled by applying the venerable principle of superposition (see Section 4.1) to its surface features. We have already noted the most important conclusion to arise from such study: the edges of the flat maria can be seen to be superimposed on the more rugged topography of the surrounding highlands, thus showing clearly that mare lavas are younger than the adjacent highlands (see Figure 12.8). Furthermore, careful observations of the maria themselves reveal that some of them are composed not of single large lava flows, but of several smaller flows that apparently formed at different times. Because the maria lack the abundant large craters of the highlands, it is also clear that many more large meteorites collided with the lunar surface *before* the maria were formed than did afterward.

By the application of similar reasoning it is also possible to determine the relative ages of many large lunar craters, both those of the highlands and the less abundant ones that occur on the maria. Often two craters, or the widely scattered debris thrown from them, can be seen to be superimposed, so that their relative ages are easily determined. More commonly it is necessary to infer the relative ages of craters by their degree of modification by later meteorite impacts (see Figure 12.9). By carefully combining all such evidence of relative age, lunar scientists have been able to compile, from telescopic photographs alone, detailed "geologic" maps of the lunar surface showing the relative ages not only of such major features as the maria and highlands, but of many individual craters and smaller lava flows as well (Figure 12.17).

From such studies a basic fourfold division of major events in lunar

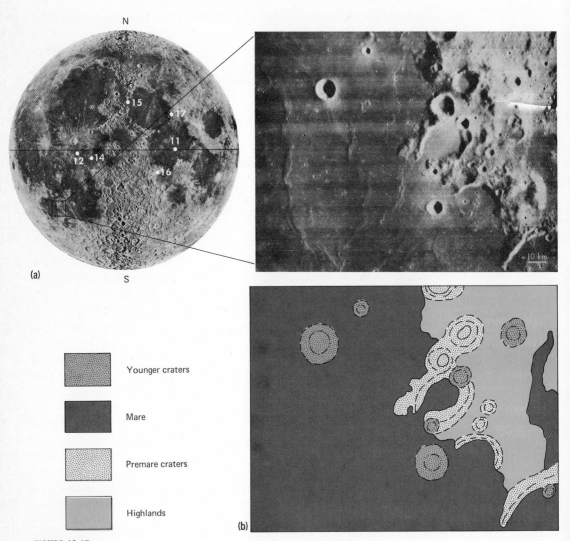

FIGURE 12.17
Lunar surface photograph (a), and a geologic map (b) compiled from it. The region shown, a part of the mare margin and adjacent highlands, contains materials formed during the four principal events of lunar history: highland rocks (oldest); premare craters; mare rocks; and young craters.
(a: NASA; b: modified from F. A. Mutch, "Geology of the Moon," 1970, after Titley.)

history has emerged: first came the solidification and differentiation of the highland rocks; then massive, large-scale cratering of the highlands; next filling of some of the largest highland craters with the mare basalts; and finally, smaller-scale cratering of both the mare and highland surfaces (Figures 12.17 and 12.18).

Radiometric Ages and Thermal History   The major events of lunar history are clearly recorded in its surface features, yet until actual moon rocks were available for radiometric dating there was no reliable means of determining the absolute ages of these events. Earlier estimates of the age of mare surfaces had been made from the abundance of their small craters. By determining the average number of meteorites reaching the earth today and making corrections for the moon's smaller mass and other factors, lunar scientists could calculate the time required to produce a given density of craters, assuming the rate of meteorite infall to have remained constant. Such estimates first suggested that the mare surfaces are very old in terms of earth history, having formed perhaps 3.0 billion years ago. This conclusion has been amply confirmed by radiometric dating (Figure 12.19).

All mare basalts so far dated have ages in the range of 3.0–4.0 billion years. Thus even these relatively late features in lunar chronology formed near the time of origin of the oldest known earth rocks. Dates from the lunar highlands are only slightly older—around 4.1 billion years—indicating that most of the rocks now exposed on the moon's surface originated in a relatively short interval early in its history. Except for continuous cratering, they have remained relatively unchanged for 3.0 billion years, or since early Precambrian time on the scale of earth history (Figure 12.19).

In addition to providing absolute ages for the highlands and maria,

FIGURE 12.18
Schematic diagram of the four principal events of lunar history: (1) formation of highland rocks (oldest); (2) premare highland cratering; (3) partial highland flooding by mare basalts; and (4) younger cratering.

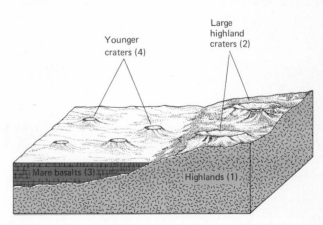

Younger craters (4)

Large highland craters (2)

Mare basalts (3)

Highlands (1)

lunar rocks also raise fundamental questions about the moon's early differentiation and thermal history. Like the rocks of the earth's crust, lunar rocks are quite *unlike* the olivine- and pyroxene-dominated meteorites which are presumably similar to the solar materials from which they consolidated. Lunar surface basalts probably formed from a partial melting of such underlying ultrabasic materials, just as do earth basalts, but there is as yet no satisfactory explanation of the consistent differences in composition between the mare and premare basalts, nor for the fact that both appear to be concentrated on only the earthside of the moon.

A further puzzle is the source of the heat which produced these

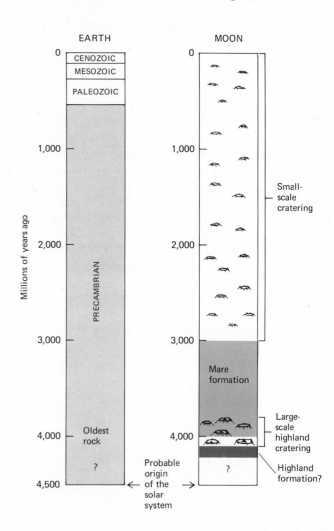

FIGURE 12.19

Comparative histories of the earth and moon. The principal events of lunar history, including highland formation, highland cratering, and mare formation, are revealed by radiometric dating to be very old, having taken place during the early Precambrian interval of earth history.

early internal meltings and then shut down to leave the moon relatively inactive for the past 3.0 billion years. It is unlikely to have been the impact energy of the large meteorites that caused the lava filling of the mare basins, for radiometric dates show that lavas from individual maria formed not all at once, but over several hundred million years of lunar history. Likewise, if the interior were once heated by radioactive decay it should *still* be hot, for there is no way for most of the heat to escape the insulating effect of the moon's outer rock layers. Yet the present moonquakes occurring halfway to the center of the moon show that its interior is now cool and rigid.

Still other puzzles are associated with the light-colored anorthosites that apparently dominate the highlands. Anorthosites are relatively rare rocks on earth, but several regions in which they occur have been intensively studied and their origin is well understood. They form when basaltic magmas are cooled very slowly deep beneath the earth's surface, rather than being extruded as lava to cool relatively rapidly. When cooled slowly, the first crystals to form are heavy magnesium- and iron-rich silicates that tend to sink to the bottom of the liquid magma, leaving the remaining fluid enriched in lighter calcium and silicon. These elements then crystallize to form an upper layer of feldspar-rich anorthosite. On the earth this process occurs, rather infrequently, in deep-seated magma chambers, some of which are now revealed at the surface by subsequent erosion. If the lunar highlands consist largely of anorthosite, as now seems probable, then a wholesale melting and differentiation of a thick outer layer of ultrabasic materials must have occurred very early in lunar history. Once again, the source of the enormous heat required for this melting is unknown.

Finally, there is the question of the large-scale meteorite impacts that produced the intensive early cratering of the highland surface. These impacts must have been concentrated in a relatively short interval of lunal history—after the consolidation of the highland rocks more than 4.0 billion years ago but before the origin of the mare basalts which fill many of the enormous craters and have maximum ages of almost 4.0 billion years. Fragments from the debris of one of the largest lava filled mare craters suggest that the material was very hot, even partially melted, for a puzzlingly long interval after it was ejected. Small meteorite fragments in the debris also indicate that the impacting object was made up mostly of iron. Beyond that, little is known of this major phase of lunar history.

Origin of the Moon Still other puzzles surround the ultimate origin of the moon and its long-term interactions with the earth. The moon almost certainly consolidated about 4.5 billion years ago from smaller masses of cosmic matter, along with the sun, earth, and planets. It may even have accreted from fragments initially in orbit around the primitive earth (Figure 12.20a). If it did, however, it is difficult to explain the differences in density between the two bodies and the concentration of most of the heavier materials in the earth's core. On the

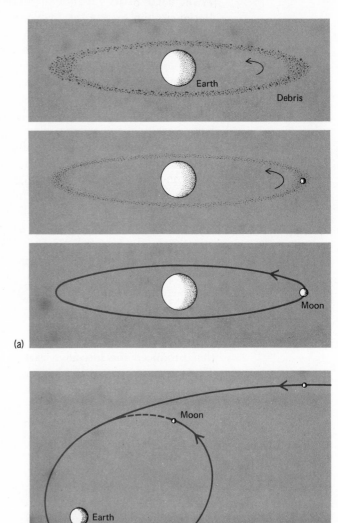

(a)

(b)

FIGURE 12.20
Hypotheses of lunar origin. (a) Accretion from debris in orbit around the primitive earth. (b) Capture of an independent body by the earth's gravitational field. There are serious objections to both hypotheses (see text).

other hand, there are also suggestions that the moon may not have been a satellite of the earth for all of its history. These come from considerations of the motions of the two bodies and the interactions of their gravitational fields.

The gravitational tides caused on the earth by the moon have the effect of gradually slowing down the earth's rate of rotation by transferring some of its rotational energy to the moon's energy of revolution about the earth. This causes the moon to become progressively more distant from the earth as the earth's rate of rotation decreases. Furthermore, the rate at which the moon is now receding from the earth can be calculated using historical records of ancient lunar eclipses. When the rate so obtained is projected into the past, it suggests that the moon was very close to the earth only 2.0 billion years ago and thus could not have been a satellite of the earth for all of its history.

Using such reasoning, some have suggested that the moon was originally a separate body, perhaps a fifth "terrestrial" or "inner" planet (along with Mercury, Venus, the earth, and Mars) that was somehow captured by the earth's gravity (Figure 12.20b). There are, however, many difficulties with this idea. Other calculations suggest that if a body as large as the moon approached the earth, it would most probably not be captured but merely have its path deflected. Furthermore, if the moon had been very near the earth as recently as 2.0 billion years ago, it would have produced enormous tidal effects and frictional melting of most of the earth's crustal rocks, events for which there is no geological evidence. In short, the question of the origin of the earth-moon system remains a fundamental problem of planetary science.

SUMMARY OUTLINE

Meteorites: Samples of the Solar System

12.1 *Meteors and meteorites:* every year millions of frozen cometary gas accretions and heavier solid fragments from space are captured by the earth's gravity, but most are vaporized as they fall through the atmosphere; about 150 meteorites reach the land surface each year, but only about 4 are recovered for study.

12.2 *Meteorite composition and origin:* most meteorites are composed of small glassy spheres of olivine and pyroxene that may represent materials melted by the initial heating of the sun; the smaller quantities of iron and other meteorites may have had a similar origin or may be fragments of larger, differentiated planetary bodies.

The Moon

12.3 *The lunar surface: maria and highlands*—the most conspicuous features of the lunar earthside are flat, dark maria and light, rugged highlands; only highlands occur on the long-hidden farside; *craters* are circular depressions of many sizes and shapes that cover the maria and highlands; some were probably caused by internal volcanoes, but most are the result of meteorite impacts; *other surface features* include small faults and foldlike wrinkle ridges, rays of debris from crater-producing impacts, and long, sinuous channels called rilles.

12.4 *Lunar materials: lunar soil*—rocks of the lunar surface are almost everywhere covered by a thick layer of pebble-, sand-, and dust-sized debris produced by meteorite impacts; *lunar rocks* are of three principal types—iron-rich basalts that form the maria and aluminum-rich basalts and light-colored anorthosites that occur in the highlands; *the moon's interior* shows faint, localized moonquakes at great depths and an earthlike outer layering, but presently lacks a magnetism-producing fluid iron core.

12.5 *Lunar history: surface chronology*—superposition of lunar surface features shows the principle events of lunar history to have been, from oldest to youngest, highland formation, massive cratering, mare formation, and finally less intensive cratering; *radiometric ages and thermal history*—highland differentiation, cratering, and mare filling took place between about 4.1 and 3.0 billion years ago, relatively soon after the moon's origin; the source of the heat for highland differentiation and basalt flooding is a major mystery; *origin of the moon*—the moon may have originally consolidated from materials in orbit around the primitive earth, or it may have been subsequently captured from elsewhere by the earth's gravity.

ADDITIONAL READINGS

Bowker, D. C., and J. K. Hughes: *Lunar Orbiter Photographic Atlas of the Moon*, National Aeronautics and Space Administration, Washington, 1971. A definitive collection of lunar surface photographs.

Goldreich, P.: "Tides and the Earth-Moon System," *Scientific American*, vol. 226, no. 4, pp. 43–52, 1972. A popular review of the problem of the moon's origin.

Hinners, N. W.: "The New Moon; A View," *Reviews of Geophysics and Space Physics*, vol. 9, no. 3, pp. 447–522, 1971. A comprehensive review article, advanced but readable.

Mason, B.: "The Lunar Rocks," *Scientific American*, vol. 225, no. 4, pp. 48–58, 1971. A popular review of the first several groups of lunar samples.

Mason, B.: "Meteorites," *American Scientist*, vol. 55, pp. 429–55, 1967. An authoritative and clearly written review article.

Mulehlberger, W. R. and E. W. Wolfe: "The Challenge of *Apollo 17*," *American Scientist*, vol. 61, pp. 660–669, 1973. A popular review of the last manned lunar landing.

Mutch, T. A.: *Geology of the Moon, A Stratigraphic View*, Princeton University Press, Princeton, N.J., 1972. A detailed but readable introduction to lunar surface chronology, with excellent illustrations.

Wood, J. A.: "The Lunar Soil," *Scientific American*, vol. 223, no. 2, pp. 14–23, 1970. A popular review article.

*Wood, J. A.: *Meteorites and the Origin of Planets*, McGraw-Hill Book Company, New York, 1968. A readable, intermediate-level survey.

*Available in paperback.

Glossary

Aa (lava) A relatively viscous, blocky surfaced basalt flow.

Alluvial fans Fan-shaped deposits of sediment formed at the base of highland terrain in arid regions.

Alluvial valleys Flat-floored valleys covered by thin deposits of alluvium.

Alluvium Sediment deposited from a stream.

Aluminum oxides Earthy masses of small crystals that commonly have resulted from intense chemical weathering; sometimes found as larger gem stones such as ruby (Al_2O_3).

Aluminum silicates Isolated tetrahedra silicates in which aluminum is the principal metallic element (e.g., Kyanite, Al_2SiO_5).

Amphibole (mineral group) Double chain silicates commonly rich in iron and magnesium ions.

Angular unconformity An unconformity in which the older rocks have been folded and eroded prior to deposition of the younger rocks.

Anhydrite A sulfate mineral; $CaSO_4$.

Anorthosites Coarse-grained plagioclase-rich rocks; the dominant rock of the lunar highlands.

Anticlines Arch-like folds having limbs that are dipping away from each other.

Aquifer A body of rock that holds and yields significant volumes of water.

Aragonite A carbonate mineral, $CaCO_3$.

Artesian wells Free-flowing water wells.

Asthenosphere A weak (plastic) zone of mantle rock lying beneath the lithosphere.

Atom The smallest particle of an element possessing the essential characteristics of the element.

Atomic Number The number of protons in each atom of an element.

Attitudes Symbols used on maps to show the strike and dip of beds.

Axial dipole magnetic fields A magnetic field caused by two opposing poles at opposite **ends of** an axis.

Bajada The joining together of many alluvial fans to make a continuous apron-like feature of sediment.

Barrier islands Small islands resulting from redistributed coastal sediments.

Batholith A large exposure of intrusive igneous rocks; results only from extensive erosion.

Bauxite An aluminum-rich weathering product formed in tropical climates.

Beach A zone within which fragments are actively moved on shore, off shore, and parallel to the shore.

Bedding The characteristic layering of sedimentary accumulations; stratification.

Bed load Particles bounced or rolled along a stream bottom during transport.

Bergschrund An air gap formed between the moving glacier and the bedrock at the head of a valley glacier; a region of intense frost-wedging.

Berm The portion of a beach that is above water level.

Biogenic chemical sediments Crystalline sediments produced by the life processes of water-dwelling plants and animals.

Biogenic sedimentary structures Features of a sedimentary rock produced by organisms following deposition (e.g., worm burrows, etc.).

Body waves Seismic waves (P and S) that travel through the interior (body) of the earth.

Breccia Large angular fragments forming a rock; may be of either volcanic or sedimentary origin.

Calcite A common carbonate mineral; $CaCO_3$ —the principal constituent of limestone.

Caldera A roughly circular feature resulting from collapse around the summit of a volcano.

Capillary waves Small irregularities on a water surface resulting from wind.

Carbonate group Minerals composed in part of the CO_3^{-2} ion group.

Carbonate rocks Rocks composed of carbonate minerals; limestone and dolomite are most common.

Catastrophism   The eighteenth century view that the earth had repeatedly undergone incredibly violent upheavals of the sort not presently observable; Georges Cuvier was the chief proponent.

Channel deposits   Coarse alluvial deposits.

Chemical sediments   Accumulations of chemically precipitated crystalline material; characterized by granular texture.

Chemical weathering   Chemical decomposition of rock-forming minerals.

Chert   Sedimentary rock composed of exceedingly fine-grained quartz and opal.

Chondrites   Stoney   meteorites   containing small glassy spheres.

Cinder cone   A relatively small volcano formed by pyroclastic eruptions from a single vent.

Cirque   A bowl-like feature of glacial origin at the head of a glaciated valley.

Clastic (texture)   A rock texture resulting from the accumulation of eroded fragments; rounded grains and pore (void) spaces between grains are usually apparent.

Clay minerals   Sheet silicates that result from chemical weathering processes; very small crystal sizes.

Cleavage   The tendency to fracture along planar surfaces; common in many minerals.

Cliff glaciers   Valley glaciers of very small size.

Composite volcanoes   Volcanoes consisting of interlayered flows and fragmental ejecta (tuff and breccia); usually andesitic in composition.

Compound   A substance composed of two or more elements.

Concordant   Igneous bodies emplaced parallel to the layering of an older rock are concordant intrusives.

Conglomerate   Sedimentary rock composed of sand- and gravel-sized fragments.

Contact metamorphism   Metamorphism from heat effects near the margin of a pluton.

Continental drift   The theory that the continents of the earth have moved laterally about with respect to each other.

Continental shelves   The submerged edges of continents.

Continental slopes   The relatively steep slopes separating continental shelves and deep ocean floors.

Convection currents   Fluid motions caused by density differences; hypothesized as the cause of plate motions in plate tectonic theory.

Coordination of atoms or ions   Refers to the number of one size particle (atom or ion) that can pack around another size particle.

Core (of the earth)   The earth's innermost zone; separated from the overlying mantle by a major seismic discontinuity.

Core-mantle discontinuity   The seismic discontinuity separating the earth's mantle and core.

Correlate   The demonstration of equivalencies between separated rock units; usually attempts are made to show *age* equivalencies.

Cosmic-ray induced nuclides   Radioactive nuclides of short half-life produced by collisons between cosmic-rays and atmospheric gases.

Covalent bonding   Atoms which are held together by sharing each other's outermost electrons.

Craters (lunar)   Circular depressions on the moon's surface.

Creep   Slow downslope movements of soil and rock.

Cross bedding   Within each major bed there are inclined layers; produced by directional currents of wind or water.

Crust   The outer, rocky shell of the earth; generally less than 60 km thick.

Crystalline   Possessing a crystalline space lattice structure.

Crystalline space lattice   A long-range three-dimensional pattern of repeated atoms or ions that characterizes all minerals.

Curie temperature   The temperature at which heated magnetized solids lose their magnetism.

Current charts   Charts showing the general pattern of surface currents in the ocean.

Currents (ocean) Motions in oceans that transport water from one location to another.

Deflation The removal of small particles of sediment by wind action.

Delta A large deposit of sediment at the mouth of a river.

Desert pavement A pavement-like surface of large fragments produced by deflation; found in some desert regions.

Deserts Regions having little moisture or vegetation.

Detrital sediments Accumulations of eroded particles; characterized by clastic texture.

Dike A discordant intrusive body.

Dip The direction and amount (in degrees) of a bed's downward inclination as measured from the horizontal.

Dip-slip (faults) Fault movements of an up–down nature.

Disconformity An unconformity where strata above and below it are essentially parallel; there is a significant age difference between them however.

Discordant Intrusives that are cross-cutting to the layering or fabric of an older rock.

Dissolved load Ions, involved in stream transport, that result from weathered rocks.

Dolomite Sedimentary rock or mineral composed primarily of the mineral dolomite [$CaMg(CO_3)_2$]; the rock and the mineral have the same name.

Dome volcanoes Formed by the emplacement of very viscous magma in the throat of a volcanic vent.

Double refraction The ability possessed by some minerals to convert one light ray into two light rays that are polarized perpendicular to each other.

Drift All glacially derived sediment, including stratified outwash deposits.

Drifts (in the oceans) Water that is moving slower than the ocean currents.

Drumlins Mounds of morainal material that have been overridden and streamlined by moving ice.

Dynamic metamorphism Metamorphism characterized by extensive shearing of mineral assemblages.

Earthquakes Natural movements of the earth usually caused by large-scale faulting.

Earthside The 60 percent of the moon's surface visible from earth.

Eddy A large region of relatively stable and immobile surface water in the center of an ocean gyre.

Elastic rebound theory A model describing faulting as the mechanism for generating earthquakes.

Electron An important sub-atomic particle possessing very little mass and an electric charge of −1.

Element A pure substance that cannot be broken down by ordinary chemical reactions.

Empirical formula The simplest combining proportions of elements to make a compound (e.g., SiO_2 or Al_2O_3).

Engineering geology The application of geologic and engineering principles to construction projects and other developments that affect man.

Eolian sediments Sedimentary accumulations of wind-blown fragments.

Ephemeral streams Streams that flow only after rainfalls.

Epicenter The point on the earth's surface located directly above an earthquake focus.

Epochs The time subdivisions of periods on the geologic timescale.

Equatorial countercurrent A one-directional ocean current flowing eastward near the equator.

Eras Major units of geologic time (e.g., Paleozoic Era).

Erosion The dynamic process of transportation and deposition of rock debris.

Erratics Glacially deposited boulders located far from their source.

Eskers Winding ridges of glacial deposits.

Estuaries Lagoon-like elongate bodies of water caused by the "drowning" of stream valleys by ocean water.

Evaporite sediments Nonbiogenic chemical sediments resulting from the evaporation of waters containing soluble elements (e.g., rock-salt, gypsum, etc.).

Exfoliation The formation of planar or curved layers of rock by jointing and weathering.

Extrusive igneous Volcanic activity.

Farside The 40 percent of the moon's surface not visible from earth.

Fault-block mountains Mountain structures originating from large scale dip-slip faulting, usually normal-slip.

Faults Fractures in rock where movement of one side with respect to the other has occurred.

Feldspar (mineral group) Framework silicates in which ions of potassium, calcium and sodium are necessary because some alumina tetrahedra exist in the structure (e.g., $KAISi_3O_8$); Orthoclase and plagioclase are the most common varieties.

Firn Roughly spherical ice particles that develop from snow.

Flood plains Flat valleys partially filled by stream sediments.

Fluvial environment The sedimentary environment of flowing streams.

Focus The point (or region) of an earthquake's origin. Earthquake foci range from very shallow to several hundred kilometers in depth.

Fold mountains Mountain structures formed largely by large scale folding of sedimentary rocks.

Folds The result of primarily plastic rock deformation; typically either arch-like or trough-like.

Foliated texture A rock texture showing minerals oriented in nearly parallel, sheetlike planes.

Formation The basic unit of rock that is depicted on a geologic map.

Fossil Evidence of the form of a once-living organism (e.g., molds, casts, or actual shells).

Free oscillations Seismically-induced vibrations of the earth as a whole; caused only by large earthquakes.

Frost wedging The principal agent of mechanical weathering; expansion of water into ice forces apart water-filled cracks in rocks.

Gabbro Plutonic igneous rock composed of pyroxene and plagioclase; a mafic rock.

Garnets Isolated tetrahedra mineral group containing a variety of metallic ions (Mg, Al, Ca, Fe, etc.).

Generating area (for waves) The region where strong winds produce wave motions.

Geologic map A map showing the distribution of different rocks as they would look if soil were removed.

Geologic structure sections Cross-sectional view of the rock formations beneath the earth's surface.

Geosyncline A large elongate basin within which great thicknesses of sedimentary rocks have accumulated.

Glacial ages Time episodes of worldwide glacial advances.

Glaciers Moving bodies of ice on the land.

Gneiss A coarsely foliated metamorphic rock; shows banding of mineral segregations.

Goethite An iron oxide mineral; FeO(OH).

Grabens Down-faulted blocks of the earth's crust; trough-like basins.

Graded bedding Type of bedding in which each individual bed shows a progressive particle size decrease going from bottom to top.

Graded stream A hypothetical stream that is neither downcutting nor depositing; a stream in equilibrium between erosion and deposition.

Granite Plutonic igneous rock rich in quartz and feldspar; a silicic rock.

Granite gneiss   A metamorphic gneiss having a mineral composition similar to granite.

Granular texture   A rock texture where crystals of approximately equal size have intergrown.

Gravel   Fragments of rock larger than 2 mm in diameter.

Gravimeters   Sensitive instruments that measure slight variations in the force of the earth's gravity at different locations.

Gravity   The property of all bodies of matter that causes them to be attracted to each other.

Gravity anomalies   Slight variations from the average gravitational force of the earth.

Gravity waves   Large wind waves capable of movement even after the wind stops blowing.

Groundmass   The fine-grained portion of a porphyritic texture.

Ground water   Water confined in the pore spaces of rock and soil.

Guyots   Flat-topped submarine mountains.

Gypsum   A sulfate mineral; $CaSO_4 \cdot 2H_2O$.

Gyres   Large circular patterns of surface currents in the oceans.

Half-life   The average life of a radioactive nuclide; the amount of time required for half of an initial amount of a nuclide to decay.

Halide (minerals)   Minerals composed of metallic ions combined with ions of the halogen group (e.g., Cl^{-1}, Br^{-1}, F^{-1}, etc.).

Halite   A halide mineral; NaCl.

Hanging valleys   Tributary valleys that drop off abruptly into their main valley; frequently occupied by waterfalls.

Hematite   An iron oxide mineral: Fe_2O_3.

Highlands (lunar)   Light colored, high relief portions of the moon's surface.

Horns   Sharp-peaked mountains resulting from glacial erosion.

Horsts   Up-faulted blocks of the earth's crust.

Hydrosphere   World-encompassing sphere of water; oceans, lakes, streams, etc.

Hydrothermal ore deposits   Valuable bodies of minerals that were deposited from hot, water-rich gases and liquids; associated with igneous processes.

Hypothesis   A tentative statement of how two or more phenomena might be related; an educated guess.

Icebergs   Large floating masses, broken off from glaciers, that have drifted into the ocean.

Ice sheets   Massive bodies of glacial ice such as those found in Antarctica and Greenland.

Igneous rocks   Rocks formed by crystallization from a hot molten mass called magma.

Intensity (earthquake)   The Modified Mercalli Intensity Scale describes the amount of damage done at *any* geographical location. It ranges from I to XII.

Intrusive igneous   Pertains to the crystallization of magma beneath the earth's surface.

Ion   A particle of atomic size that possesses an electric charge by virtue of gaining or losing electrons.

Ion groups   Combinations of ions that possess an overall charge (SiO_4^{-4}, CO_3^{-2}, etc.).

Ionic bonding   Ions held together by attraction between their opposite electrical charges.

Ionic diffusion   The migration of ions under high temperature and pressure within a solid lattice structure.

Iron meteorites   Meteorites composed predominately of iron, but typically containing some nickel.

Iron oxides   A group of minerals resulting from iron and oxygen combining in different proportions (e.g., Fe_2O_3, Fe_3O_4, etc.).

Island arcs   Curving chains of volcanic islands associated with oceanic trenches.

Isoclinal (folds)   Folds having both limbs dipping parallel to each other.

Isomorphism   When a mineral's composition varies within fixed limits while its structure remains constant.

Isoseismal map   A map delineating geographic localities that experienced different modified Mercalli Intensities during an earthquake.

Isostasy The concept that rocks of the earth will attain a gravitational equilibrium with each other if given sufficient time; this also suggests that lower density rocks "float" on higher density rocks.

Isotope See **Nuclide**

Joints Fractures in rock that lack significant displacement.

Kames Mound-like glacial deposits.

Karst topography Regions showing distinctive topographic features resulting from dissolution of limestone by ground water.

Kettles Depressions formed by melting of large blocks of ice that were incorporated with morainal debris.

Laccolith A concordant intrusive body that domes up the overlying sedimentary rocks.

Landslides Rapid downslope movements of soil or rock by mass wasting.

Laterite A soil type rich in iron oxides.

Lava Magma that has erupted onto the earth's surface.

Law A statement of how phenomena might be related that is almost certainly correct; supported by massive evidence and logic.

Law of superposition Law stating that in an undeformed sequence of sedimentary strata the oldest layer is on the bottom and is overlain by successively younger layers.

Lime muds Mud-sized grains of calcite derived from organic remains.

Lime sands Sand-sized fragments of calcite derived from organic remains.

Limestone Sedimentary rock composed predominantly of calcite; $CaCO_3$.

Lithosphere The rigid outer skin of the earth; composed of the entire crust and 60 to 80 km of the upper mantle.

Longshore transport The movement of a beach parallel to the shoreline.

Low-velocity zone A region of sharp velocity decrease in the earth's upper mantle; called the asthenosphere.

Maar volcano A large explosion crater mantled with pyroclastic debris.

Mafic Refers to rocks or minerals rich in magnesium and iron; relatively low in silicon and oxygen (e.g., basalt and pyroxene).

Magma Hot molten rock material, usually high in silicate composition.

Magnetite An iron oxide mineral; Fe_3O_4.

Magnetometers Sensitive instruments used to measure the earth's magnetic field.

Magnitude (earthquake) The Richter Magnitude Scale describes the maximum amount of ground movement measured at a fixed distance from an earthquake's focus. This scale ranges from 1 to 10.

Main magnetic field (of the earth) The principal component of earth magnetism; originates deep in the earth.

Mantle (of the earth) The thick, rigid zone of the earth's interior that lies between the crust and the core; about 80 percent of the earth's volume is in the mantle.

Marble Metamorphosed limestone.

Mare basalts Plagioclase- and pyroxene-rich rocks of the lunar maria.

Maria Dark, relatively smooth areas of the moon's surface.

Marine sediments Sediments that accumulate on the ocean floor—either shallow or deep.

Marine terrace A relatively flat feature of surf erosion.

Mass Number (atomic) The number of protons plus neutrons in each atom of an element, often known as "atomic weight."

Mass wasting Movements of rock or soil by the direct action of gravity (e.g., a rock fall).

Matter The stuff of the universe that occupies space and has mass.

Mechanical weathering Physical disintegration of rock.

Metallogenic provinces Geographic regions

characterized by concentrations of valuable metallic ore bodies.

Metamorphic rocks   Formed from pre-existing igneous or sedimentary rocks by profound modifications of texture and/or composition while remaining essentially solid; heat, pressure, and chemically active fluids are the principal agents.

Metamorphism   The process that produces metamorphic rocks; causes changes in the composition and texture of igneous or sedimentary rock.

Meteors   Extra-terrestrial materials that vaporize upon entering the earth's atmosphere.

Meteorites   Extra-terrestrial material that reaches the earth's surface.

Mica (mineral group)   Sheet silicates; muscovite and biotite are the most common varieties.

Migmatite   Rock showing an intimate mixture of granite with metamorphic material.

Mineral   A naturally-occurring inorganic solid of fixed chemical composition (e.g., quartz: SiO_2).

Model (scientific)   A mental construct of how nature might function; it may be mathematical, mechanical, or verbal.

Moho discontinuity   The seismic discontinuity at the base of the earth's crust (hence top of mantle).

Mohorovicic discontinuity   The seismic discontinuity separating the crust and the mantle; called the "moho."

Moraine   Glacial deposits having ridge-like, mound-like, or other distinctive forms.

Mountain ranges   Elongate zones of topographically high terrain.

Mud   The accumulation of clay-sized particles; grains are less than 1/256 mm in diameter.

Native element minerals   Minerals composed of uncombined elements (e.g., gold, diamond, etc.).

Natural slope angle   The angle of slope that is maintained by natural erosional processes.

Neap tides   Tides in their lowest ranges; neap tides occur twice each month.

Neptunism   An eighteenth century view that all rocks were deposited from a world-encompassing ocean; promulgated by A. G. Werner.

Neutron   An important sub-atomic particle possessing no electric charge and one unit of atomic mass; found only in an atom's nucleus.

Noble gases   Inert gases that tend not to combine with other elements to form compounds; each atom contains eight outermost electrons.

Nonbiogenic chemical sediments   Crystalline sediments formed by inorganic processes (e.g., rock salt).

Nonconformity   An unconformity in which the older rocks are non-stratified (e.g., granite).

Normal-slip faulting   Dip-slip faulting that locally lengthens the earth's crust.

Nucleus (atomic)   The massive central region of an atom or ion in which all the protons and neutrons are located.

Nuclide   Different weights (caused by varying numbers of neutrons) of a given element are called nuclides (e.g., U^{235} and U^{238} are different nuclides of uranium). The term isotope is interchangeable with nuclide.

Oblique-slip (faults)   Fault movement with components of both dip-slip and strike-slip.

Ocean-floor spreading   The separation of lithospheric plates along mid-oceanic ridges.

Oceanic ridge-rise system   A world-encircling belt of mountains and broad rises found on the ocean floor.

Oceanic trenches   Long, narrow zones of very deep ocean floor; usually parallel to a continental margin.

Olivine (mineral group)   Fe–Mg-bearing isolated tetrahedra minerals.

Ooliths   Rounded, sand-sized particles of calcite showing a concentric pattern of growth.

Organic evolution   The developmental changes in the living world through time.

Organic reefs Interlocking skeletons of living and dead marine life; commonly rich in algae and corals.

Orthoclase A feldspar mineral; $KAlSi_3O_8$.

Outwash plains Stratified deposits resulting from glacial meltwater.

Overbank deposits Fine grained (clay and sand) alluvial deposits that settle out following the flood stage of a stream.

Oxidation number The number of electrons gained, lost, or shared by an element when it forms an ion or a compound.

Pahoehoe (lava) A relatively fluid, gas-rich, basalt flow; often ropy in surface appearance.

Paleomagnetism The study of the nature of the earth's magnetic field as it existed in the geologic past.

Paradigm A model or "ground rule" that serves as a foundation to thinking.

Parallel bedding Individual layers of sediment that are parallel to each other and distinguished by slight differences in color or particle size.

Partial melting Because feldspar melts at a lower temperature than mafic minerals such as olivine and pyroxene, local concentration of silicic magma could be produced by heating and partial melting of mafic (mantle) rock.

Pedalfer A soil type rich in aluminum and iron.

Pediment A relatively flat, eroded bedrock surface extending laterally from the base of a mountain front.

Pedocal A soil type rich in calcium carbonate.

Peneplane The most mature surface of stream erosion; flat, having minimum relief.

Perennial stream Streams that flow continuously.

Periodic table A grouping of chemical elements indicating those that have similar properties.

Periods Basic units of geologic time; the time subdivisions of eras.

Phase diagram A diagram depicting the conditions under which matter will occur in different composition or states (e.g., bases, solids, liquids).

Phenocryst The larger crystals in a porphyritic texture.

Physical sedimentary structures Features of a sedimentary rock formed at the time of its deposition (e.g., cross bedding, graded bedding, etc.).

Piedmont The region extending laterally from the base of a mountain front; comprises a pediment and a bajada in desert regions.

Plagioclase An isomorphic feldspar mineral group; ranges between $NaAlSi_3O_8$ and $CaAl_2Si_2O_8$.

Plates Large slabs of lithosphere that are the principal units of plate tectonics; the earth has seven major plates and several minor ones.

Plate tectonics A model for the upper mantle and crust of the earth that envisions interactions of large, rigid lithospheric plates of rock as the cause of most deformation and volcanism.

Playa lakes Ephemeral lakes formed in arid regions.

Plunging fold A fold where the axis is inclined downward from the horizontal; usually has a "hairpin-like" outcrop pattern of the limbs.

Plutonic Pertains to the environment of rock origin deep within the earth; deep intrusive igneous rocks are plutonic in origin.

Plutons The general name for relatively massive, intrusive igneous bodies.

Poorly sorted sediment Accumulation of many different grain sizes (e.g., clay, sand, and gravel).

Porphyritic texture A volcanic rock texture in which larger crystals, called phenocrysts, are surrounded by smaller crystals or glass.

Premare basalts Plagioclase-rich basalts that are older than the mare basalts.

Primary nuclides Radioactive nuclides that have persisted since the earth first formed.

Primary waves (P waves) Seismic waves that

cause compressional motion (forward and back) in the earth; the fastest moving seismic waves.

Proton An important sub-atomic particle possessing an electric charge of $+1$ and one unit of atomic mass; found only in an atom's nucleus.

Pyrite A sulfide mineral (fool's gold); FeS_2.

Pyroclastic debris Volcanic ejecta blasted into the air by explosive eruptions (tuff and breccia are common examples).

Pyroxene (mineral group) Single chain silicates; commonly rich in iron and magnesium ions.

Quartz A framework silicate; SiO_2.

Quartzite Metamorphosed quartz-rich sandstone.

Radioactive age (of a mineral) The length of time since a radioactive parent nuclide and the daughter nuclide were held together in a closed system.

Radioactivity The spontaneous breakdown of the nucleus of an element to become a different element.

Rays (lunar) Thin coatings of light-colored debris radiating from young lunar impact craters.

Regional metamorphism Metamorphism of a large region in response to deformational pressures and elevated temperatures.

Rejuvenation The process when a stream begins to erode more actively; usually results from either uplifting or climatic changes causing increased amounts of rainfall.

Residual field The small field of earth magnetism that remains after subtracting the large dipole field.

Reversals in magnetic polarity The earth's main magnetic field has changed direction many times in the past (e.g., North magnetic pole became South magnetic pole).

Reverse-slip faulting Dip-slip faulting that locally shortens the earth's crust.

Ridges (in plate tectonics) Zones separating two lithospheric plates where new crust is being formed.

Rifts Regions where the earth's crust is pulling apart; typically formed near the center of mid-ocean ridges.

Rilles Long, narrow, trough-like valleys that occur near and within large craters on the moon.

Rises (in plate tectonics) New crust zones having more subdued topography than ridges.

Roches moutonnées Asymmetrically rounded bedrock knobs resulting from glacial erosion.

Rock An aggregate of mineral material; volcanic glass is also rock, even though non-crystalline.

Roof pendants Bodies of pre-igneous rock engulfed by plutonic material; they "hang" into the younger igneous mass.

Sand Grains ranging in diameter from 1/16 mm to 2 mm.

Sand dunes Accumulations of mobile wind-blown sand into distinctive dune forms.

Sandstone Sedimentary rock composed primarily of sand grains cemented and compacted together.

Schist A metamorphic rock having a foliated texture commonly rich in micas.

Scientific method A term sometimes used to describe the process of hypothesis testing by experimentation.

Scientific revolution When one major scientific paradigm is replaced by another.

Sea Complex ocean waves found in a generating area.

Sea ice Thin layers of ice formed from ocean water in cold polar climates.

Secondary waves (S waves) Seismic waves that cause transverse motion (sideways shearing) in the earth.

Sedimentary rocks Formed on the earth's surface by the settling out of fragments of preexisting rocks or dissolved substances; they are layered.

Seismic discontinuities Surfaces or narrow zones in the earth's interior across which there are rather abrupt changes in seismic wave velocities.

Seismic sea waves Large destructive waves that are generated by submarine faulting or landsliding. They sometimes strike coastlines with wave heights as great as 50 or 60 meters. (Also called Tsunamis.)

Seismic waves Elastic vibrations transmitted through the earth; earthquake waves.

Seismograms Records of earthquakes produced on seismographs.

Seismographs Instruments used to record the passing of seismic waves.

Semidiurnal (tides) Tides having a regular pattern of two high tides and two low tides each day.

Shale Sedimentary rock composed of clay-sized fragments; sometimes called mudstone.

Shear stress The application of force on a body in a manner that would cause its two sides to move past each other.

Shelf ice Continentally derived glacial ice floating on an adjacent sea.

Shells (electron) Energy levels in which electrons are surrounding the nucleus of an atom; higher energy levels are farther from the nucleus.

Shield areas Exposed cores of ancient mountainous terrain; the nuclei of continents.

Shield volcanoes Large, broad volcanoes that build up from repeated outpourings of fluid lava; usually basaltic in composition.

Silicate group Minerals containing silicon-oxygen tetrahedra as fundamental building blocks.

Silicate melt Liquid silicate material; magma.

Silicic Refers to rocks and minerals rich in silicon and oxygen (e.g., granite and quartz).

Sill A concordant intrusive body.

Slate A fine-grained, foliated, metamorphic rock.

Soil Altered rock debris, typically containing a variable organic content.

Soil profile The characteristic structure of a soil as seen in cross-section.

Sole markings Physical sedimentary structures found on the bottom (sole) of sedimentary strata that result from turbidity current deposition.

Space weathering The continuous abrasion of the moon's surface by small meteorite impacts.

Spheroidal weathering Rounded surfaces on rock resulting from greater rates of chemical attack at the corners and edges of rock masses.

Spring tides Tides in their greatest ranges; twice each month.

Stock An intrusive igneous body that is exposed for less than 40 square miles; smaller than a batholith.

Stoney-irons Meteorites composed of both silicate and metallic material.

Stoney meteorites Meteorites composed of silicate materials.

Stratification Layering (bedding) in sedimentary rocks.

Streams Ribbons of flowing water on the lands.

Striated pebbles Scratched and abraded fragments that result from glacial transport.

Strike The directional trend of a bed as seen on a horizontal erosion surface.

Strike-slip (faults) Fault movement of a horizontal nature; often called transcurrent faults.

Subduction zones Zones where lithospheric plates descend deep into the earth's mantle.

Submarine canyons Steep and deep canyons eroded through continental shelves and slopes.

Sulfate group Minerals composed in part of the SO_4^{-2} ion group.

Sulfide (minerals) Minerals composed of metallic elements combined with sulfur.

Surf Breaking water waves that rush onto the land.

Surface currents Horizontal ocean water movements in the upper hundred meters or so.

Surface waves Seismic waves that travel only in the outer layers of the earth (e.g., *L* waves).

Suspended load Solid particles undergoing stream transport in a primarily suspended state.

Swell Relatively simple ocean waves that have spread from the generating area.

Synclines Trough-like folds having limbs that are dipping toward each other.

Systems (of rock) The basic time–rock unit; a system comprises all rocks formed during a geologic period.

Talus Fragments of rock that have fallen in response to gravity alone.

Tarns Small lakes occupying the bottom of cirque basins.

Terrestrial sediments Sediments that accumulate on land; includes lake and stream deposits.

Tetrahedron A four-sided geometrical solid; each face is an equilateral triangle.

Texture (rock) The relationships between the grains of a rock that pertain to their (1) size distribution, (2) orientation, and (3) the manner in which they fit together; texture is the most genetically significant rock characteristic.

Theory A confidently held statement of how phenomena might be related; supported by abundant evidence.

Tidal wave The misnomer that is often applied to a seismic sea wave.

Tides World-encircling ocean movements caused by solar and lunar gravitational attractions.

Till Sediment deposited directly from glacial ice.

Topographic profile A side view of the earth's surface along a given line.

Transcurrent faults Strike-slip faults.

Transform faults A special class of strike-slip faults where actual displacement takes place only between two major plate features (e.g., movement occurs only in the region between two spreading ridges).

Transition zone A middle region of the earth's mantle separating the upper mantle from the lower mantle.

Tsunami Seismic sea waves.

Tuff Fine volcanic ash.

Turbidite sequence An accumulation of turbidity current deposits, usually characterized by graded beds showing sole markings.

Turbidity currents Sediment-laden currents of water that commonly deposit graded beds and sole markings upon reaching deep water.

Ultramafic (rocks) Refers to rocks that are very low in silicon–oxygen content (e.g., mantle rocks composed of olivine and pyroxene).

Unconformity A surface of contact between rock units that represents a significant time gap in the continuity of the rocks; a buried erosion surface.

Uniformitarianism The assumption that matter and energy have acted in an unchanging manner throughout all of earth history (i.e., natural laws have always existed).

U-shaped valleys Broad-floored, steep-walled mountain valleys; usually of glacial origin.

Velocity–depth curves Curves showing how *P* wave and *S* wave velocities change with depth in the earth.

Ventifacts Wind-faceted pebbles and cobbles formed by "sandblasting" in desert regions.

Volcanic Igneous activity involving the eruption of magma (lava) onto the earth's surface.

Volcano A cone or domed-shaped feature resulting from successive eruptions of lava or other volcanic ejecta from a localized vent.

V-shaped valleys Steep walled, narrow bottomed stream cut valleys.

Water table The upper surface of ground-water saturated rock.

Wave base The depth of water motion in a wave; about half the wave length.

Waves Temporary disturbances of a medium (e.g., ocean water) that transmit energy but do not permanently displace the medium.

Weathering   Chemical and mechanical breakdown of rock near the earth's surface.

Well-sorted sediment   Accumulation of grains that are nearly the same size (e.g., all clay).

Wind waves   Surface waves produced by wind blowing across a body of water.

Wrinkle ridges   Small-scale, foldlike structures on lunar maria.

Xenolith   A "foreign" rock incorporated in a younger intrusive igneous rock.

Index

A

Aa lava flow, 71, 72
Achondrites, 393–94
Aerial photography, 174, 186
Airy, G. B., 222, 223, 224
Airy hypothesis, 222, 223, 224
Alaska, glaciers in, 334
Alluvial fans, 118, 119, 328, 330
Alluvial valleys, 286, 287
Alluvium, 114, 284
Aluminum, 266, 361, 363, 368, 369
Aluminum oxides, 54
Aluminum silicates, 46
Amphiboles, 47–48, 51, 161
Andesite, 257, 258
Angular unconformities, 153
Anhydrite, 53
Anorthosite, 405, 407, 413
Antarctica, 302, 303, 304
  ice sheets, 323, 337, 352
Antarctic Ocean, gyres in, 306
Anticlines, 182, 184, 185, 186
Appalachian Mountains, 255
Aquifer, 275
Aragonite, 52, 53
Arctic Ocean, 302, 304
Argon-40, 156, 158, 159, 160
Aristotle, 11
Arkose, 99
Artesian well, 275
Asthenosphere, 259, 260
Astroblemes, 400
Atlantic Ocean, 132, 133, 134
  fluctuation of U. S. shoreline, 125
  gyres in, 305–6
  Midatlantic Ridge, 133, 134, 242–45, 250, 251, 253
  tidal patterns, 315–16
Atmospheric pressure, 207
Atomic nucleus, 31–32
Atomic number, 27, 155
Atoms, 25, 155
Attitudes, 186–88
Avalanches, 376, 381
Axial dipole magnetic field, 213

B

Baffin Island, ice sheet, 335
Bajada, 329, 330, 331
Banded iron formations, 361, 362, 363
Barchans, 334
Barrier islands, 121, 122, 321
Bar sands, 121, 122, 123
Basalt:
  lava flows, 71–73
  lunar, 405–7, 412
  oceanic, 71, 77–78, 79, 170, 171, 245, 248, 249, 250, 251, 257, 266
Basins:
  desert, 119
  ocean (*see* Ocean: basins)

Batholiths, 83, 84
Bauxite, 269, 271, 363
Beaches, 121, 122, 123, 319–22
Beach sands, 121, 122, 123
Bedding, 107, 138
  graded, 108, 111, 112, 128, 185, 186
  parallel, 108, 109, 112, 127, 128
Bed load, 284, 290
Bedrock, 171
Beds, 138
Bending, 191, 192–93
Bergschrund, 339, 343
Berm, 318, 319
Biogenic chemical sediments, 101, 102, 119
Biogenic sedimentary structures, 107
"Blowout" dunes, 334
Body waves, 198, 199–200
Bonding, to form compounds, 28–30
Bore holes, 196, 197, 207, 208, 219
Borings, animal, 112
Bottom-transported sediments, 132
Bowen, N. L., 80, 81, 82, 87
Breccia, 404–5
Bridgeman, Percy W., 4
Building materials:
  chemical sediments used for, 365, 367
  detrital sediments used for, 365–66
Burrows, animal, 112, 113, 123, 133, 269

C

Calcite, 52, 53
Calcium, 102
Calcium carbonate, 102
  on tropical shelves, 128, 129
Caldera, 76
Cambrian Period, 150, 151, 163
Canary Current, 306
Canyons, submarine, 323
Capillary waves, 309, 310
Carbonaceous chondrites, 393
Carbonate rocks, 104, 105–6
Carbonates, 52–53
Carbonate shelf environments, 128–32
Carbon-14, 156–58, 161, 164, 347
Carbon-14 cycle, 158
Carboniferous Period, 150
Catastrophism, 14, 19
Caverns, 276, 277
Cenozoic Era, 150, 164
Chain silicates, 44, 47–48, 51
Channel deposits, 115, 116
Channel shape, of stream, 278–79, 280, 281, 282
Chemical evaporites, 118–19
Chemicals, industrial, 367–68
Chemical sediments, 98, 99–100
  used for building materials, 365, 367

Chemical weathering, 264–72
Chert, 106, 107
Chondrites, 391, 392, 393
Chondrules, 392–93
Chronology, earth (*see* Earth: chronology)
Cinder cone volcanoes, 74, 75
Cirques, 339, 340, 344
Clastic texture, 98–99
Clay minerals, 48–49, 99, 100, 266, 267
Cleavage, 36, 51
Cliff glaciers, 339
Cliff retreat, 381
Climate:
  influence on erosion, 293
  major fluctuations of, 347–48
  and weathering, 269–72
Coal, 104, 364–65, 368
Coastal environments, 120–23
Coccolithophores, 134
Columnar jointing, 72
Composite volcanoes, 74–75
Compounds, 25–26
  empirical formula for, 30
  formation of, 28–30
Conglomerate, 104
Contact metomorphism, 89
Continental deformation, 256–57
Continental drift, 235–39, 254
Continental ice sheet, 125, 126
Continental margins, 322–24
Continental rocks, 170–75
Continental shelves, 124–28, 322–24
Continental slope, 322–23
Continents, 170–71
Convection currents, 260
Coordination, ionic, 38–39
Copper, 358, 359, 360, 369
Coral reefs, 131
Core of earth, 206
  composition, 209–10
  density, 207
  and earth magnetism, 209–14
  and earthquake wave transmission, 213–14
  structure, 213–14
Core-mantle discontinuity, 205, 206
Cosmic rays, 156–58
Country rock, 83
Covalent bonds, 28, 29
Crater Lake, Oregon, 76
Craters:
  earth, 400
  lunar, 397, 398, 399–400, 404, 409
Creep, 283, 284
Cretaceous Period, 148, 149, 150, 162, 163, 164
Cross bedding, 108, 110, 112, 119, 185, 186
Crustal deformation, and plate tectonics, 252–58
Crustal movements, 174, 175–86, 191–93, 219, 230